A GUIDE
TO USING CSMP—
The Continuous System
Modeling Program

A Program
for Simulating Physical Systems

Frank H. Speckhart

Associate Professor
Mechanical and Aerospace Engineering Department
University of Tennessee, Knoxville

Walter L. Green

Professor
Electrical Engineering Department
University of Tennessee, Knoxville

Prentice-Hall, Inc., Englewood Cliffs, New Jersey

Library of Congress Cataloging in Publication Data

SPECKHART, FRANK H
   A guide to using CSMP—the Continuous system
modeling program.

   Includes bibliographical references and index.
   1. CSMP (Computer Program) 2. Digital computer
simulation. 3. Engineering—Data processing.
I. Green, Walter L., joint author.
II. Title.
TA343.S64   001.6′424   75–19498
ISBN 0–13–371377–6

Printed in the United States of America.

PRENTICE-HALL INTERNATIONAL, INC., *London*
PRENTICE-HALL OF AUSTRALIA, PTY. LTD., *Sydney*
PRENTICE-HALL OF CANADA, LTD., *Toronto*
PRENTICE-HALL OF INDIA PRIVATE LIMITED, *New Delhi*
PRENTICE-HALL OF JAPAN, INC., *Tokyo*
PRENTICE-HALL OF SOUTHEAST ASIA (PTE.) LTD., *Singapore*

CONTENTS

3

ADVANCED FEATURES OF CSMP   81

4

FURTHER APPLICATION OF CSMP 193

5

CSMP III   244

PREFACE

During the last few years the digital computer has assumed a prominent role in system simulation and in the solution of nontrivial mathematical equations. Computer languages such as FORTRAN, BASIC, and PL/1 have provided users with a convenient method for communicating with the computer on a wide range of problems. In many cases, specialized computer programs have been developed that are directly tailored to a particular class of problems. A notable example is in the area of modeling physical systems. Several software packages are available, one of the most widely used being the Continuous System Modeling Program (S/360 CSMP and CSMP III) which was developed by the IBM Company.

CSMP is a program especially designed to allow users to simulate all types of physical systems with a minimum of programming difficulty. The language employs user-oriented statements for formulating numerically complicated mathematical operations such as integration and differentiation. Input data, problem parameters, and program output are handled by extremely simple statements. The power and flexibility of FORTRAN is retained with CSMP since it is used as the source language. Essentially, all capabilities of FORTRAN can be used in a CSMP program.

The basic objective of this text is to provide instruction for both the college student and the practicing engineer and scientist on the use of CSMP in modeling physical systems and solving mathematical equations. The book assumes that the reader has no background in CSMP and only an introductory knowledge of FORTRAN.

In many programming texts, the essential ingredients for writing a complete program (input, output, and structure statements) are contained in various chapters throughout the book. This means the user must read the entire text before formulating and writing meaningful programs. This book is organized in such a way that the reader need study only the first two chapters in order to solve significant problems.

The basic fundamentals of CSMP are presented with detailed explanations and are illustrated by example problems. From our experience as both students and teachers, we feel that this approach is more appealing and effective in the learning process. Thus, important concepts, program statements, and general software procedures are illustrated by actual working programs representing disciplines from a large cross section of engineering, mathematics, and related fields.

The first two chapters present introductory material and are designed to bring the reader to a point where it is possible to effectively use the program. Each example in Chapter Two concentrates on presenting one or two key concepts and then showing how these concepts are applied in a typical application. The third chapter has a somewhat different objective in that it concentrates on presenting the more advanced features of CSMP and is developed under five main headings; Integration Methods, Data Statements, Translation Control Statements, Subprograms, and Data Output. Once the reader has a basic knowledge of CSMP, the material in this chapter can be used to increase programming capability and as a convenient reference section. As in to the second chapter, all important concepts are illustrated by practical example programs. Chapter Four concentrates on specific examples dealing with the frequency response of a system, the simulation of digital control systems, and the simulation of logic functions.

All of the material prior to Chapter Five is written for S/360 CSMP. With the exception of output capability, there is very little difference in using S/360 CSMP and CSMP III. Consequently, the first four chapters also apply to CSMP III. The additional capabilities and slight programming changes are outlined in Chapter Five.

Practice problems are included at the close of all major chapters. Any serious potential user of CSMP should work these problems since it is practically impossible to master a program language without actually writing programs.

The material contained in the Appendix is quite useful in that it gives a complete summary of all functions and signal sources available in CSMP as well as a list of definitions, restrictions, diagnostic messages, and reserved words.

The authors wish to express their appreciation to Mr. George Miles and Mr. Wayne Toppins of the University of Tennessee Computing Center for their assistance and helpful comments. Also, we are grateful to our wives for their continuous encouragement.

F.H.S., W.L.G.

1

INTRODUCTION

In the years preceding the development and wide acceptance of the digital computer, solutions for system simulations and differential equations were commonly programmed on the analog computer. Inherent with the analog computer was the necessity for the user to give careful consideration to both magnitude and time scaling of the problem variables. Complex operations such as square roots, squaring, and trigonometric functions usually required special purpose hardware. The solution to higher-order equations or to several simultaneous equations resulted in a complicated maze of patch-panel wiring. Troubleshooting for a misplaced wiring connection on the patch board was a tedious and time-consuming task.

Following the introduction of FORTRAN and other high-level programming languages, digital computer methods were developed that made possible the numerical solution to problems that were formerly solved on the analog computer. Mathematicians, researchers, and engineers were confronted with developing and debugging their own programs or relying on the skill of a professional programmer. Relying on a programmer to develop software often results in either the user not adequately describing the problem or the programmer not correctly interpreting the user's request.

If, however, users accept the challenge of developing their own software, they are faced with the problem of devoting so much effort to writing programs that little time is left for the main responsibilities of their positions.

An Application-Oriented Program

In response to the need for a program language that did not require extensive knowledge of FORTRAN methods and numerical techniques, IBM developed an

1

application-oriented program known as *Digital Simulation Language–90* (*DSL-90*) for the 7090 digital computer series.[1] This program was later modified and adapted for use with the IBM System/360. The name of the simulation was changed to *Continuous System Modeling Program* (*System/360 CSMP*).[2-4] More recently the scope of CSMP has been extended to include greater flexibility by incorporating CRT graphic-display capability, multiple printer-plots on a single page, line-printer charts, overstrikes for graytoning and contour definition, and several new function blocks.[5-6] This version of the program is known as *CSMP III*. Since there is very little difference between S/360 CSMP and CSMP III, both forms are simply referred to as CSMP. This book will use both forms.

CSMP was particularly written to solve either a system of ordinary differential equations or analog block diagrams as encountered in system theory. In developing the program, emphasis was placed on simplified input data statements, output statements, and on program control statements that almost directly describe the mathematical equations or physical variables of the problem. In effect, CSMP allows the user to concentrate on the details of the physical system rather than the usual concerns of numerical analysis and programming.

Background Required For Using CSMP

A person entirely unacquainted with computer programming can learn to solve significant and rather complex problems with CSMP with less than two hours of preparation. An introductory knowledge of FORTRAN is a helpful but not an essential prerequisite. Using CSMP is like using any other programming language in that the scope of problems which one can solve increases with the depth of preparation and the frequency of application.

Using the CSMP program is similar in many respects to using the electronic analog computer. Many of the special CSMP statements perform the same function as typical analog computer components. As previously noted, there is never a need to be concerned with amplitude and time-scaling. CSMP digital simulation has many other advantages over the use of an analog machine in that it (1) is more accurate; (2) is easier to program; and (3) has much greater capability in handling nonlinear and time-variant problems.

Anyone working in a field of science who has a need to simulate a system or solve ordinary differential equations will quickly recognize the power of CSMP. Typically, CSMP can be used to model the dynamic behavior of an automobile, to determine the effect of adding nonlinear rubber isolaters in controlling the vibration of a machine, to study the effects of changing the deadband in a controller of a nonlinear feedback-control system, to simulate the cardiovascular system, and to predict the future performance of the stock market from a *yet to be derived* nonlinear, time-varying, multiple input/output model. Apparently, the number of potential applications for CSMP is limited only by the needs and imagination of its users.[7-11]

The Power of CSMP

The flexibility and ease of using CSMP cannot be described in a few short sentences. However, as a simple introduction, consider the following nonlinear, time-varying, differential equation:

$$\frac{d^2y}{dt^2} + yt\frac{dy}{dt} + y = t^2$$

$$y(0) = 1.0$$

$$\dot{y}(0) = 2.0$$

The program below illustrates the ease of solving this equation.

```
Y2DOT = TIME*TIME − Y − Y*TIME*YDOT
YDOT = INTGRL (2.0, Y2DOT)
Y = INTGRL (1.0, YDOT)
TIMER FINTIM = 3.0, PRDEL = 0.03
PRINT Y
END
STOP
ENDJOB
```

With the exception of control cards, the entire program listing consists of only eight statements. The output from the computer will be a column listing giving 100 discrete values of the dependent variable Y calculated uniformly over three units of problem time. As a comparison, the reader might reflect on the effort required to solve this problem by numerical methods using FORTRAN or by simulation using an analog computer.

As another simple example, suppose the step response of the following block diagram representation of a system is desired.

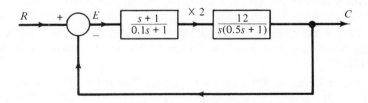

This system can be simulated using only ten CSMP statements.

As these illustrations point out, only a few program statements are required for problem simulations. Solutions using FORTRAN or an analog computer would certainly require much more complicated programming. However, FORTRAN and analog simulation are powerful in their own right. The purpose of this text is to give the reader a background in CSMP and leave the ranking of programming methods to someone else.

CSMP is most useful for small and medium-size simulations. For extremely large and complex problems the simulation may be limited by program-size restrictions. Also, because of the flexibility in programming, CSMP simulations may require slightly more computer time than custom-developed programs.

REFERENCES

1. SYN, W.M. and R.N. LINEBARGER, "DSL/90–A Digital Simulation Program for Continuous System Modeling." *1966 Spring Joint Computer Conference*, April 26–28, 1968.

2. *System/360 Continuous System Modeling Program*, User's Manual GH20-0367-4; Program Number 360A-CX-16X. IBM Corporation, Technical Publications, White Plains, N. Y.

3. *System/360 Continuous System Modeling Program (360A-CX-16X)*, Application Description, H20-0240-1. IBM Corporation, Technical Publications, White Plains, N. Y.

4. BRENNAN, R.D. and M.Y. SILBERBERG, "The System/360 Continuous System Modeling Program." *SIMULATION* 11, No. 6, December 1968, 301–308.

5. *Continuous System Modeling Program III (CSMP III)*, Application Description, program number 5734-XS9. IBM Corporation, Data Products Division, White Plains, N. Y.

6. *Continuous System Modeling Program III (CSMP III)*, Program Reference Manual SH19-7001-2, Program Number 5734-XS9. IBM Corporation, Data Processing Division, White Plains, N. Y.

7. CHUBB, B.A., "Application of a Continuous System Modeling Program to Control System Design." *Proceedings 11th Joint Automatic Control Conference*, June 24, 1970.

8. EDEN, M.S., "The Use of CSMP Digital Simulation Language in Manual Flight Control Analysis." *Proceedings 1970 Summer Computer Simulation Conference*, June 10–12, 1970.

9. BRENNAN, R.D., C.T. DeWIT, W.A. WILLIAMS, and QUATTRIN, "The Utility of a Digital Simulation Language for Ecological Modeling." *Oecologia* 4, May 1970, 113–132.

10. WINTON, H.J. and R.N. LINEBARGER, "Digital Simulation of Human Temperature Control." *Proceedings 1970 Summer Computer Simulation Conference*, June 10–12, 1970.

11. WOLF, J. and W.L. GREEN, "Simulation Study of Predictive Control." *Proceedings of the National Electronics Conference*, Vol. 27, Oct., 1972.

2

FUNDAMENTALS
OF SYSTEM/360
CSMP†

The objective of this chapter is to present those fundamentals of CSMP that will allow the beginning user to write productive programs. Extensive discussions of format, program structure, and the functions of the simulation language as separate sections within the chapter are purposely avoided. This is because the authors have found through experience that the basic fundamentals can be more readily grasped by presenting examples that illustrate practical applications of the program.

The discussions of symbols, constants, operators, structure, and format provide a working description of the simulation language. These discussions are followed by eleven examples that have been carefully selected to enable the reader to develop a fundamental knowledge of the simulation language as he progresses through the material. Problems at the end of the chapter should be considered as part of the text, for it is practically impossible for one to ever develop programming skill without personally formulating and writing programs.

A General Overview of CSMP

CSMP is an application-oriented program in that it is specially written for scientists, engineers, and analysts who are involved in work that requires the solution of ordinary differential equations or in simulating a system that has been modeled as a block diagram. Long hours of tedious program development and program debugging are not required for effecting the solution of rather complicated problems.

†Even though this chapter specifically deals with S/360 CSMP, the same programming fundamentals also apply to CSMP III. With the noted exception of the Example 2.2, all programs in this chapter will run on a CSMP III system.

The utility and ease of using CSMP stems mainly from (1) simplified program statements; (2) flexibility of program structure; and (3) a basic set of preprogrammed function blocks.

Program statements can be broken down into three categories: *data statements*, *structure statements*, and *control statements*. Data statements pertain to entities such as initial conditions for integration or numerical values for parameters and constants of the problem. Structure statements are the heart of the program in that the inner-relationship of the problem variables are here defined. Control statements are used to specify the problem run-time, the integration increment, the format of the output data resulting from the problem solution, and other specific options relating to translation and execution of the program. In reality, the user is not compelled to make a decision in terms of which category a problem statement belongs. These categories are defined simply to describe the types of statements that are utilized.

The program structure of CSMP is composed of three segments: *Initial*, *Dynamic*, and *Terminal*. Generally, data statements will appear in the Initial segment. In addition, calculations that are required to be performed only one time during a simulation (i.e., the volume of a cylinder) can be conveniently placed in this segment. The Dynamic segment is usually composed of structure statements that describe the dynamics of the system or explicitly describe a set of differential equations. The Terminal segment is the last segment in the program and is usually made up of control statements.

Many problems simulated by CSMP will not require the explicit structure described above. Examples that use the Initial, Dynamic, and Terminal segmentation as well as examples in which this segmentation can be omitted are presented in this chapter.

The flexibility of CSMP can be considerably extended to include the power of FORTRAN conditional logic and branching. This extension is accomplished through the *sort* and *nosort* options of the program. If the user does not specify sort and nosort sections, a sort subprogram within CSMP will automatically place the problem statements in the correct order for the simulation. Automatic sorting relieves the user from the task of keeping up with the proper order of program statements as required by the FORTRAN language. However, occasions often arise whereby conditional logic and branching are necessary in a simulation. The NOSORT label card provides this additional flexibility by establishing sections in the program in which the statements are executed by ordinary FORTRAN rules. A more detailed discussion on the structure of CSMP, including the sort and nosort options, will be given later.

Perhaps one of the greatest assets of CSMP is the availability of thirty-four *functional blocks*.[1] A functional block plays a role similar to a FORTRAN subroutine in that the user specifies a function but, in this case, the subroutine that specifies the properties of the function is pre-programmed in the CSMP package. The different types of functional blocks are (1) mathematical functions;

(2) system macros; (3) switching functions; (4) function generators; (5) signal sources; (6) logic functions; and (7) FORTRAN functions. A complete list of the functional blocks along with appropriate descriptions of what the functions accomplish is given in Appendix I. Illustrations using many of these functions will be presented in this chapter.

Most simulations require some form of integration. For this purpose, CSMP gives the user a choice of seven pre-programmed methods. If the user does not specify an integration method, the program will automatically use Runge-Kutta with variable integration-step size (RKS).

The user interested in installation and system requirements for running CSMP should consult appropriate IBM manuals. [2-3]

The following paragraphs describe the use of symbols, constants, and operators as normally employed in programming languages and particularly as employed in CSMP. Next, a fairly comprehensive description of the structure of CSMP is presented. Many readers may find it advantageous to proceed to Example 2.1 and return to the following material at a later time.

Symbols and Constants

All symbols used in CSMP programs must begin with an alphabetic letter (A through Z) and contain not more than six characters. Only alphabetic and numeric characters with no embedded-blanks are allowed. Some examples of valid and invalid symbols are:

Allowed	Not Allowed	Type of Error
Q1	1Q	First character is a number
B12345	B A	Embedded-blank
ABCDEF	ABCDEFG	Too many characters
I1J2	*AB	Invalid first character
K	A(A	Invalid second character

There are certain words reserved for CSMP and FORTRAN that cannot be used as symbols. These reserved words are listed in Appendix II. Unlike FORTRAN, the CSMP language does not require that symbols having a first character of I, J, K, L, M, or N be automatically treated as integers. All symbols, unless otherwise specified, will be treated as floating-point numbers.† The method of designating a symbol as an integer is discussed in Chap. 3.

Floating-point (real) constants can be written two ways. The most common method involves numbers with only a decimal point.

†A floating-point number contains a decimal. For more information see Reference 4.

234.256756

0.005945

− 45.7

+ 100.0

This type of number is limited to a total of twelve characters and not more than seven significant decimal digits. For either very large or small numbers, the format involving the letter E should be used. This type constant can assume values between approximately 10^{-75} through 10^{+75}. Examples of E-format numbers, which can be used in a CSMP program, are listed below.

E-format	Equivalent to
3.254E4	3.254×10^4
−678.1E7	-6.781×10^9
0.0231E-5	2.31×10^{-7}
1764.7E11	1.7647×10^{14}

Integers are written without a decimal point and can contain up to ten digits.

The use of subscripted variables [e.g., X(4), Y(4, 5, 7)] is permitted in the CSMP language with certain restrictions. These restrictions and guidelines for using subscripted variables are outlined in Chap. 3. Double precision is only available in CSMP III. This is covered in detail in Chap. 5.

Operators

Operators used for the basic arithmetic operations are exactly the same as FORTRAN. These operators are tabulated below.

Symbol	Function	Symbol	Function
+	addition	**	exponentiation
−	subtraction	=	replacement
*	multiplication	( )	grouping of variables and/or constants
/	division		

The order in which calculations are performed is also exactly the same as FORTRAN.

Operation	*Hierarchy*
Evaluation of functions	1st (highest)
Exponentiation (∗∗)	2nd
Multiplication and division (∗ and /)	3rd
Addition and subtraction (+ and −)	4th (lowest)

Operators of the same hierarchy, with the exception of exponentiation, are performed from left to right and exponentiation operations are performed right to left. Expressions within parentheses are always performed first. The order of some arithmetic calculations is shown in the following examples:

$$A/B*C = \frac{A}{B}C \qquad\qquad A/(B*C) = \frac{A}{B*C}$$

$$A**B**C = A^{(B^C)} \qquad\qquad (A**B)**C = (A^B)^C = A^{B*C}$$

Many programming errors can be attributed to mistakes in interpreting the correct hierarchy of operations. For this reason, it is recommended that sufficient sets of parentheses be used to insure that the arithmetic operations are performed in the desired order.

Format

As with FORTRAN, only columns 1 through 72 are read as part of the program. Consequently, columns 73 through 80 are not processed and can be used for any type of identification. Unlike FORTRAN, structure statements can begin in any column. Some typical CSMP statements printed on computer cards are shown in Fig. 2.1.

Three consecutive periods (. . .) at the end of the information contained on a computer card indicates that the statement is continued to the next card.

Y = 3.569∗SIN(3.14159∗TIME/180.0) + 54.632∗EXP(21.3∗TIME) ...
    −2.32∗X∗Z/(3.14159∗P)

⎡ Three consecutive periods
  allows continuation to the
  next card. ⎤

A statement may be continued on as many as eight cards for a total of nine cards to express a single statement. Special care should be taken *not* to separate a constant or a symbol on consecutive cards. This type of error is very difficult to find since no diagnostic messages are printed.

An asterisk in the first column denotes a comment card. Comment cards have no effect on the execution of the program. They can be placed anywhere in the program and will be printed with the program listing. Blank cards may be inserted

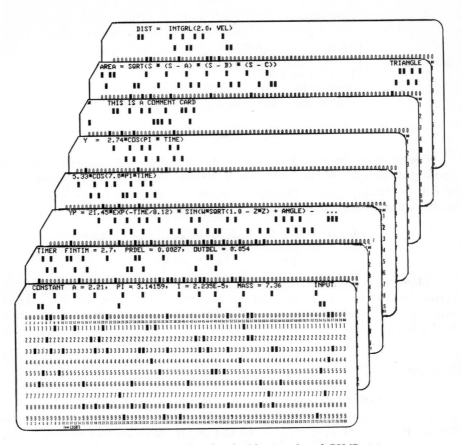

Fig. 2.1 Computer cards printed with examples of CSMP statements.

in the computer card deck to add additional lines of spacing in the listing of the program.

Structure of CSMP

The solution of most problems does not require a detailed understanding of the structure of CSMP. However, in complicated simulations where it is required to control the order of the execution of the program, an understanding of the structure of the CSMP program is necessary.

Basically the CSMP program can be divided into three segments:

INITIAL
DYNAMIC
TERMINAL

Each of the three segments can be further divided by repeated use of sort and nosort sections. The program simulation statements are contained in the body of these sections. A diagram showing the basic structure of a CSMP program is shown in Fig. 2.2.

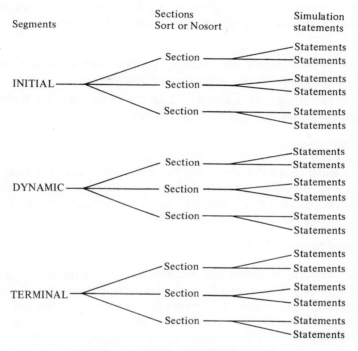

Fig. 2.2   Structure of the CSMP program.

The following paragraphs give a description of the INITIAL, DYNAMIC, and TERMINAL segments and explain their application. It should be noted that the three segments must appear in the order presented and each segment can be used only once.

Initial Segment

The Initial segment is the first to appear in the program. It is used exclusively for calculations that need to be performed only once. The use of the Initial segment can reduce computer time since all statements listed in this segment will be executed only once. All calculations for initial conditions must be included in this segment. It is not always necessary or even desirable to use the Initial segment, consequently, this segment is optional. To specify the use of the Initial segment, a card must be inserted with the label INITIAL at the beginning of the segment.

Dynamic Segment

The CSMP program uses an iterative procedure for problem solution. Since iterations only occur in the Dynamic segment, all simulation statements that are used to describe the dynamic response must be included in this segment. In effect, the Dynamic segment can be thought of as a subprogram that is executed at each iteration step. The Dynamic segment can be specified by two methods as listed below.

1. If the Initial segment is not specified, the Dynamic segment is automatically incorporated into the program. No labeling is necessary.
2. If the Initial segment is specified, a card with the label DYNAMIC must precede the statements in the Dynamic segment.

Terminal Segment

The Terminal segment is the last of the three segments and is used for those calculations that should be performed at the completion of the simulation. This segment is optional and is only executed when a statement with the label TERMINAL is placed after the Dynamic segment.

Sort and Nosort Sections

One of the most important and valuable features of the CSMP program is the sorting capability. The sorting feature chooses the correct order for the execution of the structure statements. This means that in a section where the statements are sorted (called a *sort section*), the order in which the statements are placed in the program deck has absolutely no effect on the order of statement execution. The sort section cannot be used for all types of operations. For example, the following FORTRAN conditional logic and branching statements, which control the order of execution, obviously conflict with the CSMP sorting procedure.

$$\text{IF(X.GT.4.5) GO TO 6}$$
$$\text{GO TO 7}$$
$$\text{6   IF(Y) 3, 4, 5}$$

Consequently, these types of FORTRAN statements cannot be used in a sort section. An expression such as

$$X = X + 1.0$$

in which the same variable appears on both sides of the equal sign is another type of operation that cannot be used in a sort section. Also excluded from sort sections are FORTRAN WRITE and FORMAT instructions such as the following:

$$\text{WRITE(6, 100) X, Y, Z}$$
$$\text{100 FORMAT(3F30.5)}$$

In nosort sections, no sorting is performed and therefore all statements are executed in the exact order in which they appear in the program. FORTRAN conditional logic and branching and all other similar statements that specify a particular order of control must be included in a nosort section.

All statements in the Initial and Dynamic segments not labeled by a NOSORT card are automatically placed in sort sections while statements in the Terminal segment are automatically placed in a nosort section. To specify a change in section in any of the three segments, the use of a separate card with either a SORT or NOSORT label is required. An example of a particular CSMP program having a complex structure is illustrated in Fig. 2.3.

In this example the first group of statements in the Initial segment are automatically sorted up to the point where the NOSORT statement is included. After the NOSORT card, all remaining statements in the Initial segment are contained in a nosort section. The first NOSORT label in the Dynamic segment places the first group of statements in a nosort section. The next group of statements after the SORT label are sorted. A NOSORT label then changes the last group of statements in the Dynamic segment back to a nosort section. In the Terminal segment,

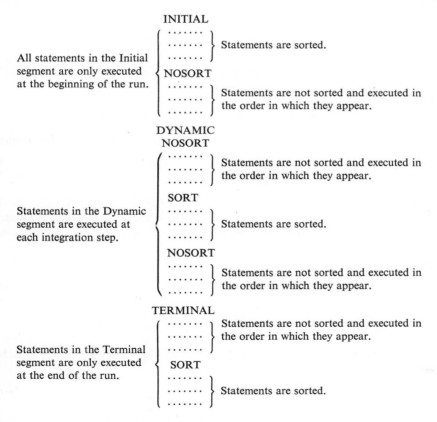

Fig. 2.3 Example of a CSMP program having a complex structure.

the first group of statements are automatically placed in a nosort section. The following SORT label changes the last group of statements to a sort section.

Note that when a section contains both sort and nosort segments, the segments will be executed in the order they appear.

Solving Problems with CSMP

The following eleven example problems illustrate the use of CSMP to simulate a wide range of engineering and mathematical problems. By studying the examples and working the problems at the end of the chapter, the reader will have a good working knowledge of CSMP.

Example 2.1 *Spring-Mass-Damper System*

Some of the very basic concepts of CSMP can be illustrated by simulating the motion of the linear spring-mass-damper system that is shown in Fig. 2.4.

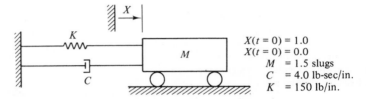

$$X(t = 0) = 1.0$$
$$X(t = 0) = 0.0$$
$$M = 1.5 \text{ slugs}$$
$$C = 4.0 \text{ lb-sec/in.}$$
$$K = 150 \text{ lb/in.}$$

Fig. 2.4 Spring-mass-damper system.

Applying Newton's second law leads directly to the following second-order equation of motion.

$$M\frac{d^2X}{dt^2} + C\frac{dX}{dt} + KX = 0 \qquad (2.1)$$

When dealing with differential equations higher than first-order, it is helpful to reduce all equations to a set of first-order expressions. For some problems, this can require a major effort. However, for Eq. 2.1, it is only necessary to introduce new symbols and rearrange variables.

Let

$$\text{VEL} = \frac{dX}{dt} \qquad \text{(velocity)}$$

and

$$ACC = \frac{d(\text{VEL})}{dt} \qquad \text{(acceleration)}$$

The resulting two first-order equations for velocity and acceleration along with the initial conditions are:

$$\frac{dX}{dt} = \text{VEL} \qquad (2.2)$$

$$\frac{d(\text{VEL})}{dt} = -(C*\text{VEL} + K*X)/M \qquad (2.3)$$

Initial conditions: $X(t = 0) = 1.0$, $\text{VEL}(t = 0) = 0.0$

With this brief explanation, a CSMP program containing only the basic features needed to numerically solve Eqs. (2.2) and (2.3) is shown in Fig. 2.5. In this example program it is not necessary to specify the Initial, Dynamic, or Terminal segments. All statements are automatically placed in the Dynamic segment and sorted. Thus the order in which the structure statements are arranged in the deck has no effect on the execution of the program. The first two statements of the program are comment cards and are identified by an asterisk in the first column. As previously mentioned, comment cards do not affect the execution of the program. The third card is a CONSTANT statement used to assign values to variables.

```
*  M = MASS,    C = DAMPING CONSTANT,   K = SPRING CONSTANT, X = DISPLACEMENT    1
*  VEL = VELOCITY,   ACC = ACCELERATICN,   XO = INITIAL DISPLACEMENT             2
      CONSTANT M = 1.5, C = 4.0, K = 150.0, XO = 1.0                             3
      ACC = (-K*X - C*VEL)/M                                                     4
      X = INTGRL(XO,VEL)                                                         5
      VEL = INTGRL(0.0,ACC)                                                      6
      PRINT X, VEL, ACC                                                          7
TITLE    SIMULATICN CF A SPRING-MASS-CAMPER SYSTEM                               8
      TIMER FINTIM = 2.0, PRDEL = 0.C5                                           9
END                                                                            10
STOP                                                                           11
END JOB                                                                        12
```

Fig. 2.5 Program for simulating spring-mass-damper system.

CONSTANT. The word CONSTANT must be the first label on the card and must be followed by at least one blank space. Otherwise, after this first blank space, blanks are not considered. Successive variables can be set equal to constants (both real and integer) by putting the variable name on the left side of the equal sign. A comma following a numeric value permits another assignment. Notice that a comma should not be placed after the last number. A CONSTANT statement must be continued to following cards by the use of three consecutive periods(...). The format of the CONSTANT card used in this example is

CONSTANT M = 1.5, C = 4.0, K = 150.0, XO = 1.0

The fourth card is an ordinary FORTRAN statement used to calculate the acceleration (ACC). Statements 5 and 6 use a special mathematical function (INTGRL) that is unique to CSMP. The INTGRL statement represents the mathematical function of integration as defined by Table 2.1.

Table 2.1

Formulation for Integration

CSMP *Form*	*Mathematical Function*
Definition: Integration	$Y = \int_0^t X\,dt + IC$
Y = INTGRL(IC, X)	
IC = Y(0)	Equivalent Laplace transform
	$X(s) \longrightarrow \boxed{\dfrac{1}{s}} \longrightarrow Y(s)$

In the definition presented in Table 2.1, the integration is with respect to the independent variable t. The symbol TIME is a reserved word in the CSMP program and is used to represent the independent variable. The integrand can be an algebraic expression, but the initial condition (IC) of the INTGRL function must be a constant or symbol. If the INTGRL function is included as part of a statement, it must be the rightmost part of the expression.

Allowed	*Not Allowed*
Y = INTGRL(7.0, X-Z)	Y = INTGRL(3.0 + 4.0, X-Z)
Y = 10.0 + INTGRL(Q,X)	Y = INTGRL(Q,X) + 10.0
Y = 3.0*INTGRL(Q,X)	Y = INTGRL(Q,X)*3.0

INTGRL is only one of several mathematical functions available to the CSMP user. A listing and a definition for all functions are contained in Appendix I.

The fifth and sixth statements of Fig. 2.5 represent the integration of Eqs (2.2) and (2.3). In statement 5, the velocity VEL is integrated to obtain the displacement X with the initial condition of X0. In the same manner, the sixth statement will perform the integration of the acceleration ACC to obtain the velocity VEL with the initial velocity equal to zero. The seventh card is a PRINT statement which has the following general definition.

PRINT. The PRINT card is used to specify the variables that will be printed at each specific interval during the simulation. All variables following the PRINT label plus the independent variable TIME will be printed and correctly labeled. A comma must be inserted between successive variable names and at least one blank space must follow the PRINT label. The PRINT label should appear only once in the program. If more than one PRINT statement is included, only the last will be executed. If necessary, three periods (...) can be used to continue to following cards. An example of the PRINT statement is:

PRINT X, VEL, ACC

A column format is automatically used for output when there are less than nine dependent variables printed. An example of this is shown in Fig. 2.6 which is the output of the program of Fig. 2.5. Up to forty-nine dependent variables can be printed. When more than eight are specified, an equation-form output format is used as shown in Fig. 2.7. Notice that the output for both Figs. 2.6 and 2.7 is E-format with five significant digits. The user has no other choice of output format when using the CSMP PRINT statement. The eighth card in the program is a TITLE statement.

TITLE. The TITLE card allows the user to specify a heading that will appear at the top of each page of printed output. At least one blank space must follow the label TITLE and the first character must be a number or an alphabetic letter. Continuation to successive cards is not permitted, but up to five TITLE cards can

---

```
SIMULATION OF A SPRING-MASS-DAMPER SYSTEM      RKS      INTEGRATION

TIME            X              VEL              ACC
0.0          1.0000E 00     0.0            -1.0000E 02
5.0000E-02   8.8283E-01    -4.4884E 00     -7.6313E 01
1.0000E-01   5.7793E-01    -7.3878E 00     -3.8093E 01
1.5000E-01   1.7863E-01    -8.2319E 00      4.0884E 00
2.0000E-01  -2.1177E-01    -7.C839E 00      4.0068E 01
2.5000E-01  -5.0491E-01    -4.4556E 00      6.2373E 01
3.0000E-01  -6.4574E-01    -1.1340E 00      6.7598E 01
3.5000E-01  -6.2098E-01     2.0328E 00      5.6677E 01
4.0000E-01  -4.5698E-01     4.3385E 00      3.4129E 01
4.5000E-01  -2.C871E-01     5.362CE 00      6.5725E 00
5.0000E-01   5.6407E-02     5.0288E 00     -1.9051E 01
5.5000E-01   2.7551E-01     3.5845E 00     -3.7109E 01
6.0000E-01   4.0411E-01     1.4990E 00     -4.4409E 01
6.5000E-01   4.2404E-01    -6.6988E-01     -4.0618E 01
7.0000E-01   3.4430E-01    -2.4145E 00     -2.7991E 01
7.5000E-01   1.9559E-01    -3.3879E 00     -1.0524E 01
8.0000E-01   2.0609E-02    -3.4633E 00      7.1746E 00
8.5000E-01  -1.3725E-01    -2.7355E 00      2.1020E 01
9.0000E-01  -2.4395E-01    -1.4716E 00      2.8319E 01
9.5000E-01  -2.8142E-01    -2.81C0E-02      2.8217E 01
1.0000E 00  -2.4S71E-01     1.2417E 00      2.1660E 01
1.0500E 00  -1.6472E-01     2.0683E 00      1.0957E 01
1.1000E 00  -5.2589E-02     2.3178E 00     -9.2178E-01
1.1500E 00   5.7602E-02     2.C048E 00     -1.1106E 01
1.2000E 00   1.4084E-01     1.2714E 00     -1.7474E 01
1.2500E 00   1.8140E-01     3.3817E-01     -1.9042E 01
1.3000E 00   1.7533E-01    -5.5613E-01     -1.6050E 01
1.3500E 00   1.2983E-01    -1.2113E 00     -9.7523E 00
1.4000E 00   6.0245E-02    -1.5071E 00     -2.0055E 00
1.4500E 00  -1.4458E-02    -1.42C6E 00      5.2340E 00
1.5000E 00  -7.6524E-02    -1.C192E 00      1.0370E 01
1.5500E 00  -1.1330E-01    -4.3434E-01      1.2489E 01
1.6000E 00  -1.1952E-01     1.7708E-01      1.1480E 01
1.6500E 00  -9.7571E-02     6.7160E-01      7.9662E 00
1.7000E 00  -5.5995E-02     9.5046E-01      3.0650E 00
1.7500E 00  -6.7754E-03     S.7667E-01     -1.9269E 00
1.8000E 00   3.7855E-02     7.7575E-01     -5.8542E 00
1.8500E 00   6.8238E-02     4.2210E-01     -7.9494E 00
1.9000E 00   7.9188E-02     1.5851E-02     -7.9611E 00
1.9500E 00   7.0622E-02    -3.4333E-01     -6.1466E 00
2.0000E 00   4.6938E-02    -5.7898E-01     -3.1499E 00
```

Fig. 2.6 Column-type output for spring-mass-damper system.

```
P                                                        SIMP      INTEGRATION

ME =  0.0         X  = 1.0000E 03    Y  = 1.0000E 06    Z  =  4.5776E-05   Q  = 0.0
                  R  = 1.0000E 00    V  = 0.0           T  =  0.0          W  = 0.0
                  E1 = 1.0000E 03    T1 = 5.0000E 02    DD =  1.0000E 09   AB = 0.0

MF =  1.2500E 01  X  = 2.2535E 03    Y  = 5.0782E 06    Z  = -1.2535E 02   Q  = 2.8169E 04
                  R  = 4.4376E-01    V  = 1.1868E 01    T  =  5.2665E 00   W  = 7.9347E 08
                  E1 = 1.6692E 02    T1 = 5.1250E 02    DD =  1.1444E 10   AB = -3.3524E 06

ME =  2.5000E 01  X  = 5.0781E 03    Y  = 2.5787E 07    Z  = -4.0781E 02   Q  = 1.2695E 05
                  R  = 1.9692E-01    V  = 2.4423E 01    T  =  4.8094E 00   W  = 1.6117E 10
                  E1 = 1.9531E 02    T1 = 5.2500E 02    DD =  1.3095E 11   AB = -5.0577E 07

MF =  3.7500E 01  X  = 1.1443E 04    Y  = 1.3094E 08    Z  = -1.0443E 03   Q  = 4.2911E 05
                  R  = 8.7390E-02    V  = 3.7111E 01    T  =  3.2431E 00   W  = 1.8414E 11
                  E1 = 2.9722E 02    T1 = 5.3750E 02    DD =  1.4984E 12   AB = -4.4347E 08

MF =  5.0000E 01  X  = 2.5786E 04    Y  = 6.6494E 08    Z  = -2.4786E 03   Q  = 1.2893E 06
                  R  = 3.8780E-02    V  = 4.9768E 01    T  =  1.9300E 00   W  = 1.6623E 12
                  E1 = 5.0561E 02    T1 = 5.5000E 02    DD =  1.7146E 13   AB = -3.1809E 09
```

Fig. 2.7 Example of equation-type output.

be used. Each TITLE card provides one line of heading on each page of output. An example of a TITLE statement is

TITLE SIMULATION OF A SPRING-MASS-DAMPER SYSTEM

The ninth card is called a TIMER statement.

TIMER. The TIMER card is used to specify the variables that control the run-time, print increment, integration interval (step-size), and minimum allowable integration interval. In Fig. 2.5, only two TIMER variables are used, FINTIM, and PRDEL. An example of the TIMER statement is

$$\text{TIMER} \quad \text{FINTIM} = 2.5, \text{PRDEL} = 0.05$$

At least one blank   A comma must
space must follow   separate listings
the label TIMER

FINTIM. FINTIM is a symbol that appears on the TIMER card and determines the value of TIME (independent variable) at which the run is terminated. FINTIM is set equal to the desired simulation time and must be included on the TIMER card.

PRDEL. This TIMER variable controls the increment for the output of the PRINT statement. If the value for PRDEL is not specified, it is automatically set equal to FINTIM/100. If a value of OUTDEL is also included on a TIMER card, the output increment of the PRINT statement can change. At this point, there is no need to explain the relationship between PRDEL and OUTDEL, since it will be covered later in this chapter.

The three other variables that can appear on the TIMER card are defined below. A detailed explanation is included in Example 2.2 and in Chap. 3.

OUTDEL. Print increment for the print-plot output

DELT. Integration interval

DELMIN. Minimum integration interval

The format for specifying the above variables on a TIMER card is similar to a CONSTANT card. At least one blank must follow the label TIMER, and successive listings of variables must be separated by commas. The order in which the variables are listed on the TIMER card is not important.

The last three statements (END, STOP, and ENDJOB) must be included to signify the end of the program. These cards must appear in the order shown in Fig. 2.5 and the ENDJOB statement must begin in the first column. For more complicated simulations, the END, STOP, and ENDJOB cards can be used to control the execution of the program. A detailed explanation will follow when advanced topics are considered in Chap. 3. For most simulations the user needs only to use these cards to signify the end of the program.

Example 2.2   *Simulation of a Block Diagram*

The previous example illustrated one of the principal attributes of CSMP, the solution of a differential equation. Another important application is determining the response of a system that is modeled as a block diagram. Block diagram representation of a system is often used in control system analysis and design. However, the concept of a block diagram is not restricted to control systems alone. In fact, this application extends to such areas as physiology, transportation, energy conservation, and economics.

The basic premise of the block diagram, or signal-flow graph, stems from the application of Laplace transforms or operator notation. For example, a system might be represented by the second-order, linear, time-invariant differential equation

$$\frac{d^2x\,(t)}{dt^2} + A\frac{dx\,(t)}{dt} + Bx(t) = ku(t) \tag{2.4}$$

If $dx\,(0)/dt = x(0) = 0$, application of Laplace transforms yields

$$\frac{X(s)}{U(s)} = \frac{K}{s^2 + As + B} \tag{2.5}$$

In the sense of the block diagram, the above relationship can be represented by the configuration shown in Fig. 2.8.

$$U(s) \longrightarrow \boxed{\dfrac{K}{s^2 + As + B}} \xrightarrow{\ X(s)\ }$$

Fig. 2.8 Block diagram representation of a second-order differential equation.

Laplace transforms and block diagrams offer the convenience of analyzing a system without the necessity of working directly with the system equations. In effect, the Laplace transform changes a linear differential equation to an algebraic expression.

Suppose we consider the simplified schematic diagram of a position control system shown in Fig. 2.9. Neglecting viscous friction and the electrical time-constant of the motor, this position control system can also be represented by the block diagram of Fig. 2.10. Generally, the block diagram is much easier to work with than the schematic.

As an example, suppose the parameters in Fig. 2.10 have the values of $K_m = 4$,

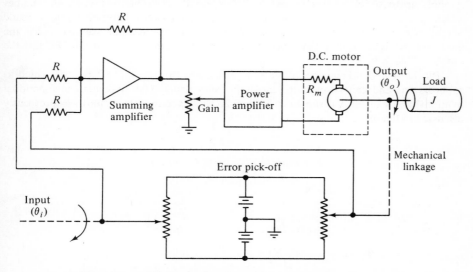

Fig. 2.9 Simplified schematic diagram of a position control system.

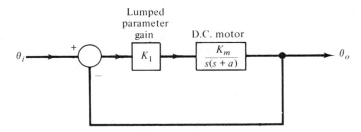

Fig. 2.10 Block diagram of the position control system.

$K_1 = 1.5625$, and $a = 2$. We desire to simulate the response of the system to a unit-step input.

To solve this problem, we first restructure the system diagram so that it is compatible to CSMP modeling. A suitable diagram for this purpose is presented in Fig. 2.11.

Two basic dynamic blocks are present in this representation and can be broken out as separate entities as follows:

$$X2 \longrightarrow \boxed{\dfrac{1}{\left(\dfrac{s}{a} + 1\right)}} \xrightarrow{\ X1\ } \qquad X1 \longrightarrow \boxed{\dfrac{1}{s}} \xrightarrow{\ \text{OUTPUT}\ }$$

Block A   Block B

There is no real significance attached to the order of the blocks, block A could have appeared as the block nearest the output and preceded by block B. The arrangement of the blocks, for linear systems, is a matter of user preference. The symbols X1, X2, X3, ERROR, etc., are intermediate variables. They could have been designated by other symbols up to six characters in length. It is normally desirable to use variable names which describe physical variables in the problem. A significant point that should be made is that each block must have an input and an output variable. For example, X2 is the input and X1 is the output for block A; whereas, X1 is the input and OUTPUT is the output of block B.

Blocks A and B have special meaning in CSMP. In a sense they are represented as subprograms and can be executed by simple statements. The mathematical functional relationship for block A is shown in Table 2.2. The $1/s$ term (integration) was previously discussed in Example 2.1.

Table 2.2

Formulation for First Order Lag

General Form	*Function*
Definition: 1ST ORDER LAG (Real pole)	$P\dfrac{dY}{dt} + Y = X$
Y = REALPL(IC, P, X)	Laplace form: $\dfrac{1}{Ps + 1}$
IC = Y(0)	$X(s) \longrightarrow \boxed{\dfrac{1}{Ps + 1}} \xrightarrow{\ Y(s)\ }$

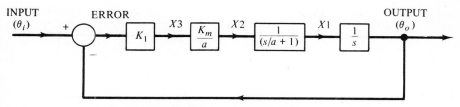

Fig. 2.11 A CSMP representation of the system in Fig. 2.10.

These formulations help explain why the original system was subdivided as given in Fig. 2.11. Note that the general form for REALPL is $1/(Ps + 1)$. This means that when we encounter terms of the form $1/(s + a)$ we must divide the numerator and denominator by a. This changes the form to $(1/a)/(s/a + 1)$ and the $(1/a)$ term in the numerator can either be associated with another constant or represented by a single block as follows.

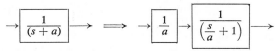

A program listing for the CSMP simulation of this problem is presented in Fig. 2.12.† For convenience, the CSMP block representation corresponding to this program is

```
*    CSMP PROGRAM FOR EXAMPLE 2 - BLOCK DIAGRAM OF SECOND ORDER SYSTEM
*    NOTE THAT THE ASTERISK IN COLUMN ONE SIGNIFIES A COMMENT STATEMENT


TITLE PRINTED LISTING OF "OUTPUT" (SIMULATION) AND "OUTEXT" (ANALYTICAL)
*    THE TITLE IS ALWAYS GIVEN AT THE TOP OF THE "PRINTED LISTING" PAGE(S)
     INPUT   = STEP(0.0)
     ERROR   = INPUT - OUTPUT
     OUTPUT  = INTGRL(0.0,X1)
     X1      = REALPL(0.0,1.0/2.0,X2)
     X2      = 2.0*X3
     X3      = 1.5625*ERROR


* THE FOLLOWING STATEMENTS ARE NOT PART OF THE CSMP SIMULATION. THEY
*ARE USED TO CALCULATE THE EXACT SOLUTION OF "OUTPUT" FOR COMPARISON
     OUTEXT  = 1.0 - ((EXP(-WN*ZETA*TIME))/SQRT(1.0 - ZETA*ZETA))*...
     SIN(WN*SQRT(1.0 -ZETA*ZETA)*TIME + THETA)
     WN      = 2.5
     ZETA    = 0.4
     THETA   = ATAN(SQRT(1.0 - ZETA*ZETA)/ZETA)
* END OF SPECIAL CALCULATION SECTION FOR "OUTPUT" COMPARISON


TIMER FINTIM = 5.2, OUTDEL = 0.16, PRDEL = 0.04
PRTPLT  OUTPUT(ERROR), ERROR(,0.5)
PRINT OUTPUT, OUTEXT
LABEL STEP RESPONSE FOR SECOND ORDER SYSTEM
END
STOP
ENDJOB
```

Fig. 2.12 CSMP listing for Example 2.2.

†The symbol OUTPUT cannot be used in a CSMP III program since it is a reserve word that has a special meaning. To run this program on CSMP III, the symbol, OUTPUT, must be replaced by another symbol.

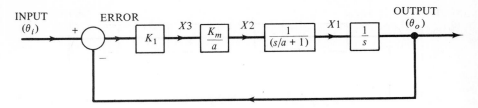

Fig. 2.13   Programming block diagram for Example 2.2.

repeated in Fig. 2.13. One can easily relate the input-output variables of the blocks in Fig. 2.13 to the statements given in the program listing. Since the arrangement of the cards is not unique, one may start with any block and proceed to describe the system.

Several new terms have been introduced in this problem that were not used in Example 2.1. These terms include PRTPLT, OUTDEL, STEP and LABEL. A general definition of these terms will be given before examining the computer output for Example 2.2.

PRTPLT. An example of a PRTPLT output is given in Fig. 2.14. PRTPLT statements are used to specify those variables that the user desires to be printer-plotted. Continuation cards are not allowed, but up to ten separate PRTPLT statements containing 100 variables can be used. A card giving a typical PRTPLT request might appear as

PRTPLT X1, OUTPUT, ERROR

In this case, separate printer plots will be made for X1, OUTPUT, and ERROR. Another useful feature of the PRTPLT request is the capability of printing up to a maximum of three additional problem variables on the same page as the printer-plotted variable. These additional variables appear on the far right side of the page in column format. The statement

PRTPLT X1(ERROR, OUTPUT), X2, OUTPUT

means that separate printer plots will be made for X1, X2, and OUTPUT. In addition, the values of ERROR and OUTPUT will be printed on the right side of the pages that give the printer-plot of X1.

The range and scale of a printer-plotted variable can be controlled as follows:

PRTPLT	SIGX(0.2, 3.0)	statement (1)
PRTPLT	SIGX(, 4.0)	statement (2)
PRTPLT	SIGX(-1.2,)	statement (3)
PRTPLT	SIGX(0.4, , XOUT)	statement (4)

The information given by each statement is:

Statement (1)—The printer-plot of SIGX will not show values less than 0.2 or greater than 3.0. Consequently, the scale of the printer-plot will be adjusted for a maximum of 3.0 and a minimum of 0.2.

Statement (2)—The printer-plot of SIGX does not have a specified lower bound but the upper bound is 4.0.

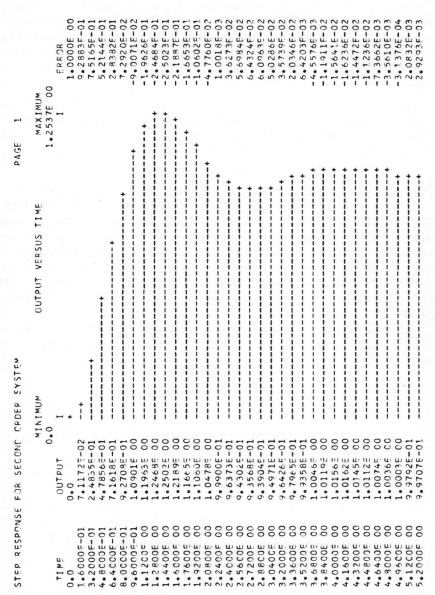

Fig. 2.14 Printer-plot for step response of Example 2.2.

Statement (3)—SIGX will not have printer-plotted values less than −1.2. The upper bound is unspecified.

Statement (4)—The printer-plot of SIGX has a lower bound of 0.4 and the upper bound unspecified. Also, XOUT will be listed on the right side of the SIGX printer-plot. Finally, printer-plots can only be made using the independent variable (usually TIME) along the abscissa and the requested dependent variable as the ordinate. In other words, it is not possible to plot two dependent variables simultaneously as in a phase-plane plot.

OUTDEL. OUTDEL controls the output increment of the independent variable for printer-plots. If printer-plot points are desired every 0.04 time-units, we use OUTDEL = 0.04. The OUTDEL specification is included on the TIMER card. A typical example of an OUTDEL assignment is

$$\text{TIMER FINTIM} = 10.0, \text{ OUTDEL} = 0.02$$

If printer-plots are requested but OUTDEL is not specified, the program will use an OUTDEL = PRDEL. If PRDEL is not given in the program, OUTDEL is set to FINTIM/100. When OUTDEL and PRDEL are both specified, the smaller of the two will be adjusted to be a submultiple of the larger. A useful guide in selecting the OUTDEL interval is that each page of OUTPUT will contain fifty lines. Thus, if the user desires the printer-plot to be contained on one page only, the OUTDEL value should be OUTDEL = FINTIM/50. In general, the OUTDEL selection, based on the number of pages for the plot, is given by OUTDEL = FINTIM/(50 × number of output pages).

STEP. STEP is one of several signal sources available in CSMP. This particular signal provides a unit-step forcing-function and is described in Table 2.3.

Table 2.3

Formulation for a Unit Step Signal

General Form		*Function*
Definition:	Unit-step input or forcing function.	$Y = 0 \quad t < T$ $Y = 1 \quad t \geq T$
	$Y = \text{STEP}(T)$	Y (step plot: 1.0, T, t)

A unit-step input applied at $t = 0$ is expressed by

$$\text{XIN} = \text{STEP}(0.0)$$

If the input forcing-function is a step of weight 3 applied at $t = 2.4$, the expression is written as

$$\text{XIN} = 3.0*\text{STEP}(2.4)$$

LABEL. LABEL cards are used for placing an open-format title or heading at the top of each page of printer-plotted output. The first non-blank character following LABEL must be alphabetic or numeric. After the first non-blank character, special symbols (=, /, *, etc.) may be used within the heading. As with PRTPLT cards, continuation cards are not permitted, but up to ten LABEL cards may be used for each simulation. Consider the following illustration.

```
PRTPLT ROVER, SIGMA
LABEL RESPONSE OF SYSTEM MODEL — PROBLEM 1
PRTPLT X1
LABEL STATE X1 WITH INITIAL CONDITION X2 = 2.0
```

Since the first LABEL statement is associated with the first PRTPLT statement, the second with the second, and so on, the heading, RESPONSE OF SYSTEM MODEL–PROBLEM 1, will appear at the top of each page of printer-plotted output for the variables ROVER and SIGMA. The printer-plot of X1 will have a heading of STATE X1 WITH INITIAL CONDITION X2 = 2.0. If the number of PRTPLT cards exceeds the number of LABEL cards, the excess print-plots will not have headings.

Refer to Fig. 2.12 and consider the following portion of the listing:

```
TIMER FINTIM = 8.0, OUTDEL = 0.16, PRDEL = 0.04
PRTPLT OUTPUT (ERROR), ERROR (, 0.5)
LABEL STEP RESPONSE FOR SECOND ORDER SYSTEM
PRINT OUTPUT, OUTEXT
```

The TIMER card gives the information that the simulation will run for 8.0 sec, printer-plotted output will occur every 0.16 sec, and printed output every 0.04 sec. We note that with a FINTIM of 8.0 and OUTDEL of 0.16, there will be one page for each printer-plotted variable (number of pages = FINTIM/(OUTDEL *50)).

The PRTPLT card specifies printer plots for OUTPUT and ERROR. The plot for OUTPUT will have ERROR given in tabular form at the right side of the page. The plot of ERROR will have an upper bound of 0.5. The heading, STEP RESPONSE FOR SECOND ORDER SYSTEM will appear at the top of each page of printer-plotted output. The printer-plot for OUTPUT is given in Fig. 2.14 while Fig. 2.15 shows the plot for ERROR.

The PRINT card designates that OUTPUT and OUTEXT will be printed out. As indicated in the program listing of Fig. 2.12, the exact analytical solution (OUTEXT) is calculated. Notice also in Fig. 2.16 the very close agreement between the CSMP numerical solution (OUTPUT) and the exact analytical solution. This is an indication of the accuracy that can be expected.

Example 2.3 *Solution of the Van der Pol Equation*

This example illustrates the ease of solving a nonlinear differential equation for various parameter values. The particular example is the well-known Van der Pol equation, which is shown below with the selected initial conditions.

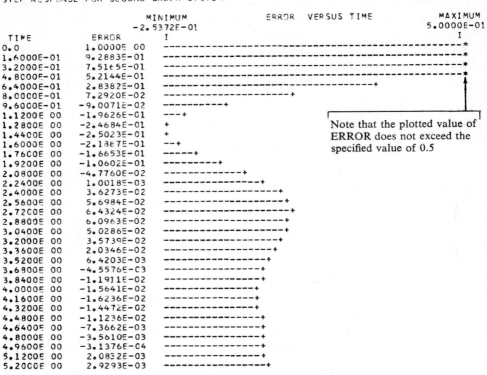

STEP RESPONSE FOR SECOND ORDER SYSTEM   PAGE   1

Fig. 2.15   System error response for Example 2.2.

$$\ddot{X} + E(X^2 - 1)\dot{X} + X = 0 \tag{2.6}$$

$$X(0) = 2.0$$

$$\dot{X}(0) = 0.0$$

A helpful first step is to change the second-order differential equation to an equivalent set of first-order equations, easily accomplished as

$$\dot{X} = XD \tag{2.7}$$

$$\ddot{X} = XDD = -E(X^2 - 1)XD - X \tag{2.8}$$

The program for solving Eqs. (2.7) and (2.8) is straight forward, as shown in Fig. 2.17. An ordinary FORTRAN statement is used to calculate XDD and two INTGRL functions are used to integrate XDD and XD to obtain XD and X, respectively.

In this example the following PARAMETER card is used to make a sequence of simulation runs for five different values of the constant E.

PARAMETER E = (0.05, 0.5, 2.0, 10.0, 50.0)

One separate run is made for each value of E. Figs. 2.18 and 2.19 show the printer-plot outputs for E = 2.0 and E = 50.0. Note that the parameter value E is listed at the beginning of each plot and on each page of PRINT output.

PRINTED LISTING OF "OUTPUT" (SIMULATION) AND "OUTEXT" (ANALYTICAL)

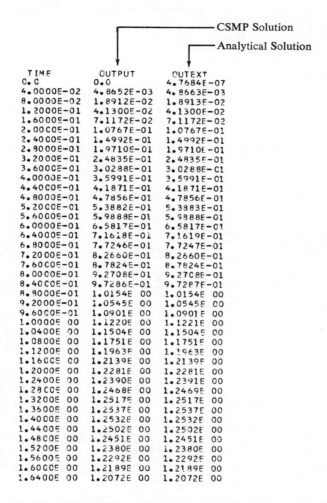

Fig. 2.16   Diagram illustrating printer output for Example 2.2.

When making a sequence of runs using a PARAMETER card, all printer-plots for the same variable will have a common scale. This means that some plots may not cover the full range (example: Fig. 2.19); however, a common scale has the advantage that a direct comparison can be made between plots of sequential runs.

In Example 2.1, a description of the CONSTANT statement was given. The application of both the INCON and PARAMETER cards is identical to the CONSTANT statement. They are completely interchangeable but are lettered differently as a matter of convenience to the user. The description given below is, therefore, equally valid for all three cards.

```
LABEL    SOLUTION OF THE VAN DER POL EQUATION

*    THE FOLLOWING PARAMETER STATEMENT RECYCLES THE PROGRAM FOR THE FIVE
*    DIFFERENT VALUES OF E.
     PARAMETER  E = (0.05, 0.5, 2.0, 10.0, 50.0)

     INCON XI =   2.0,   XDI = 0.0
     XDD = -E*(X*X - 1.0)*XD - X
     X = INTGRL(XI,XD)
     XD = INTGRL(XDI,XCC)
     PRTPLT X (XD,XDD)
     TIMER FINTIM = 10.0,   OUTDEL = 0.25
END
STOP
ENDJOB
```

Fig. 2.17 Program to solve Van der Pol equation.

```
SOLUTION OF THE VAN DER POL EQUATION                          PAGE  1

                       MINIMUM           X   VERSUS TIME     MAXIMUM
                       -2.0197E 00       E   = 2.0000E 00    2.0174E 00
   TIME      X         I                                             I    XD          XDD
0.0          2.0000E 00 -------------------------------------------+     0.0         -2.0000E 00
2.5000E-01   1.9597E 00 -------------------------------------------+     -2.6104E-01 -4.7673E-01
5.0000E-01   1.8839E 00 -------------------------------------------+     -3.3346E-01 -1.8382E-01
7.5000E-01   1.7955E 00 -------------------------------------------+     -3.7194E-01 -1.4120E-01
1.0000E 00   1.6980E 00 ------------------------------------------+      -4.0872E-01 -1.5853E-01
1.2500E 00   1.5905E 00 -----------------------------------------+       -4.5342E-01 -2.0331E-01
1.5000E 00   1.4701E 00 ----------------------------------------+        -5.1291E-01 -2.7892E-01
1.7500E 00   1.3320E 00 ---------------------------------------+         -5.9721E-01 -4.0725E-01
2.0000E 00   1.1679E 00 -------------------------------------+           -7.2507E-01 -6.4001E-01
2.2500E 00   9.6270E-01 -----------------------------------+             -9.3568E-01 -1.0997E 00
2.5000E 00   6.8613E-01 --------------------------------+                -1.3174E 00 -2.0806E 00
2.7500E 00   2.7397E-01 ---------------------------+                     -2.0640E 00 -4.0920E 00
3.0000E 00   -3.9365E-01 -------------------+                            -3.3366E 00 -5.2454E 00
3.2500E 00   -1.3112E 00 --------+                                       -3.4977E 00  6.3417E 00
3.5000E 00   -1.9025E 00 -+                                              -1.1781E 00  8.0742E 00
3.7500E 00   -2.0197E 00 +                                               -1.5696E-02  2.1164E 00
4.0000E 00   -1.9815E 00 +                                               2.5613E-01  4.8245E-01
4.2500E 00   -1.9069E 00 -+                                              3.2795E-01  1.7775E-01
4.5000E 00   -1.8201E 00 --+                                             3.6465E-01  1.3333E-01
4.7500E 00   -1.7247E 00 ---+                                            3.9919E-01  1.4817E-01
5.0000E 00   -1.6199E 00 ----+                                           4.4076E-01  1.8815E-01
5.2500E 00   -1.5033E 00 ------+                                         4.9547E-01  2.5489E-01
5.5000E 00   -1.3704E 00 --------+                                       5.7187E-01  3.6616E-01
5.7500E 00   -1.2143E 00 ---------+                                      6.8561E-01  5.6372E-01
6.0000E 00   -1.0219E 00 -----------+                                    8.6868E-01  9.4486E-01
6.2500E 00   -7.6847E-01 ---------------+                                1.1921E 00  1.7647E 00
6.5000E 00   -4.0125E-01 ------------------+                             1.8165E 00  3.4494E 00
6.7500E 00   1.8610E-01 ------------------------+                        2.9768E 00  5.5613E 00
7.0000E 00   1.0705E 00 ---------------------------------+               3.7796E 00 -2.1747E 00
7.2500E 00   1.8063E 00 ------------------------------------------+      1.7696E 00 -9.8139E 00
7.5000E 00   2.0137E 00 --------------------------------------------+    1.8361E 00 -3.1357E 00
7.7500E 00   1.9970E 00 --------------------------------------------+    -2.1868E-01 -6.9013E-01
8.0000E 00   1.9278E 00 ------------------------------------------+      -3.1531E-01 -2.1472E-01
8.2500E 00   1.8435E 00 ------------------------------------------+      -3.5592E-01 -1.3607E-01
8.5000E 00   1.7504E 00 ----------------------------------------+        -3.8979E-01 -1.4150E-01
8.7500E 00   1.6482E 00 ---------------------------------------+         -4.2895E-01 -1.7555E-01
9.0000E 00   1.5349E 00 -------------------------------------+           -4.7959E-01 -2.3427E-01
9.2500E 00   1.4068E 00 ------------------------------------+            -5.4923E-01 -3.3123E-01
9.5000E 00   1.2577E 00 ----------------------------------+              -6.5109E-01 -5.0014E-01
9.7500E 00   1.0765E 00 --------------------------------+                -8.1150E-01 -8.1869E-01
1.0000E 01   8.4246E-01 ------------------------------+                  -1.0878E 00 -1.4740E 00
```

Fig. 2.18 Printer-plot output for solution of Van der Pol equation E = 2.0.

PARAMETER, CONSTANT, INCON. These data statements are used for assigning numerical values. Only the first five letters are required to signify their use. Thus it is not necessary to write PARAMETER in full since PARAM is acceptable. This also applies to many other CSMP statements. For example, PRTPLOT may be used rather than PRTPLT.

Regarding PARAMETER, CONSTANT, and INCON; at least one blank must follow the label card while the remaining entries must have the general format

```
SOLUTION OF THE VAN DER POL EQUATION                              PAGE   1

                        MINIMUM        X     VERSUS TIME      MAXIMUM
                       -2.0197E 00     E    = 5.0000E 01      2.0174E 00
   TIME        X        I                                      I      XD          XDD
 0.0         2.0000E 00 -----------------------------------------------+   0.0        -2.0000E 00
 2.5000E-01  1.9967E 00 -----------------------------------------------+  -1.3368E-02 -2.5845E-04
 5.0000E-01  1.9934E 00 -----------------------------------------------+  -1.3156E-02 -3.7365E-02
 7.5000E-01  1.9900E 00 -----------------------------------------------+  -1.3320E-02 -1.8507E-02
 1.0000E 00  1.9866E 00 -----------------------------------------------+  -1.3462E-02 -3.1214E-03
 1.2500E 00  1.9833E 00 -----------------------------------------------+  -1.3293E-02 -3.3574E-02
 1.5000E 00  1.9799E 00 -----------------------------------------------+  -1.3544E-02 -2.4862E-03
 1.7500E 00  1.9765E 00 -----------------------------------------------+  -1.3564E-02 -5.3825E-03
 2.0000E 00  1.9731E 00 -----------------------------------------------+  -1.3620E-02 -2.9764E-03
 2.2500E 00  1.9696E 00 -----------------------------------------------+  -1.3643E-02 -5.4636E-03
 2.5000E 00  1.9662E 00 -----------------------------------------------+  -1.3708E-02 -1.8091E-03
 2.7500E 00  1.9628E 00 -----------------------------------------------+  -1.3756E-02 -8.8120E-04
 3.0000E 00  1.9593E 00 -----------------------------------------------+  -1.3802E-02 -1.8787E-04
 3.2500E 00  1.9559E 00 -----------------------------------------------+  -1.3831E-02 -2.0018E-03
 3.5000E 00  1.9524E 00 -----------------------------------------------+  -1.3870E-02 -2.4099E-03
 3.7500E 00  1.9489E 00 -----------------------------------------------+  -1.3918E-02 -1.6050E-03
 4.0000E 00  1.9454E 00 -----------------------------------------------+  -1.3961E-02 -1.5526E-03
 4.2500E 00  1.9419E 00 -----------------------------------------------+  -1.3978E-02 -5.2452E-03
 4.5000E 00  1.9384E 00 -----------------------------------------------+  -1.4058E-02 -2.2030E-04
 4.7500E 00  1.9349E 00 -----------------------------------------------+  -1.4073E-02 -4.2686E-03
 5.0000E 00  1.9313E 00 -----------------------------------------------+  -1.4147E-02 -2.8706E-04
 5.2500E 00  1.9278E 00 -----------------------------------------------+  -1.4190E-02 -4.9400E-04
 5.5000E 00  1.9242E 00 -----------------------------------------------+  -1.4235E-02 -5.8270E-04
 5.7500E 00  1.9206E 00 -----------------------------------------------+  -1.4125E-02 -2.1666E-02
 6.0000E 00  1.9171E 00 -----------------------------------------------+  -1.4331E-02 -2.6703E-04
 6.2500E 00  1.9135E 00 -----------------------------------------------+  -1.4322E-02 -7.7343E-03
 6.5000E 00  1.9099E 00 -----------------------------------------------+  -1.4426E-02 -2.0790E-04
 6.7500E 00  1.9062E 00 -----------------------------------------------+  -1.4474E-02 -2.0504E-04
 7.0000E 00  1.9026E 00 -----------------------------------------------+  -1.4523E-02 -2.0504E-04
 7.2500E 00  1.8990E 00 -----------------------------------------------+  -1.4512E-02 -7.9346E-03
 7.5000E 00  1.8953E 00 -----------------------------------------------+  -1.4552E-02 -9.2669E-03
 7.7500E 00  1.8917E 00 -----------------------------------------------+  -1.4672E-02 -2.1172E-04
 8.0000E 00  1.8880E 00 -----------------------------------------------+  -1.4677E-02 -6.0024E-03
 8.2500E 00  1.8843E 00 -----------------------------------------------+  -1.4740E-02 -4.5366E-03
 8.5000E 00  1.8806E 00 -----------------------------------------------+  -1.4809E-02 -2.4033E-03
 8.7500E 00  1.8769E 00 -----------------------------------------------+  -1.4879E-02 -2.2888E-04
 9.0000E 00  1.8731E 00 -----------------------------------------------+  -1.4893E-02 -5.0983E-03
 9.2500E 00  1.8694E 00 -----------------------------------------------+  -1.4968E-02 -2.4109E-03
 9.5000E 00  1.8656E 00 -----------------------------------------------+  -1.5030E-02 -1.5574E-03
 9.7500E 00  1.8618E 00 -----------------------------------------------+  -1.5072E-02 -3.0708E-03
 1.0000E 01  1.8581E 00 -----------------------------------------------+  -1.5145E-02 -1.0271E-03
```

Fig. 2.19 Printer-plot output for solution of Van der Pol equation E = 50.

of variable name, equal sign, and numerical assignment. Additional assignments are made on the same card by placing a comma after the numerical value. Any number of assignments may appear on a card. Up to eight continuation cards are allowable by using three consecutive periods at the end of the card. Valid applications are

> PARAMETER AX = 6.0, MW = 2.0, TAU = 1.0, . . .
> RHO = 0.1, GAMMA = 6.2E − 03
> PARAM EDGE = 3.2, START = 6.4, LAST = −0.5
> INCON ICX1 = 0.115, ICX2 = 0.0
> CONST ALPHA = 1.35, BETA = −6.134
> CONSTANT X = 5.6, Y = −45.78

where all of these cards are used in the same program. The results of the simulation would have been unchanged by using

> CONSTANT AX = 6.0, MW = 2.0, TAU = 1.0, . . .
> RHO = 0.1, GAMMA = 6.2E − 03, ICX1 = 0.115, . . .
> ICX2 = 0.0, ALPHA = 1.35, BETA = −6.134, . . .
> X = 5.6, Y = −45.78, EDGE = 3.2, START = 6.4, . . .
> LAST = −0.5

In the above, CONSTANT could be replaced by either INCON or PARAMETER.

Calculations are not allowed in a data statement. An example of an invalid statement is shown below.

CONSTANT X = 5.8, Y = 4.2/3.14

                                        └────Illegal operation

As illustrated in Example 2.3, an attractive feature of these data statements is their capability of making sequential runs. A typical statement is

INCON ICX1 = (0.0, 2.0, −1.2, 6.0), ICX2 = 2.3

The program will make four simulation runs in which ICX1 is assigned the values appearing in the parentheses in the order given from left to right. The value of ICX2 = 2.3 is used for each run. As a slight modification one may use

PARAMETER ICX1 = (0.0, 3∗0.2, 1.4, 2∗2.0)

which will result in seven simulation runs with the values of ICX1 being 0.0, 0.2, 0.4, 0.6, 1.4, 2.0, 4.0. A maximum of fifty runs can be made in any sequence.

Only one variable can be used to make sequential runs. Examples of *invalid* data statements which attempt to make sequential runs using two variables for two different programs are

PARAMETER X = (2.0, 5.0, 9.0), Y = (3.0, 4.0, 7.0)
CONSTANT X = (2.0, 5.0, 9.0)
PARAMETER Y = (3.0, 4.0, 7.0)

Example 2.4   *Transient Temperature Response*

In some simulations it is necessary to use functional relationships that are contained in tabular or graphic form. Fig. 2.20 shows an example of a graphic relationship between

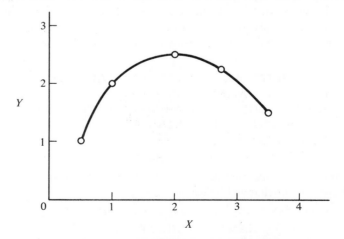

Fig. 2.20 Graphic relationship between X and Y.

X and *Y* that can be represented by using a FUNCTION statement. Points on the curve of Fig. 2.20 are entered on a FUNCTION card as illustrated below.

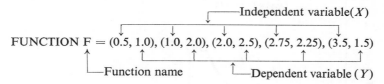

The symbol F represents the chosen name of this particular function but any other valid symbol can be used. The independent variable for each point is always listed first starting with the smallest algebraic value. This variable must monotonically increase to assure that the slope is neither infinite nor the function multivalued. Increment size of the independent variable may be unequal as shown in the above example. Parentheses can be used to group the coordinates of each point but are not required. Commas must separate *all* numbers. The list of values can be extended to additional cards by the use of three consecutive periods (. . .). There are no specific restrictions on the number of points per function.

The two types of statements which use the data contained on the previous FUNCTION card are

$$Y = AFGEN(F, X)$$
$$Y = NLFGEN(F, X)$$

where Y is the dependent variable (ordinate) and X is the independent variable (abscissa).

The AFGEN statement provides linear interpolation between consecutive data points. In effect, the curve is represented by a series of straight-line segments. The more closely spaced the points, the more accurate the representation of the function.

NLFGEN uses a Lagrange quadratic interpolation between points. Consequently, the NLFGEN element cannot accurately represent functions containing abrupt changes. An example of this is shown by Fig. 2.21.

Notice that the error in using the NLFGEN statement is greatest where the data points are widely separated and where the curve changes abruptly.

Both AFGEN and NLFGEN functions can be treated as ordinary variables and, consequently, included as part of structure statements. An example is

$$Z = 3.5*AFGEN(F, X)/(3.8 - NLFGEN(L, TIME))$$

If the value of the input variable is outside the specified range of the function, a diagnostic message is printed and the simulation proceeds without interruption. The value returned by AFGEN or NLFGEN is either the first or last dependent variable in the FUNCTION statement. For example, if the input value of X in the graphical relationship of Fig. 2.20 were 4.0, the output would be 1.5. In a similar manner, the output would be 1.0 for all input values less than 0.5.

This example illustrates the simulation of the transient temperature response of a small hot copper cylinder which is quenched in water. Special emphasis is placed on the use of the FUNCTION statement in conjunction with the AFGEN function.

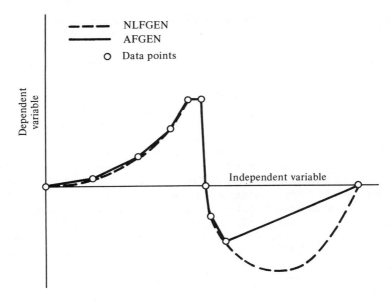

- – – – NLFGEN
- ———— AFGEN
- O Data points

Fig. 2.21 Comparison of curve fit using AFGEN and NLFGEN functions.

Figure 2.22 defines the physical setup of the problem.

Applying the basic law of heat conduction and assuming uniform cylinder temperature and constant water temperature yields

$$\text{TDOT} = \dot{T} = \text{H}*\text{AREA}*(\text{TA} - \text{T})/(\text{RO}*\text{C}*\text{V}) \tag{2.9}$$

$$\text{T} = \int_0^t \dot{T} \, dt + \text{TINITL} \tag{2.10}$$

where
- T = temperature of cylinder
- TA = temperature of water, 530°R
- TINITL = initial temperature of cylinder, 1960°R
- AREA = surface area of cylinder = $\pi(D^2/2 + DL)$
- RO = density of copper, 559 lb/ft³
- C = specific heat of copper, 0.0915 Btu/lb-°R
- V = volume of cylinder = $\pi D^2 L/4$
- D = diameter of cylinder, 0.02 ft
- L = length of cylinder, 0.1 ft
- H = overall heat transfer coefficient, a function of temperature as shown in Table 2.4.

Data for the heat transfer coefficient is taken from Table 2.4 and entered into the program on a FUNCTION statement called HBOIL.

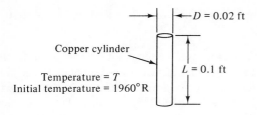

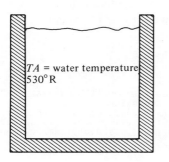

Fig. 2.22 Schematic showing the quenching of a hot copper cylinder in water.

Table 2.4

Boiling Heat Transfer Coefficient as a Function of Temperature

Temperature (°R)	Heat Transfer Coefficient (Btu/ft²-hr-°R)
672.1	114
673.0	278
687.0	667
702.0	3680
737.0	13200
822.0	3680
1082	880
1672	880
2672	1140

Notice that this program contains all three segments: Initial, Dynamic, and Terminal. The use of the three segments is not required for this particular program, but is included in order to show a logical classification of the structure statements. The Initial segment contains both the FUNCTION statement and the CONSTANT card which are used to specify the parameters of the system, Since the calculations used to determine AREA and V are only required once, they are included in the Initial segment.

In the Dynamic segment, an AFGEN statement is used to specify H, a FORTRAN type statement is used to calculate TDOT, and the INTGRL function is used to integrate TDOT to obtain T.

The run is terminated either by reaching the finish time (FINTIM) or by the control of the FINISH card.

FINISH. The FINISH label allows the user to terminate a run prior to the time as specified by FINTIM. A run can be ended when any dependent variable first crosses or reaches a given bound. In the program of Fig. 2.23, the run will terminate when T reaches or crosses the value of 682°R, or when TIME reaches FINTIM. If more than one FINISH card is included, only the last card is considered. A single FINISH statement can contain a maximum of ten terminating conditions. Continuation to successive cards is permitted.

```
LABEL      QUENCHING OF A HOT COPPER CYLINDER
INITIAL
       CONSTANT RO = 559.0, C = 0.0915, TA = 530.0, L = 0.1, D = 0.02,...
       TINITL = 1960.0, PI = 3.14159
*    THE FOLLOWING VALUES LISTED ON THE FUNCTION CARD ARE THE HEAT TRANSFER
*    COEFFICIENT AS A FUNCTION OF TEMPERATURE
FUNCTION  HBOIL = (672.1,114.0),(673.0,278.0),(687.0,667.0),          ...
                  (702.0,3680.0),(737.0,13200.0),(822.0,3680.0),      ...
                  (1082.0,880.0),(1672.0,880.0),(2672.0,1140.0)
       AREA = PI*(D*D/2.0 + D*L)
       V = PI*D*D*L/4.0
DYNAMIC
       H = AFGEN(HBOIL,T)
       TDOT = H*AREA*(TA - T)/(RO*C*V)
       T = INTGRL(TINITL,TDOT)
TERMINAL
       TIMER  FINTIM = 0.006,  OUTDEL = 0.00001
       FINISH  T = 682.0
       PRTPLT T  (H)
END
STOP
ENDJOB
```

Fig. 2.23 Program to calculate the transient temperature of a copper cylinder.

The FINISH statement

$$\text{FINISH} \quad X = -3.0, Y = Z$$

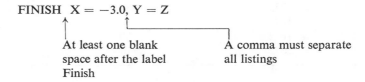

At least one blank space after the label Finish

A comma must separate all listings

shows two conditions for ending the run. The first condition (X = − 3.0) will terminate the run when X first reaches or crosses − 3.0. The second requirement (Y = Z) will end the simulation when Y = Z, or when Y − Z first changes sign. The run will be terminated when either of the conditions are met.

Since the FINISH conditions are checked at each integration interval, the run may be terminated at a time that is not a multiple of PRDEL or OUTDEL. If this happens, printing will occur at the time when the FINISH condition is met.

Figure 2.24 shows the PRTPLT output for the time history of the temperature

```
QUENCHING OF A HOT COPPER CYLINDER                                    PAGE   1

                        MINIMUM           T    VERSUS TIME        MAXIMUM
                        6.7907E 02                                1.9600E 03
 TIME        T          I                                         I       H
 0           1.9600E 03 -------------------------------------------------+   9.5488E 02
 .0000E-05   1.9029E 03 -----------------------------------------------+     9.4003E 02
 .0000E-05   1.8489E 03 ----------------------------------------------+      9.2599E 02
 .0000E-05   1.7978E 03 -------------------------------------------- --+     9.1270E 02
 .0000E-05   1.7493E 03 -------------------------------------------+        9.0010E 02
 .0000E-05   1.7033E 03 -----------------------------------------+          8.8814E 02
 .0000E-05   1.6596E 03 ----------------------------------------+           8.8000E 02
 .0000E-05   1.6176E 03 --------------------------------------+             8.8000E 02
 .0000E-05   1.5772E 03 ------------------------------------+               8.8000E 02
 .0000E-05   1.5383E 03 ----------------------------------+                 8.8000E 02
 .0000E-04   1.5009E 03 --------------------------------+                   8.8000E 02
 .1000E-04   1.4648E 03 ------------------------------+                     8.8000E 02
 .2000E-04   1.4301E 03 ----------------------------+                       8.8000E 02
 .3000E-04   1.3967E 03 --------------------------+                         8.8000E 02
 .4000E-04   1.3645E 03 ------------------------+                           8.8000E 02
 .5000E-04   1.3335E 03 ----------------------+                             8.8000E 02
 .6000E-04   1.3036E 03 ---------------------+                              8.8000E 02
 .7000E-04   1.2749E 03 -------------------+                                8.8000E 02
 .8000E-04   1.2472E 03 -----------------+                                  8.8000E 02
 .9000E-04   1.2206E 03 ----------------+                                   8.8000E 02
 .0000E-04   1.1949E 03 --------------+                                     8.8000E 02
 .1000E-04   1.1702E 03 ------------+                                       8.8000E 02
 .2000E-04   1.1465E 03 -----------+                                        8.8000E 02
 .3000E-04   1.1236E 03 ---------+                                          8.8000E 02
 .4000E-04   1.1015E 03 --------+                                           8.8000E 02
 .5000E-04   1.0802E 03 ------+                                             8.9888E 02
 .6000E-04   1.0566E 03 -----+                                             1.1537E 03
 .7000E-04   1.0279E 03 ----+                                              1.4629E 03
 .8000E-04   9.9394E 02 ---+                                               1.8283E 03
 .9000E-04   9.5509E 02 --------+                                          2.2468E 03
 .0000E-04   9.1216E 02 ------+                                            2.7090E 03
 .1000E-04   8.6658E 02 -----+                                            3.1999E 03
 .2000E-04   8.2011E 02 ---+                                              3.8917E 03
 .3000E-04   7.3674E 02 --+                                              1.3130E 04
 .4000E-04   6.9375E 02 +                                                2.0237E 03
 .5000E-04   6.8613E 02 +                                                6.4290E 02
 .6000E-04   6.8224E 02 +                                                5.3473E 02
 .7000E-04   6.7907E 02 +                                                4.4664E 02
```

Fig. 2.24 Printer-plot output for temperature of copper cylinder.

of the copper cylinder. Notice that the FINISH card controlled the termination of the program at TIME = 0.00037 hr.

Example 2.5   *Three-Stage Saturn Vehicle*

For some types of simulations it is necessary to use conditional branching statements to control the sequence of calculations. An example of this type problem is the simulated launch of a three-stage Saturn rocket in which each stage requires different equations to describe the physical parameters. The FORTRAN IF statement can readily be used to insure that calculations are performed in proper order.

Figure 2.25 gives the significant system parameters and also shows the forces acting on the vehicle.

Applying Newton's second law and neglecting all minor effects and disturbances, the equations of motion for a rocket traveling in a radial direction from the center of the earth are

$$m\dot{V} = \dot{m}V_o - mg - \tfrac{1}{2}\rho C_d A V^2 \tag{2.11}$$

and

$$\text{ALT} = \int_o^t V\,dt \tag{2.12}$$

Program Symbols
(Fig. 2.26)

where m = mass of total rocket in slugs   MASS

V = velocity in ft/sec   VEL

$\dot{V}$ = acceleration in ft/sec^2 ACC

$\dot{m}$ = mass flow rate of propellant in slug/sec (constant for each stage)   MDOT

V_o = outflow velocity of propellant relative to rocket in ft/sec   VOUT

g = acceleration of gravity which is given by   G

$$g = 32.17\left[\frac{r_e}{r_e + \text{ALT}}\right]^2$$

r_e = radius of earth in feet   RADERH

ALT = altitude of rocket in feet   ALT

ρ = density of air in slug/ft^3 RO
An approximate value of ρ is given by the

$$\rho = 0.00238\,e^{-\text{ALT}/24000}$$

$C_d A$ = cross-sectional area of the rocket times the drag coefficient in ft^2 DRAG

While Eqs. (2.11) and (2.12) are valid for each stage, the parameters m, $\dot{m}$, V_o, and $C_d A$ will change for each burn.

Figure 2.26 shows the CSMP program listing for the simulation of the launch. All three segments (Initial, Dynamic, Terminal) are included in this program. The use of the three segments is good programming practice since it separates the different types of program statements.

The Initial segment is used to specify the constants in the program and makes a one-time calculation to find the radius of the earth RADERH in feet.

Since the parameters in Eq. (2.11) depend on the particular burn stage, FORTRAN IF statements are utilized to transfer control to the appropriate group of structure statements. To allow the use of logic branching statements, the first part of the Dynamic segment is changed to a nosort section.

By examining the program structure, one can make the following observations.

If the flight time TIME is less than the burn time of the first stage BURNT1, control is transferred to statement 1. The instantaneous mass MASS of the total moving rocket is calculated and the appropriate values for the propellant-mass-flow rate MDOT, drag-coefficient times cross-sectional area DRAG, and propellant-outflow velocity VOUT for the first stage are specified. Control is then transferred to statement 3 which is a FOR-

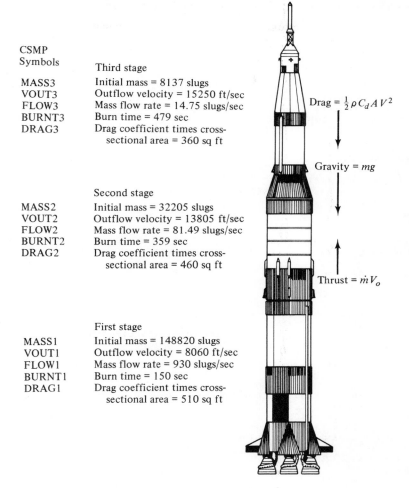

CSMP
Symbols

Third stage

MASS3   Initial mass = 8137 slugs
VOUT3   Outflow velocity = 15250 ft/sec
FLOW3   Mass flow rate = 14.75 slugs/sec
BURNT3   Burn time = 479 sec
DRAG3   Drag coefficient times cross-
      sectional area = 360 sq ft

$\text{Drag} = \frac{1}{2}\rho C_d A V^2$

Gravity = mg

Second stage

MASS2   Initial mass = 32205 slugs
VOUT2   Outflow velocity = 13805 ft/sec
FLOW2   Mass flow rate = 81.49 slugs/sec
BURNT2   Burn time = 359 sec
DRAG2   Drag coefficient times cross-
      sectional area = 460 sq ft

Thrust = $\dot{m}V_o$

First stage

MASS1   Initial mass = 148820 slugs
VOUT1   Outflow velocity = 8060 ft/sec
FLOW1   Mass flow rate = 930 slugs/sec
BURNT1   Burn time = 150 sec
DRAG1   Drag coefficient times cross-
      sectional area = 510 sq ft

Fig. 2.25 Saturn three-stage rocket.

TRAN CONTINUE statement. After the CONTINUE statement, the Dynamic segment is changed to a sort section. Calculations are then made for acceleration of gravity G, density of air RO, and acceleration of the rocket ACC. INTGRL statements are used to find the velocity VEL and altitude of the rocket ALT. The final two statements in the Dynamic segment are used to calculate the velocity in miles per hour MPH and the altitude in miles MILES. Control is then transferred back to the beginning of the Dynamic segment where the iterative procedure is repeated.

     The final portion of the Dynamic segment was changed to a sort section to eliminate the possibility of having calculations performed in the wrong order.

     The value of FINTIM is set equal to the total burn time of all three stages and will consequently terminate the program at the end of the third-stage burn.

```
LABEL    SIMULATION OF THE LAUNCH OF A SATURN V THREE STAGE ROCKET
INITIAL
        CONSTANT MASS1 = 148820.0, MASS2 = 32205.0, MASS3 = 8137.0,    ...
        FLOW1 = 930.0, FLOW2 = 81.49, FLOW3 = 14.75, VOUT1 = 8060.0,   ...
        VOUT2 = 13805.0, VOUT3 = 15250.0, DRAG1 = 510.0, DRAG2 = 460.0,...
        DRAG3 = 360.0, BURNT1 = 150.0, BURNT2 = 359.0
        RADERH = 3960.0*5280.0
DYNAMIC
   NOSORT
        IF(TIME.LE.BURNT1) GO TO 1
        IF(TIME.LT.(BURNT1 + BURNT2)) GO TO 2
*    CALCULATIONS FOR THIRD STAGE
        MASS = MASS3 - FLOW3*(TIME - BURNT1 - BURNT2)
        MDOT = FLOW3
        DRAG = DRAG3
        VOUT = VOUT3
        GO TO 3
*    CALCULATIONS FOR FIRST STAGE
     1 MASS = MASS1 + MASS2 + MASS3 - FLOW1*TIME
        MDOT = FLOW1
        DRAG = DRAG1
        VOUT = VOUT1
        GO TO 3
*    CALCULATIONS FOR SECOND STAGE
     2 MASS = MASS2 + MASS3 - FLOW2*(TIME - BURNT1)
        MDOT = FLOW2
        DRAG = DRAG2
        VOUT = VOUT2
     3 CONTINUE
   SORT
        G = 32.17*((RADERH/(RADERH + ALT))**2)                    GRAVITY
        RO = 0.00238*EXP(-ALT/24000.0)                           DENSITY
        ACC = (MDOT*VOUT - MASS*G - 0.5*DRAG*RO*VEL*VEL)/MASS
        VEL = INTGRL(0.0,ACC)
        ALT = INTGRL(0.0,VEL)
        MPH = VEL*60.0/88.0
        MILES = ALT/5280.0
TERMINAL
        TIMER FINTIM = 988.0, PRDEL = 4.0, OUTDEL = 20.0
        PRINT MILES, MPH, ACC,G, RO
        PRTPLT MPH  (MILES,ACC)
*    THE FOLLOWING FOUR STATEMENTS ARE USED TO MAKE CALCULATIONS AFTER THE
*    RUN IS COMPLETED AND TO PRINT THE RESULTS.
        FEET = MILES*5280.0
        WEIGHT = MASS*32.17
  100 FORMAT(2E20.5)
        WRITE(6,100) FEET, WEIGHT
END
STOP
ENDJOB
```

Fig. 2.26   Program to simulate the launch of a Saturn rocket.

Ordinary FORTRAN FORMAT and WRITE statements can be used in nosort sections. An example of this is shown by the last two statements in the Terminal segment. Since the statements in the Terminal segment are only executed at the end of the program, the final altitude in feet FEET and the final earth-weight of the rocket WEIGHT can be calculated and will be the last output of the program. A detailed discussion of the use of FORTRAN output statements is included in Chap. 3 in the section on data output.

The PRTPLT output showing the vehicle velocity profile is given in Fig. 2.27.

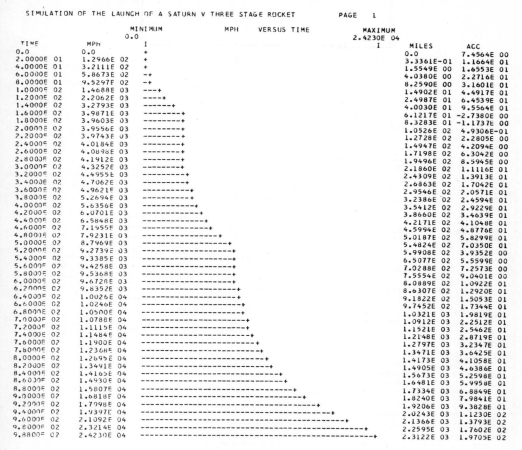

Fig. 2.27   Printer-plot output for launch of Saturn rocket.

Example 2.6   *Control System with Compensation*

A problem often encountered by the control engineer is to compensate a system so that both dynamic and steady state design specifications are met. This problem illustrates how a simple compensator can be added to a system in order to improve the transient response to a unit-step input while increasing the static-loop gain.

Consider the closed-loop configuration in Fig. 2.28. The system step response with a loop gain of 300/24 and without compensation is given in Fig. 2.29. We note that the response has excessive oscillation and a peak overshoot of 89%.

The system requirements are to increase the static-loop gain to 300/12 and design a compensator so that the overshoot for a step input is approximately 20% and the time to the first peak is equal to or less than 1 sec.

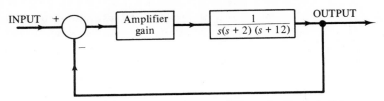

Fig. 2.28 Uncompensated system diagram for Example 2.6.

```
                            MINIMUM            OUTPUT VERSUS TIME          MAXIMUM
                              0.0                                         1.8894E 00
                                                                              I
   TIME              OUTPUT            I
0.0                  0.0              +
6.0000E-02           8.8246E-03       +
1.2000E-01           5.8440E-02       -+
1.8000E-01           1.6470E-01       ----+
2.4000E-01           3.2761E-01       --------+
3.0000E-01           5.3782E-01       --------------+
3.6000E-01           7.8032E-01       --------------------+
4.2000E-01           1.0368E 00       ---------------------------+
4.8000E-01           1.2876E 00       ----------------------------------+
5.4000E-01           1.5136E 00       -----------------------------------------+
6.0000E-01           1.6977E 00       -----------------------------------------------+
6.6000E-01           1.8261E 00       --------------------------------------------------+
7.2000E-01           1.8894E 00       ----------------------------------------------------+
7.8000E-01           1.8834E 00       ----------------------------------------------------+
8.4000E-01           1.8094E 00       -------------------------------------------------+
9.0000E-01           1.6735E 00       -----------------------------------------------+
9.6000E-01           1.4871E 00       ----------------------------------------+
1.0200E 00           1.2649E 00       ----------------------------------+
1.0800E 00           1.0245E 00       ---------------------------+
1.1400E 00           7.8462E-01       --------------------+
1.2000E 00           5.6374E-01       --------------+
1.2600E 00           3.7874E-01       ---------+
1.3200E 00           2.4353E-01       ------+
1.3800E 00           1.6805E-01       ----+
1.4400E 00           1.5754E-01       ----+
1.5000E 00           2.1215E-01       -----+
1.5600E 00           3.2703E-01       --------+
1.6200E 00           4.9270E-01       ------------+
1.6800E 00           6.9587E-01       ------------------+
1.7400E 00           9.2048E-01       ------------------------+
1.8000E 00           1.1490E 00       ------------------------------+
1.8600E 00           1.3637E 00       -------------------------------------+
1.9200E 00           1.5482E 00       ------------------------------------------+
1.9800E 00           1.6886E 00       ----------------------------------------------+
2.0400E 00           1.7744E 00       -------------------------------------------------+
2.1000E 00           1.7996E 00       -------------------------------------------------+
2.1600E 00           1.7628E 00       -----------------------------------------------+
2.2200E 00           1.6675E 00       -----------------------------------------+
2.2800E 00           1.5217E 00       -----------------------------------------+
2.3400E 00           1.3370E 00       ----------------------------------+
2.4000E 00           1.1282E 00       ----------------------------+
2.4600E 00           9.1155E-01       -----------------------+
2.5200E 00           7.0391E-01       ------------------+
2.5800E 00           5.2118E-01       --------------+
2.6400E 00           3.7723E-01       ---------+
2.7000E 00           2.8279E-01       -------+
2.7600E 00           2.4464E-01       ------+
2.8200E 00           2.6518E-01       -------+
2.8800E 00           3.4222E-01       ---------+
2.9400E 00           4.6926E-01       -----------+
3.0000E 00           6.3599E-01       ----------------+
```

Fig. 2.29 Uncompensated system step response for Example 2-6.

Using classical design procedures, one can define a proposed compensator with a transfer function given by

$$G_c(s) = \frac{\overset{\text{(lead)}}{(s + 2.2)}\overset{\text{(lag)}}{(s + 0.2)}}{(s + 8.0)(s + 0.0275)} \tag{2.13}$$

The diagram in Fig. 2.30 includes this compensator as well as a block arrangement suitable for CSMP simulation. Figure 2.31 gives the program listing using the variable names as specified in Fig. 2.30. When compared with Example 2.2, one notes that the essential difference is the addition of the compensator, $G_c(s)$. The CSMP statement and formulation for a simple zero and pole (taken together) are summarized in Table 2.5.

Table 2.5

Formulation for a Lead-Lag Structure

General Form	*Function*
Definition: Pole-zero (often called lead-lag compensator)	$\tau_2 \dot{Y} + Y = \tau_1 \dot{X} + X$
$Y = \text{LEDLAG}\,(\tau_1, \tau_2, X)$	Laplace form: $\dfrac{\tau_1 s + 1}{\tau_2 s + 1}$

$$X(s) \longrightarrow \boxed{\dfrac{\tau_1 s + 1}{\tau_2 s + 1}} \longrightarrow Y(s)$$

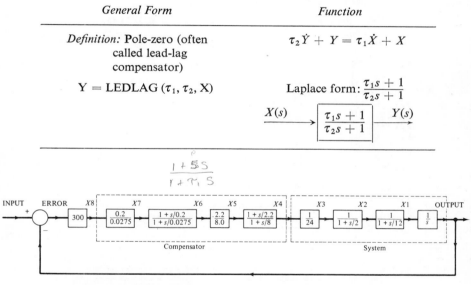

Fig. 2.30 Diagram for programming compensated system of Example 2.6.

From a practical point of view, the LEDLAG formulation (as illustrated in this example) can be used for either a lead or lag compensator as defined in classical control theory.

The step response of the system with the compensator is shown in Fig. 2.32. We see that the problem specifications are met. One important reason for employing CSMP for a problem of this nature is that the graphic methods used in designing compensators do not generally yield a convenient method for determining the system step response. This is particularly true for higher-order systems.

```
INPUT = STEP(0.0)
ERROR = INPUT - OUTPUT
X8 = 300.0*ERROR
X7 = (0.2/0.0275)*X8
X6 = LEDLAG(1.0/0.2,1.0/0.0275,X7)
X5 = (2.2/8.0)*X6
X4 = LEDLAG(1.0/2.2,1.0/8.0,X5)
X3 = (1.0/24.0)*X4
X2 = REALPL(0.0,1.0/2.0,X3)
X1 = REALPL(0.0,1.0/12.0,X2)
OUTPUT = INTGRL(0.0,X1)

TIMER FINTIM = 6.0, OUTDEL = 0.12
PRTPLT OUTPUT
LABEL       RESPONSE FOR COMPENSATED SYSTEM - EXAMPLE 2-6
END
STOP
ENDJOB
```

Fig. 2.31 CSMP program listing for Example 2.6.

```
 RESPONSE FOR COMPENSATED SYSTEM - EXAMPLE 2-6                          PAGE    1

                            MINIMUM              OUTPUT VERSUS TIME          MAXIMUM
                            0.0                                              1.2068E 00
   TIME           OUTPUT       I                                               I
 0.0            0.0            +
 1.2000E-01     4.9649E-02     --+
 2.4000E-01     2.4100E-01     ---------+
 3.6000E-01     5.1065E-01     ---------------------------+
 4.8000E-01     7.7689E-01     ------------------------------------------+
 6.0000E-01     9.8866E-01     --------------------------------------------------+
 7.2000E-01     1.1266E 00     -------------------------------------------------------+
 8.4000E-01     1.1941E 00     ----------------------------------------------------------+
 9.6000E-01     1.2068E 00     -----------------------------------------------------------+
 1.0800E 00     1.1846E 00     ---------------------------------------------------------+
 1.2000E 00     1.1458E 00     ------------------------------------------------------+
 1.3200E 00     1.1045E 00     ---------------------------------------------------+
 1.4400E 00     1.0691E 00     ------------------------------------------------+
 1.5600E 00     1.0436E 00     ----------------------------------------------+
 1.6800E 00     1.0283E 00     ----------------------------------------------+
 1.8000E 00     1.0216E 00     ---------------------------------------------+
 1.9200E 00     1.0209E 00     ---------------------------------------------+
 2.0400E 00     1.0235E 00     ---------------------------------------------+
 2.1600E 00     1.0274E 00     ---------------------------------------------+
 2.2800E 00     1.0309E 00     ----------------------------------------------+
 2.4000E 00     1.0335E 00     ----------------------------------------------+
 2.5200E 00     1.0348E 00     ----------------------------------------------+
 2.6400E 00     1.0349E 00     ----------------------------------------------+
 2.7600E 00     1.0342E 00     ----------------------------------------------+
 2.8800E 00     1.0330E 00     ----------------------------------------------+
 3.0000E 00     1.0315E 00     ----------------------------------------------+
 3.1200E 00     1.0301E 00     ----------------------------------------------+
 3.2400E 00     1.0288E 00     ----------------------------------------------+
```

Fig. 2.32 Compensated system step response for Example 2.6.

Example 2.7 *RLC Circuit Problem*

The purpose of this example is to illustrate how CSMP can be used to simulate an electric circuit. A series RLC circuit is given in Fig. 2.33. Suppose it is desired to find the voltage across the capacitor in response to a pulse-train input of unit amplitude. There

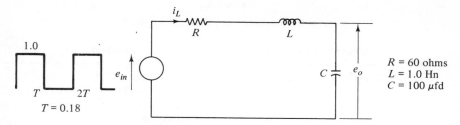

Fig. 2.33 Series RLC circuit.

are several approaches that can be used to determine the solution. Three methods will be considered here.

Method 1—Transfer Function. The transfer function between $e_o(t)$ and $e_{in}(t)$ can be expressed as:

$$\frac{E_o(s)}{E_{in}(s)} = \frac{\frac{1}{LC}}{s^2 + \left(\frac{R}{L}\right)s + \frac{1}{LC}} \tag{2.14}$$

This can easily be simulated by CSMP provided we know the nature of the roots of the characteristic equation. Two possibilities exist in that the roots can either be real or complex. If the roots are real we can solve the problem by applying REALPL twice. A function (CMPXPL) which requires only one program statement can be used for either real or complex roots. The solution using this function will be presented later in this example.

Method 2—State Variable Representation. The following equations can be written for the previous circuit.

$$e_{in}(t) = Ri(t) + L\frac{di\,(t)}{dt} + e_o(t) \tag{2.15}$$

$$i(t) = \frac{C\,de_o\,(t)}{dt} \tag{2.16}$$

Rearranging yields

$$\dot{e}_o(t) = \frac{i(t)}{C} \tag{2.17}$$

$$\dot{i}(t) = -\frac{e_o(t)}{L} - \frac{Ri(t)}{L} + \frac{e_{in}(t)}{L} \tag{2.18}$$

These last two equations form a state-variable representation of the system where the states are $e_o(t)$ and $i(t)$.

Method 3—Direct Circuit Application. Kirchoff's law, applied directly to the RLC circuit, yields

$$e_{in}(t) = Ri(t) + L\frac{di\,(t)}{dt} + \frac{1}{C}\int_o^t i(t)\,dt \tag{2.19}$$

A "graphical" representation for the solution of this equation is given later.

Pulse Train Generation. The input voltage given in Fig. 2.33 is a train of square waves which can be generated by using two standard CSMP functions. These two functions are IMPULS and PULSE and are defined in Tables 2.6 and 2.7.

Table 2.6

Formulation for an Impulse Generator†

Description	Function
$Y = IMPULS(P_1, P_2)$ 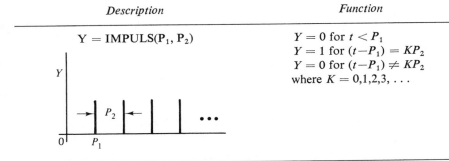	$Y = 0$ for $t < P_1$ $Y = 1$ for $(t - P_1) = KP_2$ $Y = 0$ for $(t - P_1) \neq KP_2$ where $K = 0,1,2,3,\ldots$

†This impulse function does not have the same properties as the well-known Dirac (impulse) function.

Table 2.7

Formulation for Pulse Generator

Description	Function
$Y = PULSE(P,X)$ 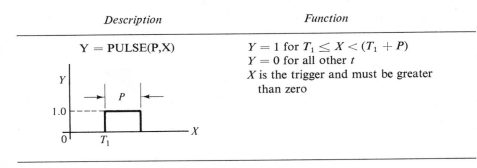	$Y = 1$ for $T_1 \leq X < (T_1 + P)$ $Y = 0$ for all other t X is the trigger and must be greater than zero

As an example, suppose the following pulse-train is desired.

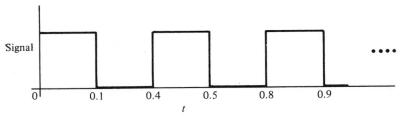

The trigger signal for the pulse will be

$$FLASH = IMPULS(0.0, 0.4)$$

where FLASH (a dummy variable) plays the role of X for the pulse function. The expression for the output pulse-train is then given by

$$\text{SIGNAL} = \text{PULSE}(0.1,\text{FLASH})$$

With this preliminary background we return to the circuit problem and give a solution using each method described above.

Solution by Method 1. As a first step in this solution one should determine the particular values of ξ and ω_n from Eq. (2.14) (the characteristic equation), which can be easily accomplished by comparing

$$s^2 + \frac{R}{L} s + \frac{1}{LC} = 0 \qquad (2.20)$$

with

$$s^2 + 2\xi\omega_n s + \omega_n^2 = 0 \qquad (2.21)$$

Recall from linear system theory that if

$\xi = 1$; the roots are real and equal (critical damped)

$\xi > 1$; the roots are real and unequal (overdamped)

$\xi < 1$; the roots are complex (underdamped)

Comparing coefficients in Eqs. (2.20) and (2.21) gives

$$\omega_n = \sqrt{1/LC} \qquad (2.22)$$

$$\xi = \frac{R}{2\sqrt{L/C}} \qquad (2.23)$$

Using the values of R, L, and C from Fig. 2.33 results in

$$\omega_n = 100.0, \qquad \xi = 0.3 \qquad (2.24)$$

which implies that the system has complex roots. Complex roots can be directly simulated by using the function defined in the Table 2.8, provided the roots are in the denominator of the transfer function. Roots in the denominator of a transfer function are called poles and thus for this case we refer to these roots as complex poles.

Table 2.8

Formulation for Complex Poles†

General Form	Function
Definition: Complex poles $Y = \text{CMPXPL}(\text{IC1},\text{IC2},\zeta,\omega_n,X)$ $\text{IC1} = Y(0)$ $\text{IC2} = \dot{Y}(0)$ For any value of ζ	$\ddot{y}(t) + 2\xi\omega_n\dot{y}(t) + \omega_n^2 y(t) = x(t)$ Equivalent Laplace form $X \longrightarrow \boxed{\dfrac{1 \times \omega_n^2}{s^2 + 2\xi\,\omega_n s + \omega_n^2}} \longrightarrow Y$

†The poles are not required to be complex to use CMPXPL. As noted in the table any value of ζ is acceptable and thus the function can also be used when the poles are real.

A diagram giving a block arrangement for simulating Eq. (2.20) is given in Fig. 2.34. The actual program listing of the simulation statements is given directly below the diagram in Fig. 2.35.

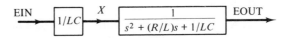

Fig. 2.34 Simulation diagram for RLC circuit (Method 1 solution).

```
*       EXAMPLE PROBLEM FOR RLC NETWORK - EXAMPLE 2-7

PARAM L = 1.0, C = 0.0001, R = 60.0
TRIG = IMPULS(0.0,0.36)
EIN = PULSE(0.18,TRIG)
X = EIN*(1.0/(L*C))
EOUT = CMPXPL(0.0,0.0,R/(2.0*SQRT(L/C)),SQRT(1.0/(L*C)),X)

TIMER FINTIM = 0.4, OUTDEL = 0.008
PRTPLT EOUT(EIN)
LABEL PULSE TRAIN RESPONSE ACROSS CAPACITOR OF RLC NETWORK
PRTPLT EIN
LABEL PLOT SHOWING THE PULSE TRAIN TO RLC NETWORK
END
STOP
ENDJOB
```

Fig. 2.35 Program listing for RLC circuit (Method 1 solution).

For convenience, the circuit parameter values are listed on a PARAMETER card. This gives the added flexibility of changing the circuit values without repunching several problem statement cards.

The voltage response across the capacitor is shown by the printer-plot of Fig. 2.36 which gives the response for a time increment involving two input pulses.

Solution by Method 2. The program listing for the solution of Eqs. (2.17) and (2.18) is given in Fig. 2.37. An advantage of using this approach is that one is not required to determine the form of the roots for the circuit characteristic equation. Formulating the problem in this manner has the added advantage of giving the solution for the current in the circuit. This current response is given in Fig. 2.38.

The voltage $e_o(t)$ is identical to the solution in method 1; therefore, another printer-plot for this variable is not given.

Solution by Method 3. The diagram in Fig. 2.39 presents one form of solution for Eq. (2.19). Anyone familiar with analog computer simulation recognizes that differentiation has a tendency to produce noise (spikes). However, this approach was selected here to stress the use of the derivative function in CSMP. This function is defined in Table 2.9.

```
                        MINIMUM            EOUT    VERSUS TIME          MAXIMUM
                        -3.7134E-01                                     1.3716E 00
IME            EOUT           I                                         I         EIN
0              0.0            ---------+                                          1.0000E 00
0C00E-03       2.6053E-01     ------------------+                                 1.0000E 00
6C00E-02       7.7806E-01     --------------------------------+                   1.0000E 00
4000E-02       1.2052E 00     ------------------------------------------+         1.0000E 00
2000E-02       1.3707E 00     -----------------------------------------------+    1.0000E 00
0000E-02       1.2945E 00     ---------------------------------------------+      1.0000E 00
8000E-02       1.1054E 00     ------------------------------------------+         1.0000E 00
6CC0E-02       9.3762E-01     -------------------------------------+              1.0000E 00
4000E-02       8.6384E-01     ----------------------------------+                 1.0000E 00
2000E-02       8.8378E-01     -----------------------------------+                1.0000E 00
0C00E-02       9.5213E-01     -------------------------------------+              1.0000E 00
8000E-02       1.0175E 00     ---------------------------------------+            1.0000E 00
6CC0E-02       1.0495E 00     ----------------------------------------+           1.0000E 00
0400E-01       1.0455E 00     ----------------------------------------+           1.0000E 00
1200E-01       1.0211E 00     ---------------------------------------+            1.0000E 00
2000E-01       9.9581E-01     --------------------------------------+             1.0000E 00
2800E-01       9.8220E-01     --------------------------------------+             1.0000E 00
3600E-01       9.8235E-01     --------------------------------------+             1.0000E 00
4400E-01       9.9094E-01     --------------------------------------+             1.0000E 00
5200E-01       1.0006E 00     ---------------------------------------+            1.0000E 00
6C00E-01       1.0063E 00     ---------------------------------------+            1.0000E 00
5800E-01       1.0068E 0C     ---------------------------------------+            1.0000E 00
7600E-01       1.0038E 00     ---------------------------------------+            1.0000E 00
8400E-01       9.2917E-01     -------------------------------------+              0.0
9200E-01       4.8989E-01     -------------------------+                          0.0
0C00E-01       -1.7989E-02    ----------+                                         0.0
0800E-01       -3.2360E-01    -+                                                  0.0
1600E-01       -3.5677E-01    +                                                   0.0
2400E-01       -2.0539E-01    ----+                                               0.0
3200E-01       -1.2740E-02    -----------+                                        0.0
4000E-01       1.1261E-01     -------------+                                      0.0
4800E-01       1.3591E-01     ----------------+                                   0.0
5600E-01       8.4858E-02     ------------+                                       0.0
4C00E-01       1.2366E-02     -----------+                                        0.0
7200E-01       -3.8451E-02    ---------+                                          0.0
8000E-01       -5.1376E-02    ---------+                                          0.0
8800E-01       -3.4623E-02    ---------+                                          0.0
9600E-01       -7.5722E-03    -----------+                                        0.0
9400E-01       1.2821E-02     -----------+                                        0.0
1200E-01       1.9267E-02     -----------+                                        0.0
2000E-01       1.3975E-02     -----------+                                        0.0
2800E-01       3.9656E-03     ----------+                                         0.0
3600E-01       -4.1406E-03    ----------+                                         0.0
4400E-01       -7.1634E-03    ----------+                                         0.0
5200E-01       -5.5832E-03    ----------+                                         0.0
5C00E-01       -1.9129E-03    ----------+                                         0.0
6800E-01       2.6182E-01     ------------------+                                 1.0000E 00
7600E-01       7.8072E-01     --------------------------------+                   1.0000E 00
8400E-01       1.2075E 00     ------------------------------------------+         1.0000E 00
9200E-01       1.3716E 00     -----------------------------------------------+    1.0000E 00
0C00E-01       1.2941E 00     ---------------------------------------------+      1.0000E 00
```

Fig. 2.36 Computer solution for voltage across the capacitor
(Method 1 solution).

Table 2.9

Formulation for Differentiation

General Form	Function
Definition: Differentiation	
Y = DERIV(IC,X)	$y(t) = \dfrac{dx\,(t)}{dt}$
IC = $\dot{X}(0)$	Laplace form:
	$X \rightarrow \boxed{\quad s \quad} \longrightarrow Y$

The program listing for the simulation is given in Fig. 2.40 and is accompanied by the error message

SIMULATION INVOLVED AN ALGEBRAIC LOOP CONTAINING THE FOLLOWING ELEMENTS

   I   X3   I

This statement means that the sort subprogram in CSMP cannot find an integration or other memory block† in the loop involving I — X3 — I. The program will not run in this case. To correct this situation, one must re-cast the problem so that every loop in the diagram contains a memory block or integration. Loops are not always obvious for a given set of equations unless one draws a block diagram or signal-flow graph.

```
*          SOLUTION TO RLC CIRCUIT USING STATE VARIABLE APPROACH
PARAM L = 1.0,  C = 0.0001,  R = 60.0
TRIG = IMPULS(0.0,0.36)
EIN = PULSE(0.18,TRIG)
EODOT = I/C
*   I IS THE CURRENT THROUGH THE INDUCTOR
IDOT = (1.0/L)*(-EO - I*R + EIN)
EO = INTGRL(0.0,EODOT)
*    EO IS THE VOLTAGE ACROSS THE CAPACITOR
I = INTGRL(0.0,IDOT)

TIMER FINTIM = 0.4,  OUTDEL = 0.008
PRTPLT EO(EIN)
LABEL  VOLTAGE ACROSS THE CAPACITOR - EXAMPLE 2-7
PRTPLT I(EIN)
LABEL  CURRENT THROUGH THE INDUCTOR - EXAMPLE 2-7
END
STOP
ENDJOB
```

Fig. 2.37   Program listing for RLC solution (Method 2).

†A discussion of memory functions is given in Chap. 3.

		MINIMUM −6.5032E−03	I VERSUS TIME	MAXIMUM 6.4814E−03	
E	I	I		I	EIN
	0.0	+------------------------+			
00E−03	5.6998E−03	+---+			1.0000E 00
COE−02	6.4814E−03	+--+			1.0000E 00
00E−02	3.8425E−03	+----------------------------------+			1.0000E 00
00E−02	3.5775E−04	+------------------------+			1.0000E 00
00E−02	−1.9717E−03	+----------------+			1.0000E 00
00E−02	−2.4636E−03	+--------------+			1.0000E 00
00E−02	−1.5812E−03	+------------------+			1.0000E 00
00E−02	−2.7314E−04	+------------------------+			1.0000E 00
00E−02	6.6814E−04	+---------------------------+			1.0000E 00
00E−02	9.2889E−04	+----------------------------+			1.0000E 00
00E−02	6.4277E−04	+---------------------------+			1.0000E 00
00E−02	1.5598E−04	+-----------------------+			1.0000E 00
00E−01	−2.2049E−04	+----------------------+			1.0000E 00
00E−01	−3.4729E−04	+---------------------+			1.0000E 00
00E−01	−2.5846E−04	+----------------------+			1.0000E 00
00E−01	−7.8947E−05	+-----------------------+			1.0000E 00
00E−01	7.0206E−05	+-----------------------+			1.0000E 00
C0E−01	1.2871E−04	+-----------------------+			1.0000E 00
00E−01	1.0291E−04	+-----------------------+			1.0000E 00
00E−01	3.7367E−05	+-----------------------+			1.0000E 00
00E−01	−2.1201E−05	+-----------------------+			1.0000E 00
00E−01	−4.7230E−05	+-----------------------+			1.0000E 00
00E−01	−4.0583E−05	+-----------------------+			1.0000E 00
00E−01	−5.6506E−03	+--+			0.0
00E−01	−6.5032E−03	+			0.0
C0E−01	−3.8977E−03	+--------+			0.0
00E−C1	−4.0707E−04	+---------------------+			0.0
00E−01	1.9498E−03	+---------------------------------+			0.0
00E−01	2.4693E−03	+------------------------------------+			0.0
00E−01	1.6012E−03	+-------------------------------+			0.0
00E−01	2.9236E−04	+----------------------+			0.0
00E−01	−6.5865E−04	+------------------+			0.0
00E−01	−9.2999E−04	+----------------+			0.0
C0E−01	−6.4989E−04	+------------------+			0.0
00E−01	−1.6339E−04	+----------------------+			0.0
00E−01	2.1648E−04	+-----------------------+			0.0
C0E−01	3.4731E−04	+------------------------+			0.0
00E−01	2.6097E−04	+-----------------------+			0.0
00E−01	8.1788E−05	+-----------------------+			0.0
00E−01	−6.8531E−05	+----------------------+			0.0
00E−01	−1.2856E−04	+----------------------+			0.0
00E−01	−1.0378E−04	+----------------------+			0.0
00E−01	−3.8439E−05	+----------------------+			0.0
00E−01	2.0526E−05	+-----------------------+			0.0
00E−C1	2.1380E−04	+-----------------------+			1.0000E 00
00E−01	5.8070E−03	+--+			1.0000E 00
00E−01	6.4711E−03	+--+			1.0000E 00
00E−01	3.7644E−03	+----------------------------------+			1.0000E 00
00E−01	2.7532E−04	+------------------+			1.0000E 00
00E−01	−2.0171E−03	+----------------+			1.0000E 00

Fig. 2.38   Current response of RLC circuit (Method 2 solution).

49

Solution for:

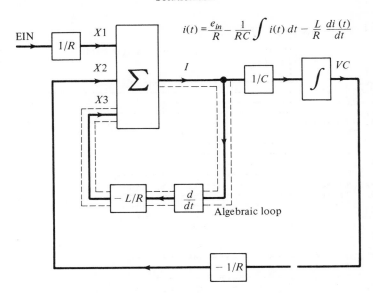

$$i(t) = \frac{e_{in}}{R} - \frac{1}{RC}\int i(t)\,dt - \frac{L}{R}\frac{di(t)}{dt}$$

Fig. 2.39   Solution diagram for RLC circuit using DERIV function.

```
CCNSTANT L = 1.0, C = 0.0001, R = 60.0

TRIG = IMPULS(0.0,0.36)
EIN = PULSE(0.18, TRIG)
X1 = (EIN)/R
X2 = -VC
*VC IS THE VOLTAGE ACROSS THE CAPACITOR
X3 = (L/R)*DERIV(0.0,I)
I  = X1 + X2 + X3
*  I IS THE CURRENT IN THE CIRCUIT
VC = (1.0/R*C)*INTGRL(0.0,I)
TIMER FINTIM = 0.4, OUTDEL = 0.008
PRTPLT VC, I
LABEL  SOLUTION FOR CIRCUIT USING DERIV - EXAMPLE 2-7
END
STOP
```

```
SIMULATION INVOLVES AN ALGEBRAIC LOOP CONTAINING THE FOLLOWING ELEMENTS
  I        X3       I

  OUTPUTS      INPUTS     PARAMS   INTEGS + MEM BLKS   FORTRAN  DATA CDS
  12(500)     35(1400)    5(400)    1+   0=   1(300)    9(600)      5
```

****PROBLEM CAN NOT BE EXECUTED****

Fig. 2.40   Program listing for RLC solution using DERIV function.

Although the simulation will not run using the DERIV function, one should not interpret this to mean a loop cannot have a derivative term. To correct the present problem a solution for the capacitor voltage that has an integrator in each loop is given in Fig. 2.41. The program listing in Fig. 2.42 follows directly from the diagram and the simulation results are identical to those obtained by previous methods.

The objective of this example has been to demonstrate how CSMP can be used for the solution of an electric circuit. However, it should be noted that other specialized programs such as ECAP (Electric Circuit Analysis Program) are generally better suited for circuit analysis and design.

Solution for:

$$\frac{di\,(t)}{dt} = \frac{e_{in}}{L} - \frac{1}{LC}\int i(t)\,dt - \frac{R\,i(t)}{L}$$

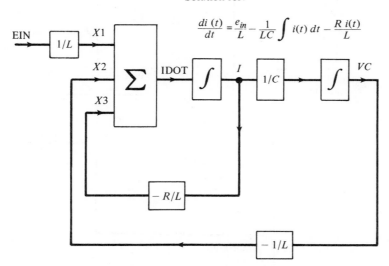

Fig. 2.41 Diagram for Method 3 solution of RLC circuit using all integrators.

```
CONSTANT L = 1.0,  C = 0.0001,  R = 60.0

TRIG = IMPULS(0.0,0.36)
EIN = PULSE(0.18, TRIG)
*VC IS THE VOLTAGE ACROSS THE CAPACITOR
*  I IS THE CURRENT IN THE CIRCUIT
X1 = (EIN)/L
X2 = -VC/L
X3 = -(R/L)*I
IDOT = X1 + X2 + X3
VC =(1.0/C)*INTGRL(0.0,I)
I = INTGRL(0.0,IDOT)
TIMER FINTIM = 0.4, OUTDEL = 0.008
PRTPLT VC, I
LABEL  RLC CIRCUIT USING INTGRL TERMS ONLY - EXAMPLE 2-7
END
STOP
ENDJOB
```

Fig. 2.42 Program listing for Method 3 using all integrator blocks.

Example 2.8   *Temperature Control for a Chemical Process*

A simplified diagram of a temperature-control system is shown in Fig. 2.43. Basically, this system regulates the temperature of a chemical material by passing it through a heat exchanger unit. An output-temperature sensor compares the process temperature to the desired set-point. Any difference between the measured and set-point signals is conditioned by a proportional-integral controller whose output makes correcting adjustments to the steam control valve.

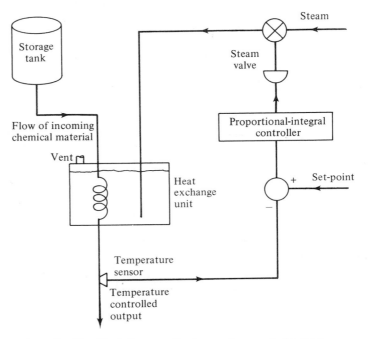

Fig. 2.43   Simplified diagram of a temperature control system.

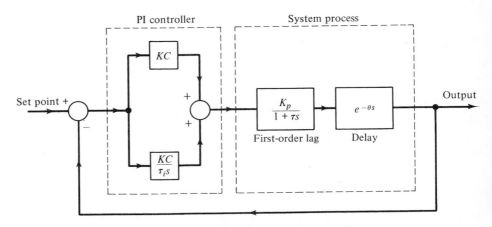

Fig. 2.44   Block diagram of temperature control system.

Figure 2.44 gives a block-diagram representation of the linearized system. If the output of the first-order lag is $f(t)$, then an ideal delay of θ units will produce $f(t - \theta)$. In terms of Laplace transforms this delay is expressed by $e^{-\theta s}$ and so represented in the block diagram. The functional relationship for the delay block is given in Table 2.10.

Table 2.10

Formulation for Ideal Delay

General Form	*Function*

Definition: Ideal Delay
  Y $=$ DELAY(N, P, X)
  P $=$ Delay time
  N $=$ Number of points sampled
        during P (N must be an
        integer)
	$Y(t) = X(t - P) \; t \geq P$
	$Y(t) = 0 \qquad\quad t < P$
	Laplace form: e^{-Ps}
	
	$X(s) \longrightarrow \boxed{\; e^{-Ps} \;} \longrightarrow Y(s)$

The only thing likely to cause confusion is the number of points sampled during P. As noted in Table 2.10, N must be an integer and should not include a decimal (i.e., 21.0 is not allowed but 21 is). As a rule the value of N should be equal to the approximate number of integration steps expected during the delay time P.†

Slight modifications of the original block diagram are necessary before directly writing the simulation statements. One suitable, but not unique, representation is given in Fig. 2.45 where the gain constants have been isolated from the dynamic elements. As stated in a previous example, each block must have input and output variable assignments.

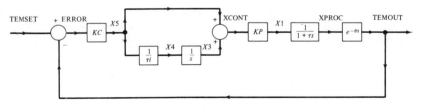

Fig. 2.45   Simulation diagram for Example 2.8.

The simulation program listing is shown in Fig. 2.46. The data statements PARAMETER, CONSTANT and INCON are used to assign numerical values to the specified variables. The use of PARAMETER KC $= (1.8, 12.0)$ will cause the program to run first with KC $= 1.8$ and second with KC $= 12.0$. A complete description of PARAMETER, INCON and CONSTANT was given in Example 2.3.

---

†Refer to Appendix I for further discussion on selecting N.

```
*    CHEMICAL PRCCESS CONTROL
PARAMETER  KC = (1.8,12.0),  THETA = 19.0
CCNSTANT KP = 6.45,  TAU = 544.0,  TAUI = 540.0
INCON  ICX3 = 0.0,   ICRLPL = 0.0
TEMSET = STEP(0.0)
ERROR = TEMSET - TEMOUT
X5 = KC*ERROR
X4 = (1.0/TAUI)*X5
X3 = INTGRL(ICX3,X4)
XCONT = X5 + X3
X1 = (KP)*XCONT
XPROC = REALPL(ICRLPL, TAU, X1)
TEMOUT = DELAY(25,THETA, XPROC)

TIMER FINTIM = 200.0, OUTDEL = 4.0, PRDEL = 2.0
PRTPLT TEMOUT
LABEL  TEMPERATURE CF PROCESS OUTPUT - EXAMPLE 2-8
PRTPLT XCONT
LABEL  OUTPUT OF PI CONTROLLER - EXAMPLE 2-8
PRINT TEMOUT
END
STOP
ENDJOB
```

Fig. 2.46   CSMP program listing for the system in Fig. 2.45.

The control statements indicate that the simulation will run for 200 sec (FINTIM). Printer-plots will be made for TEMOUT and XCONT with each plotted point occurring every 4 sec (OUTDEL). In addition, the values of TEMOUT occurring every 2 sec (PRDEL) will be printed on separate pages from those of the printer-plots of TEMOUT.

The response of TEMOUT for KC = 1.8 and 12.0 is shown in Figs. 2.47 and 2.48, respectively. Not only is the requested plot LABEL given at the top of the output but the particular value of KC also appears. Whenever repeated runs occur, the run value of the variable parameter will be given at the top of all PRTPLT and PRINT pages.

We note that the printer-plot points of TEMOUT for KC = 1.8 are practically all the same. The reason for this is that the printer-plot points are quantized to always cover fifty printer spaces across the page. If a multiple-value parameter is used, the quantization is made over the range of the maximum and minimum values of the total data set. For this example, the largest value in the data set is 34.855 and the smallest value is −29.642. It is incidental that both values occur for KC = 12.0. The quantization step is therefore

$$\text{Quantization step} = \frac{34.855 + 29.652}{50} \doteq 1.29 \text{ units} \qquad (2.25)$$

The variation of TEMOUT for KC = 1.8 is from 0 to 1.0 (see Fig. 2.47) which is less than one quantization unit. The only reason the plotted output level changes for this case is that 0.0 is almost at a quantization step point and the addition of approximately 0.106 causes a level change.

One might look upon multiple-parameter value runs as having the disadvantage of possibly giving one or more obscured plots. This may be true, but, on the other hand, the plots are scaled to a common relative amplitude which is often a desired characteristic. Furthermore, it is always possible to rerun the simulation using a single-value parameter. Figure 2.49 shows the rerun of this problem for KC = 1.8 only.

The remaining printer-plots and printed output for XCONT and TEMOUT are omitted since they give the same general type of information as previously presented.

```
                        MINIMUM              TEMOUT VERSUS TIME          MAXIMUM
                       -2.9642E 01           KC     = 1.8000E 00        3.4855E 01
    TIME         TEMOUT          I                                          I
0.0             0.0            -----------------------+
4.0000E 00      0.0            -----------------------+
8.0000E 00      0.0            -----------------------+
1.2000E 01      0.0            -----------------------+
1.6000E 01      0.0            -----------------------+
2.0000E 01      2.1342E-02     -----------------------+
2.4000E 01      1.0671E-01     -----------------------+
2.8000E 01      1.9209E-01     ------------------------+
3.2000E 01      2.7747E-01     ------------------------+
3.6000E 01      3.6285E-01     ------------------------+
4.0000E 01      4.4710E-01     ------------------------+
4.4000E 01      5.2521E-01     ------------------------+
4.8000E 01      5.9603E-01     ------------------------+
5.2000E 01      6.5957E-01     ------------------------+
5.6000E 01      7.1582E-01     ------------------------+
6.0000E 01      7.6485E-01     ------------------------+
6.4000E 01      8.0707E-01     ------------------------+
6.8000E 01      8.4309E-01     ------------------------+
7.2000E 01      8.7353E-01     ------------------------+
7.6000E 01      8.9902E-01     ------------------------+
8.0000E 01      9.2016E-01     ------------------------+
8.4000E 01      9.3757E-01     ------------------------+
8.8000E 01      9.5176E-01     ------------------------+
9.2000E 01      9.6324E-01     ------------------------+
9.6000E 01      9.7244E-01     ------------------------+
1.0000E 02      9.7974E-01     ------------------------+
1.0400E 02      9.8548E-01     ------------------------+
1.0800E 02      9.8994E-01     ------------------------+
1.1200E 02      9.9337E-01     ------------------------+
1.1600E 02      9.9596E-01     ------------------------+
1.2000E 02      9.9789E-01     ------------------------+
1.2400E 02      9.9929E-01     ------------------------+
1.2800E 02      1.0003E 00     ------------------------+
1.3200E 02      1.0010E 00     ------------------------+
1.3600E 02      1.0014E 00     ------------------------+
1.4000E 02      1.0017E 00     ------------------------+
1.4400E 02      1.0018E 00     ------------------------+
1.4800E 02      1.0018E 00     ------------------------+
1.5200E 02      1.0018E 00     ------------------------+
1.5600E 02      1.0018E 00     ------------------------+
1.6000E 02      1.0017E 00     ------------------------+
1.6400E 02      1.0015E 00     ------------------------+
1.6800E 02      1.0014E 00     ------------------------+
1.7200E 02      1.0013E 00     ------------------------+
1.7600E 02      1.0012E 00     ------------------------+
1.8000E 02      1.0011E 00     ------------------------+
1.8400E 02      1.0010E 00     ------------------------+
1.8800E 02      1.0009E 00     ------------------------+
1.9200E 02      1.0009E 00     ------------------------+
1.9600E 02      1.0008E 00     ------------------------+
2.0000E 02      1.0007E 00     ------------------------+
```

Fig. 2.47   Printer-plot of TEMOUT for Example 2.8 ($KC = 1.8$).

```
                          MINIMUM           TEMOUT VERSUS TIME          MAXIMUM
                         -2.9642E 01        KC    = 1.2000E 01         3.4855E 01
     TIME        TEMOUT        I                                           I
0.0            0.0            -------------------+
4.0000E 00     0.0            -------------------+
8.0000E 00     0.0            -------------------+
1.2000E 01     0.0            -------------------+
1.6000E 01     0.0            -------------------+
2.0000E 01     1.4228E-01     --------------------+
2.4000E 01     7.1142E-01     --------------------+
2.8000E 01     1.2806E 00     ---------------------+
3.2000E 01     1.8498E 00     ---------------------+
3.6000E 01     2.4190E 00     ----------------------+
4.0000E 01     2.9377E 00     ----------------------+
4.4000E 01     3.1830E 00     -----------------------+
4.8000E 01     3.1045E 00     -----------------------+
5.2000E 01     2.7020E 00     ----------------------+
5.6000E 01     1.9755E 00     ---------------------+
6.0000E 01     9.4448E-01     --------------------+
6.4000E 01    -2.6877E-01     ------------------+
6.8000E 01    -1.4835E 00     -----------------+
7.2000E 01    -2.5153E 00     ----------------+
7.6000E 01    -3.1798E 00     ---------------+
8.0000E 01    -3.3006E 00     ---------------+
8.4000E 01    -2.7538E 00     ----------------+
8.8000E 01    -1.5125E 00     -----------------+
9.2000E 01     3.3142E-01     -------------------+
9.6000E 01     2.6321E 00     ----------------------+
1.0000E 02     5.0847E 00     -------------------------+
1.0400E 02     7.3241E 00     --------------------------+
1.0800E 02     8.9245E 00     ----------------------------+
1.1200E 02     9.5744E 00     ----------------------------+
1.1600E 02     8.9782E 00     ----------------------------+
1.2000E 02     7.0053E 00     --------------------------+
1.2400E 02     3.7240E 00     -----------------------+
1.2800E 02    -5.5492E-01     ------------------+
1.3200E 02    -5.3826E 00     --------------+
1.3600E 02    -1.0056E 01     ----------+
1.4000E 02    -1.3805E 01     --------+
1.4400E 02    -1.5793E 01     ------+
1.4800E 02    -1.5532E 01     ------+
1.5200E 02    -1.2594E 01     ---------+
1.5600E 02    -7.0404E 00     -------------+
1.6000E 02     8.8519E-01     ------------------+
1.6400E 02     1.0234E 01     -----------------------------+
1.6800E 02     1.9771E 01     ------------------------------------+
1.7200E 02     2.8025E 01     ------------------------------------------+
1.7600E 02     3.3331E 01     ----------------------------------------------+
1.8000E 02     3.4549E 01     -----------------------------------------------+
1.8400E 02     3.0642E 01     --------------------------------------------+
1.8800E 02     2.1323E 01     -------------------------------------+
1.9200E 02     7.2223E 00     --------------------------+
1.9600E 02    -1.0552E 01     ----------+
2.0000E 02    -2.9642E 01     +
```

Fig. 2.48 Printer-plot of TEMOUT for Example 2.8 (*KC* = 12.0).

56

```
TEMPERATURE OF PROCESS OUTPUT - EXAMPLE 2-8                    PAGE   1

                    MINIMUM            TEMOUT VERSUS TIME          MAXIMUM
                      0.0                                        1.0020E 00
 TIME              TEMOUT          I                                 I
0.0               0.0             +
4.0000E 00        0.0             +
8.0000E 00        0.0             +
1.2000E 01        0.0             +
1.6000E 01        0.0             +
2.0000E 01        2.1342E-02      -+
2.4000E 01        1.0671E-01      -----+
2.8000E 01        1.9209E-01      ---------+
3.2000E 01        2.7747E-01      -------------+
3.6000E 01        3.6282E-01      -----------------+
4.0000E 01        4.4672E-01      ----------------------+
4.4000E 01        5.2483E-01      --------------------------+
4.8000E 01        5.9565E-01      -----------------------------+
5.2000E 01        6.5919E-01      --------------------------------+
5.6000E 01        7.1544E-01      ----------------------------------+
6.0000E 01        7.6451E-01      ------------------------------------+
6.4000E 01        8.0680E-01      --------------------------------------+
6.8000E 01        8.4288E-01      ---------------------------------------+
7.2000E 01        8.7339E-01      ----------------------------------------+
7.6000E 01        8.9894E-01      -----------------------------------------+
8.0000E 01        9.2015E-01      ------------------------------------------+
8.4000E 01        9.3760E-01      ------------------------------------------+
8.8000E 01        9.5184E-01      -------------------------------------------+
9.2000E 01        9.6336E-01      --------------------------------------------+
9.6000E 01        9.7259E-01      --------------------------------------------+
1.0000E 02        9.7991E-01      ---------------------------------------------+
1.0400E 02        9.8567E-01      ---------------------------------------------+
1.0800E 02        9.9013E-01      ---------------------------------------------+
1.1200E 02        9.9356E-01      ----------------------------------------------+
1.1600E 02        9.9615E-01      ----------------------------------------------+
1.2000E 02        9.9807E-01      ----------------------------------------------+
1.2400E 02        9.9946E-01      ----------------------------------------------+
1.2800E 02        1.0005E 00      -----------------------------------------------+
1.3200E 02        1.0011E 00      -----------------------------------------------+
1.3600E 02        1.0016E 00      -----------------------------------------------+
1.4000E 02        1.0018E 00      -----------------------------------------------+
1.4400E 02        1.0019E 00      -----------------------------------------------+
1.4800E 02        1.0020E 00      ------------------------------------------------+
1.5200E 02        1.0019E 00      -----------------------------------------------+
1.5600E 02        1.0018E 00      -----------------------------------------------+
1.6000E 02        1.0017E 00      -----------------------------------------------+
1.6400E 02        1.0016E 00      -----------------------------------------------+
1.6800E 02        1.0015E 00      -----------------------------------------------+
1.7200E 02        1.0014E 00      -----------------------------------------------+
1.7600E 02        1.0013E 00      -----------------------------------------------+
1.8000E 02        1.0011E 00      -----------------------------------------------+
1.8400E 02        1.0010E 00      -----------------------------------------------+
1.8800E 02        1.0010E 00      -----------------------------------------------+
1.9200E 02        1.0009E 00      -----------------------------------------------+
1.9600E 02        1.0008E 00      -----------------------------------------------+
2.0000E 02        1.0008E 00      -----------------------------------------------+
```

Fig. 2.49 Plotted output of TEMOUT with $KC = 1.8$ and full-scale plotting.

57

Example 2.9 *Simulation of Gear Train Containing Backlash*

The function generators shown in Appendix I are very useful for simulating nonlinear systems. An example of a system using a dead-space function is shown in Fig. 2.50.

This system represents a sinusoidal torque driving a flywheel through a set of gears having excessive backlash. The shaft, connecting gear 2 and the flywheel, is not rigid but has a spring constant of $K = 4000$ in-lb/radian. The contact force between gear teeth depends upon their relative position. Figure 2.51 shows the contact force as zero in the position where the teeth are not in contact and increasing in a linear manner after contact has been made.

The functional relationship between force and displacement can be simulated by a dead space (DEADSP) function generator.

Table 2.11

Formulation of Dead Space Function

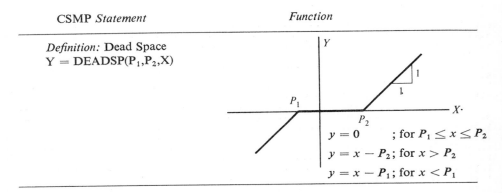

CSMP *Statement*	*Function*
Definition: Dead Space $Y = \text{DEADSP}(P_1, P_2, X)$	$y = 0$; for $P_1 \leq x \leq P_2$ $y = x - P_2$; for $x > P_2$ $y = x - P_1$; for $x < P_1$

Using the force characteristics shown on Fig. 2.51, the following DEADSP function can be used to simulate the contact force between gear teeth.

$$F = KG*DEADSP(-0.1, 0.1, R1*THETA1 - R2*THETA2)$$

where $R1*THETA1 - R2*THETA2$ = relative movement between gear teeth.

The equations for the angular accelerations of the gears and flywheel are:

$$\dot{\omega}_1 = WD1 = (T*SIN(W*TIME) - R1*F)/I1 \tag{2.26}$$

$$\dot{\omega}_2 = WD2 = (R2*F + K*(THETA3 - THETA2))/I2 \tag{2.27}$$

$$\dot{\omega}_3 = WD3 = K*(THETA2 - THETA3)/I3 \tag{2.28}$$

In the program of Fig. 2.52, the angular accelerations are calculated by ordinary FORTRAN statements. INTGRL functions are used to integrate angular accelerations and velocities to obtain angular velocities and angular displacements, respectively. All initial conditions are assumed to be zero.

It is often desirable to know the maximum and minimum values of variables during a run. The RANGE statement can be used for this purpose. In this example, it is used to list the maximums and minimums for 10 variables.

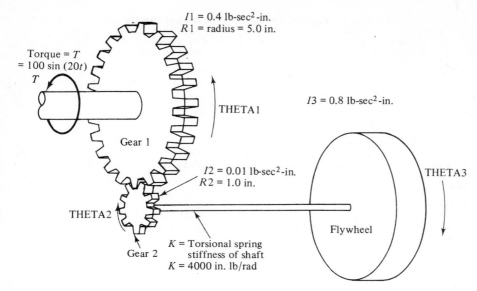

$I1 = 0.4$ lb-sec^2-in.
$R1 = $ radius $= 5.0$ in.

Torque $= T$
$= 100 \sin (20t)$

T

THETA1

$I3 = 0.8$ lb-sec^2-in.

Gear 1

$I2 = 0.01$ lb-sec^2-in.
$R2 = 1.0$ in.

THETA3

THETA2

$K = $ Torsional spring
stiffness of shaft
$K = 4000$ in. lb/rad

Gear 2

Flywheel

Fig. 2.50 Schematic of gear train system.

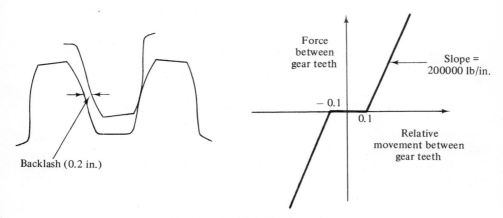

Force
between
gear teeth

Slope =
200000 lb/in.

-0.1

0.1

Relative
movement between
gear teeth

Backlash (0.2 in.)

Fig. 2.51 Backlash in gear teeth.

RANGE. The RANGE card allows the user to obtain both the maximum and minimum values of specified variables that occur during the run. Included with the maximum and minimum values are the times at which the values occurred. The times of maximum and minimum do not necessarily occur at a multiple of PRDEL or OUTDEL. An example of the output from the RANGE statement for this problem is shown in Fig. 2.53.

RANGE F, WD1, WD2, WD3, W1, W2, W3, THETA1, THETA2, THETA3

```
TITLE    SIMULATION OF THE DYNAMICS OF A GEAR TRAIN CONTAINING BACKLASH
   CONSTANT   I1 = 0.4, I2 = 0.01, I3 = 0.8, R1 = 5.0, R2 = 1.0,         ...
   KG = 200000.0, K = 4000.0, T = 100.0, W = 20.0
   F = KG * DEADSP(-0.1, 0.1, R1*THETA1 - R2*THETA2)
   WD1 = (T * SIN(W*TIME) - R1 * F)/I1
   WD2 = (R2 * F + K * (THETA3 - THETA2))/I2
   WD3 = K * (THETA2 - THETA3)/I3
   W1 = INTGRL(0.0, WD1)
   W2 = INTGRL(0.0, WD2)
   W3 = INTGRL(0.0, WD3)
   THETA1 = INTGRL(0.0, W1)
   THETA2 = INTGRL(0.0, W2)
   THETA3 = INTGRL(0.0, W3)
   PRINT F, WD1, WD2, WD3, W1, W2, W3, THETA1, THETA2, THETA3
   RANGE F, WD1, WD2, WD3, W1, W2, W3, THETA1, THETA2, THETA3
TIMER  FINTIM = 0.5, PRDEL = 0.005
END
STOP
ENDJOB
```

Fig. 2.52 Program for gear train simulation.

```
           PROBLEM DURATION 0.0            TO   5.0000E-01
```

VARIABLE	MINIMUM	TIME	MAXIMUM	TIME
F	-9.2400E 02	2.7134E-01	9.4110E 02	4.1301E-01
WD1	-1.1534E 04	4.1301E-01	1.1361E 04	2.7134E-01
WD2	-9.2923E 04	2.7134E-01	8.8692E 04	4.1301E-01
WD3	-1.7390E 02	2.2181E-01	1.8759E 02	3.8948E-01
W1	-3.0321E 00	2.1984E-01	3.2531E 00	3.8741E-01
W2	-1.8832E 01	3.6290E-01	2.0539E 01	1.8764E-01
W3	-2.7485E-01	3.1375E-01	2.7382E 00	4.6904E-01
THETA1	0.0	0.0	1.3119E-01	4.5464E-01
THETA2	-4.2604E-03	3.7764E-02	6.5328E-01	4.9929E-01
THETA3	0.0	0.0	6.3820E-01	5.0000E-01

Fig. 2.53 Example of output from a range statement.

Only one RANGE statement can be used, but this statement can contain up to 100 variables by using continuation cards.

Since more than eight variables are requested in the PRINT statement, the "equations format" as shown in Fig. 2.54 is automatically selected. Two TITLE cards are used to obtain two lines of heading on each page of output.

Example 2.10 *Beam Deflection Problem*

There exists a large class of engineering problems where it is necessary to use a trial and error solution. Two-point boundary-value problems generally fall within this class. The CSMP program has a feature for handling this type of problem and can be illustrated with the following cantilever-beam example.

This problem can be simply stated: find the deflection and stress of a thin steel cantilever beam of varying thickness with a 30 lb weight hung on the end. Figure 2.55 gives the basic configuration of the problem along with beam measurements.

The thin tapered beam will have a large deflection and consequently a large slope. For this reason, the exact equation for the radius of curvature is used in the following differential equation for beam deflection.

```
TIME = 0.0          F      = 0.0        WD1   = 0.0        WD2  = 0.0        WD3   = 0.0
                    W1     = 0.0        W2    = 0.0        W3   = 0.0        THETA1= 0.0
                    THETA2= 0.0         THETA3= 0.0

TIME = 5.0000E-03   F      = 0.0        WD1   = 2.4958E 01  WD2  = 0.0        WD3   = 0.0
                    W1     = 6.2448E-02  W2    = 0.0        W3   = 0.0        THETA1= 1.0411E-04
                    THETA2= 0.0         THETA3= 0.0

TIME = 1.0000E-02   F      = 0.0        WD1   = 4.9667E 01  WD2  = 0.0        WD3   = 0.0
                    W1     = 2.4917E-01  W2    = 0.0        W3   = 0.0        THETA1= 8.3166E-04
                    THETA2= 0.0         THETA3= 0.0

TIME = 1.5000E-02   F      = 0.0        WD1   = 7.3880E 01  WD2  = 0.0        WD3   = 0.0
                    W1     = 5.5829E-01  W2    = 0.0        W3   = 0.0        THETA1= 2.7999E-03
                    THETA2= 0.0         THETA3= 0.0

TIME = 2.0000E-02   F      = 0.0        WD1   = 9.7354E 01  WD2  = 0.0        WD3   = 0.0
                    W1     = 9.8673E-01  W2    = 0.0        W3   = 0.0        THETA1= 6.6135E-03
                    THETA2= 0.0         THETA3= 0.0

TIME = 2.5000E-02   F      = 0.0        WD1   = 1.1986E 02  WD2  = 0.0        WD3   = 0.0
                    W1     = 1.5302E 00  W2    = 0.0        W3   = 0.0        THETA1= 1.2859E-02
                    THETA2= 0.0         THETA3= 0.0

TIME = 3.0000E-02   F      = 0.0        WD1   = 1.4116E 02  WD2  = -3.5422E 03  WD3   = 4.4277E 01
                    W1     = 5.7468E-01  W2    = 1.1544E 01  W3   = 1.6570E-02  THETA1= 2.0949E-02
                    THETA2= 8.8597E-03   THETA3= 4.1982E-06

TIME = 3.5000E-02   F      = 0.0        WD1   = 1.6105E 02  WD2  = -6.7610E 03  WD3   = 8.4512E 01
                    W1     = -9.4938E-01 W2    = -4.0249E 00 W3   = 4.3920E-01  THETA1= 2.1422E-02
                    THETA2= 1.8021E-02   THETA3= 1.1190E-03

TIME = 4.0000E-02   F      = 0.0        WD1   = 1.7934E 02  WD2  = 7.6842E 02  WD3   = -9.6052E 00
                    W1     = -1.2725E 00 W2    = 4.6252E 00  W3   = 4.4856E-01  THETA1= 1.5725E-02
                    THETA2= 1.6512E-03   THETA3= 3.5723E-03

TIME = 4.5000E-02   F      = 0.0        WD1   = 1.9583E 02  WD2  = -6.6252E 02  WD3   = 8.2815E 00
                    W1     = -3.3377E-01 W2    = -3.6696E 00 W3   = 5.5224E-01  THETA1= 1.1675E-02
                    THETA2= 7.6850E-03   THETA3= 6.0287E-03

TIME = 5.0000E-02   F      = 0.0        WD1   = 2.1037E 02  WD2  = 5.5572E 02  WD3   = -6.9464E 00
                    W1     = 6.8258E-01  W2    = 4.7077E 00  W3   = 4.4753E-01  THETA1= 1.2517E-02
                    THETA2= 7.1776E-03   THETA3= 8.5669E-03

TIME = 5.5000E-02   F      = 0.0        WD1   = 2.2280E 02  WD2  = -4.4811E 02  WD3   = 5.6014E 00
                    W1     = 1.7664E 00  W2    = -3.7386E 00 W3   = 5.5310E-01  THETA1= 1.8613E-02
                    THETA2= 1.2157E-02   THETA3= 1.1037E-02

TIME = 6.0000E-02   F      = 0.0        WD1   = 2.3301E 02  WD2  = -8.3174E 03  WD3   = 1.0397E 02
                    W1     = 9.2847E-01  W2    = -2.6537E 00 W3   = 7.3739E-01  THETA1= 2.3806E-02
                    THETA2= 3.4727E-02   THETA3= 1.3933E-02

TIME = 6.5000E-02   F      = 0.0        WD1   = 2.4089E 02  WD2  = -3.2118E 03  WD3   = 4.0147E 01
                    W1     = -7.2547E-01 W2    = -8.4676E 00 W3   = 1.0940E 00  THETA1= 2.1541E-02
                    THETA2= 2.6572E-02   THETA3= 1.8542E-02
```

Fig. 2.54 Output from gear train simulation.

$$\frac{\dfrac{d^2y}{dx^2}}{\left[1 + \left(\dfrac{dy}{dx}\right)^2\right]^{3/2}} = \frac{M}{EI} \tag{2.29}$$

Symbols program
(Fig. 2.56)

where M = bending moment in beam	FORCE*(LO − X)
E = Young's Modulus (30×10^6 psi)	E
I = Moment of inertia = $\frac{1}{12}wt^3$	I
w = width of beam (1.0 in)	WIDTH
t = thickness of beam	T
y = deflection of beam	Y
$\dfrac{dy}{dx}$ = slope of beam	YP
$\dfrac{d^2y}{dx^2} = \dfrac{M}{EI}\left[1 + \left(\dfrac{dy}{dx}\right)^2\right]^{3/2}$	YPP

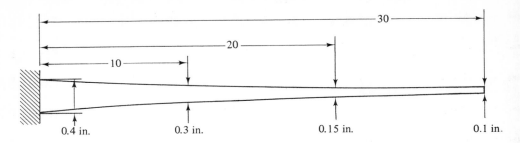

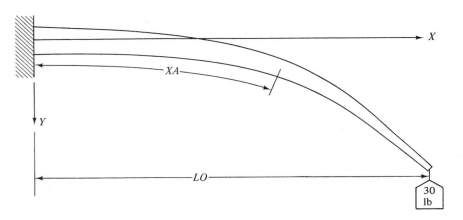

Fig. 2.55   Cantilever beam.

Notice that in Eq. (2.29), the independent variable X is the position in the horizontal direction. This is the first example in which the independent variable has not been time (TIME). This detail can easily be handled by renaming the independent variable using the control statement RENAME. A listing of the program in Fig. 2.56 shows an application of this statement.

RANAME.   The RENAME statement can be used to change the symbol for any of the six reserved names (DELT, DELMIN, FINTIM, OUTDEL, PRDEL, and TIME), as shown below:

$$\text{RENAME TIME} = \text{X}$$

In this example, the symbol X will be used in place of the reserve word TIME. The independent variable listed in all PRINT and PRTPLT outputs will be the variable X. Successive renaming on the same card must be separated by a comma. No continuation cards (...) are allowed with RENAME; however, several RE-NAME cards may be used.

The only requirement for the location of the RENAME card is that it must be placed before the TIMER statement.

```
LABEL  LARGE DEFLECTION OF A CANTILEVER BEAM
       RENAME TIME = X
INITIAL
       CONSTANT L= 30.0, WIDTH = 1.0,E = 10.0E6, FORCE = 30.0, LO = 26.0
       FUNCTION THICK = (0.0,0.4),(10.0,0.3),(20.0,0.15),(30.0,0.1)
DYNAMIC
       C = SQRT(1.0 + YP*YP)
       XA = INTGRL(0.0,C)
       T = NLFGEN(THICK,XA)
       I = WIDTH*T*T*T/12.0
       YPP = FORCE*(LO - X)*C*C*C/(E*I)
       YP = INTGRL(0.0,YPP)
       Y = INTGRL(0.0,YP)
       STRESS = FORCE*(LO - X)*T/(2.0*I)
TERMINAL
       TIMER FINTIM = 30.0
       FINISH XA = L
       IF(ABS(X - LO).LT.0.3) GO TO 1
       LO = LO + 0.2*(X - LO)
       CALL RERUN
     1 CONTINUE

*      THE FOLLOWING END CARD RESETS THE INITIAL CONDITIONS AND INITIATES
*      ANOTHER RUN USING A VALUE OF OUTDEL = 0.6.

END
       TIMER OUTDEL = 0.6
       PRTPLT Y (STRESS,XA,YP)
END
STOP
ENDJOB
```

Fig. 2.56 Program for deflection of cantilever beam.

Since the beam is curved, the X coordinate is not equal to the distance as measured along the center line of the beam. The distance from the base along the beam is given by

$$XA = \int_o^x \left[1.0 + \left(\frac{dy}{dx} \right)^2 \right]^{1/2} dx \qquad (2.30)$$

CSMP statements for the above expression are

$$C = SQRT(1.0 + YP*YP)$$
$$XA = INTGRL(0.0,\ C)$$

The value of XA must be used as the argument of the nonlinear function-generator statement to calculate the thickness of the beam.

$$T = NLFGEN(THICK,\ XA)$$

Since the beam is curved in its final deflected shape, the exact horizontal LO position of the end of the beam is not known. An initial estimate of 26 in. is made on a CONSTANT card. With this initial estimate the values of y, dy/dx, d^2y/dx^2 and stress are calculated as

$$\frac{d^2y}{dx^2} = YPP = FORCE*(LO - X)*C*C*C/(E*I) \qquad (2.31)$$

$$\frac{dy}{dx} = \int_o^x \frac{d^2y}{dx^2} dx = YP = INTGRL(0.0,\ YPP) \qquad (2.32)$$

$$y = \int_o^x \frac{dy}{dx}\, dx = \mathrm{Y} = \mathrm{INTGRL}(0.0,\, \mathrm{YP}) \qquad (2.33)$$

$$\sigma = \frac{mt}{2I} = \mathrm{STRESS} = \mathrm{FORCE}*(\mathrm{LO} - \mathrm{X})*\mathrm{T}/(2.0*\mathrm{I}) \qquad (2.34)$$

The calculations in the Dynamic segment continue until the distance integrated along the beam XA is equal to the total length L. At this point the first run is terminated by the following FINISH statement:

FINISH XA = L

The FINISH statement terminates a run when the variable XA reaches or first crosses the specified bound L. At this point the actual horizontal position of the end of the beam X is compared with the first estimate LO. If the first estimate for the horizontal location of the end of the beam is not sufficiently close (less than 0.3 in.) to the calculated value, additional runs will be required until this criterion is satisfied. The CALL RERUN feature is used to solve this trial and error problem.

CALL RERUN. The CALL RERUN statement must be placed in the Terminal segment to recycle the program through additional runs with new parameters. All constants used in succeeding runs will have the same values as specified in the Initial segment unless otherwise specified in the Terminal segment. The program listing of Fig. 2.56 illustrates the use of the CALL RERUN statement.

After the first run is terminated by the FINISH statement, the difference between the actual horizontal position of the end of the beam X is compared with the initial guess LO. If the absolute difference is greater than or equal to 0.3 in., a new estimate for LO is calculated and, the CALL RERUN statement will recycle the program using the new estimate for LO and all other constants as specified in the Initial segment.

Successive runs will occur until the absolute difference between the actual horizontal position and the estimate is within the desired range. When this occurs, the FORTRAN IF statement will transfer the execution to the CONTINUE statement which bypasses the CALL RERUN card.

In this example, the CONTINUE card is used as a FORTRAN statement. If the statement number were not included, the CONTINUE card would have been executed as a CSMP statement. A detailed explanation of the CSMP CONTINUE card is given in Chap. 3.

Notice that there is no output statement before the first END card. This means that there will be no output until the final run is completed. If output is desired at the end of each iterative run, PRINT, or PRTPLT statements should be placed before the first END card. The first END card is used after a CALL RERUN statement to reset the independent variable X to zero, and to initiate another run. The use of the END statement for making multiple runs is covered in Chap. 3. The PRTPLT output for the final run is shown in Fig. 2.57.

One of the most important statements for trial and error solution is the algo-

```
LARGE DEFLECTION OF A CANTILEVER BEAM                          PAGE   1

                        MINIMUM          Y    VERSUS  X        MAXIMUM
                          0.0                                  1.2606E 01
 X           Y           I                                          I    STRESS      XA          YP
•.0          0.0         +                                               2.9250E 04  0.0         0.0
.0000E-01    2.6519E-03  +                                               2.9452E 04  6.0001E-01  8.8725E-03
.2000E 00    1.0687E-02  +                                               2.9653E 04  1.2001E 00  1.7945E-02
.8000E 00    2.4228E-02  +                                               2.9852E 04  1.8002E 00  2.7226E-02
.4000E 00    4.3402E-02  +                                               3.0048E 04  2.4005E 00  3.6724E-02
.0000E 00    6.8342E-02  +                                               3.0243E 04  3.0010E 00  4.6448E-02
.6000E 00    9.9186E-02  +                                               3.0434E 04  3.6018E 00  5.6408E-02
.2000E 00    1.3608E-01  +                                               3.0622E 04  4.2030E 00  6.6615E-02
.8000E 00    1.7918E-01  +                                               3.0806E 04  4.8045E 00  7.7079E-02
.4000E 00    2.2863E-01  +                                               3.0985E 04  5.4066E 00  8.7813E-02
.0000E 00    2.8461E-01  -+                                              3.1159E 04  6.0092E 00  9.8829E-02
.6000E 00    3.4728E-01  -+                                              3.1326E 04  6.6124E 00  1.1014E-01
.2000E 00    4.1684E-01  -+                                              3.1486E 04  7.2164E 00  1.2176E-01
.8000E 00    4.9346E-01  -+                                              3.1638E 04  7.8213E 00  1.3371E-01
.4000E 00    5.7736E-01  --+                                             3.1780E 04  8.4272E 00  1.4600E-01
.0000E 00    6.6874E-01  --+                                             3.1912E 04  9.0341E 00  1.5865E-01
.6000E 00    7.6782E-01  ---+                                            3.2031E 04  9.6422E 00  1.7168E-01
.0200E 01    8.7484E-01  ---+                                            3.2277E 04  1.0252E 01  1.8513E-01
.0800E 01    9.9010E-01  ---+                                            3.2751E 04  1.0863E 01  1.9919E-01
.1400E 01    1.114CE 00  ----+                                           3.3279E 04  1.1475E 01  2.1399E-01
.2000E 01    1.2470E 00  ----+                                           3.3866E 04  1.2090E 01  2.2962E-01
.2600E 01    1.3897E 00  -----+                                          3.4520E 04  1.2707E 01  2.4622E-01
.3200E 01    1.5427E 00  ------+                                         3.5249E 04  1.3326E 01  2.6391E-01
.3800E 01    1.7067E 00  ------+                                         3.6066E 04  1.3948E 01  2.8289E-01
.4400E 01    1.8825E 00  -------+                                        3.6984E 04  1.4573E 01  3.0337E-01
.5000E 01    2.0711E 00  --------+                                       3.8019E 04  1.5202E 01  3.2562E-01
.5600E 01    2.2737E 00  ---------+                                      3.9194E 04  1.5835E 01  3.5001E-01
.6200E 01    2.4916E 00  ---------+                                      4.0538E 04  1.6474E 01  3.7698E-01
.6800E 01    2.7267E 00  ----------+                                     4.2089E 04  1.7118E 01  4.0717E-01
.7400E 01    2.9810E 00  -----------+                                    4.3898E 04  1.7770E 01  4.4142E-01
.8000E 01    3.2574E 00  ------------+                                   4.6041E 04  1.8430E 01  4.8093E-01
.9200E 01    3.5595E 00  -------------+                                  4.8625E 04  1.9102E 01  5.2752E-01
.9800E 01    3.8924E 00  ---------------+                                5.1818E 04  1.9788E 01  5.8402E-01
.0400E 01    4.2630E 00  ----------------+                               5.2945E 04  2.0494E 01  6.5323E-01
.1000E 01    4.6785E 00  ------------------+                             5.2539E 04  2.1224E 01  7.3370E-01
.1600E 01    5.1462E 00  --------------------+                           5.1556E 04  2.1984E 01  8.2803E-01
.2200E 01    5.6757E 00  ----------------------+                         4.9863E 04  2.2785E 01  9.4020E-01
.2800E 01    6.2791E 00  ------------------------+                       4.7302E 04  2.3636E 01  1.0757E 00
.3400E 01    6.9728E 00  ---------------------------+                    4.3680E 04  2.4553E 01  1.2422E 00
.4000E 01    7.7780E 00  ------------------------------+                 3.8772E 04  2.5553E 01  1.4495E 00
.4600E 01    8.7224E 00  ---------------------------------+              3.2331E 04  2.6677E 01  1.7075E 00
.5200E 01    1.1142E 01  ----------------------------------------+       2.4167E 04  2.7943E 01  2.0147E 00
.5800E 01    1.2606E 01  -------------------------------------------+    1.4345E 04  2.9379E 01  2.3267E 00
                                                                   +     3.6008E 03  3.0961E 01  2.5207E 00
```

Fig. 2.57   PRTPLT output for deflection of beam.

rithm used in the terminal segment to recalculate LO. The calculation that was used in this example is

$$LO = LO + 0.2*(X - LO) \qquad (2.35)$$

This statement is not meant to be the optimum algorithm for calculating the new estimate for LO. It is only a calculation that yielded desired results. Extreme care should be used in choosing algorithms for recalculating new estimates of parameters. These calculations can affect not only the number of iterative runs but can often make the difference between a convergent or divergent solution.

Example 2.11   *Stress Analysis of Cam-Follower System*

In all of the previous examples, some form of numerical integration was performed. This example illustrates the use of CSMP to make a sequence of calculations where no integration is conducted. Other computer languages such as FORTRAN and BASIC could be used for this particular type of problem but CSMP has the advantage of particular capabilities such as FUNCTION and PRTPLT.

The problem is to calculate the contact stress between a cam and roller-follower at 2° cam increments. A schematic of the physical system is shown in Fig. 2.58.

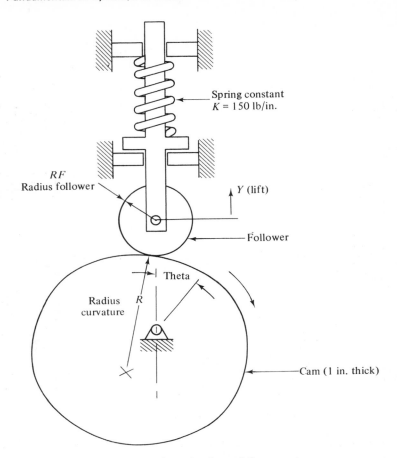

Fig. 2.58   Schematic of cam-follower system.

The stress is a function of the angular position of the cam, consequently, the independent variable is THETA. The first statement of the program of Fig. 2.59 uses the RENAME card to change the symbol for the independent variable from TIME to THETA.

The maximum compressive contact stress for a 1 in. thick steel cam and roller-follower is given by the Hertz formula.

$$\text{STRESS} = 2291\sqrt{\frac{F(R + RF)}{R*RF}} \text{ (psi)} \tag{2.36}$$

where R = radius of curvature of cam profile at the
point of contact (in.)

RF = radius of roller-follower (in.)

F = radial load (lb)

```
          RENAME TIME = THETA
LABEL     CALCULATION OF CONTRACT STRESS BETWEEN CAM AND FOLLOWER
          CONSTANT RF = 0.75,   K = 150.0,   YI = 0.25
          FUNCTION LIFT =(0.0,0.0),(30.0,0.0937),(60.0,0.375),(90.0,0.75)...
          ,(120.0,1.125),(150.0,1.406),(180.0,1.5),(210.0,1.5),          ...
          (240.0,1.357),(270.0,0.9818),(300.0,0.375),(330.0,0.1433),      ...
          (360.0,0.0)
          FUNCTION RADCUV =  (0.0,2.0),(30.0,2.4),(60.0,2.5),(90.0,2.75),...
          (120.0,3.0),(150.0,3.3),(180.0,3.5),(210.0,3.5),(240.0,2.95),   ...
          (270.0,2.8),(300.0,2.6),(330.0,2.5),(360.0,2.0)
          Y = NLFGEN(LIFT,THETA)
          F = K*(Y + YI)
          R = NLFGEN(RADCUV,THETA)
          STRESS = 2291.0*SQRT(F*(R + RF)/(R*RF))
          PRTPLT STRESS (F,R,Y)
          TIMER FINTIM = 360.0,  OUTDEL = 2.0
END
STOP
ENDJOB
```

Fig. 2.59 Program to calculate contact stress between cam and follower.

The radius of curvature of the cam surface is measured from the drawing of the cam at 30° increments and this information is included in the following FUNCTION statement.

FUNCTION RADCUV = (0.0,2,0),(30.0,2.4),(60.0,2.5),(90.0,2.75), ...
(120.0,3.0),(150.0,3.3),(180.0,3.5),(210.0,3.5),(240.0,2.95),   ...
(270.0,2.8),(300.0,2.6),(330.0,2.5),(360.0,2.0)

To simplify the calculation of force, the cam is assumed to be rotating at a speed where the inertial forces can be neglected. A static analysis yields the following expression for the force between the cam and follower.

$$F = K(Y + YI) \tag{2.37}$$

where Y = lift of follower

YI = initial compression of spring

The lift Y is also a function of THETA and is measured from the drawing at 30° increments. This information is included in the following FUNCTION statement.

FUNCTION LIFT = (0.0,0.0),(30.0,0.0937),(60.0,0.375),(90.0,0.75), ...
(120.0,1.125),(150.0,1.406),(180.0,1.5),(210.0,1.5),   ...
(240.0,1.357),(270.0,0.9818),(300.0,0.375),(330.0,0.1433),   ...

Since both the radius of curvature of the cam and the lift are represented by smooth curves, the NLFGEN function-generating element is used to calculate values for R and Y.

The TIMER card controls the increment and range of calculations. In this example, FINTIM was set equal to 360.0 which corresponds to a cam rotation of one revolution. A value of OUTDEL = 2.0 forces the calculations to be performed and print-plotted at 2° increments. One page of the printer-plot output is shown in Fig. 2.60. Note the maximum and minimum values for the print-plot variable STRESS are always included in the PRTPLT output.

```
CALCULATION OF CONTRACT STRESS BETWEEN CAM AND FOLLOWER          PAGE   1

                       MINIMUM              STRESS VERSUS THETA        MAXIMUM
                       1.8996E 04                                      4.7359E 04
    THETA        STRESS       I                                           I        F              R             Y
    0.0        1.8996E 04  +                                                   3.7500E 01    2.0000E 00    0.0
    2.0000E 00  1.9197E 04  +                                                  3.8437E 01    2.0267E 00    6.2467E-0
    4.0000E 00  1.9396E 04  +                                                  3.9374E 01    2.0533E 00    1.2493E-
    6.0000E 00  1.9592E 04  -+                                                 4.0311E 01    2.0800E 00    1.8740E-
    8.0000E 00  1.9785E 04  -+                                                 4.1248E 01    2.1067E 00    2.4987E-
    1.0000E 01  1.9975E 04  -+                                                 4.2185E 01    2.1333E 00    3.1233E-
    1.2000E 01  2.0163E 04  --+                                                4.3122E 01    2.1600E 00    3.7480E-
    1.4000E 01  2.0349E 04  --+                                                4.4059E 01    2.1667E 00    4.3727E-
    1.6000E 01  2.0533E 04  --+                                                4.4996E 01    2.2133E 00    4.9973E-
    1.8000E 01  2.0714E 04  ---+                                               4.5933E 01    2.2400E 00    5.6220E-
    2.0000E 01  2.0894E 04  ---+                                               4.6870E 01    2.2667E 00    6.2467E-
    2.2000E 01  2.1071E 04  ---+                                               4.7807E 01    2.2933E 00    6.8713E-
    2.4000E 01  2.1246E 04  ---+                                               4.8744E 01    2.3200E 00    7.4960E-
    2.6000E 01  2.1420E 04  ----+                                              4.9681E 01    2.3467E 00    8.1207E-
    2.8000E 01  2.1591E 04  ----+                                              5.0618E 01    2.3733E 00    8.7453E-
    3.0000E 01  2.1761E 04  ----+                                              5.1555E 01    2.4000E 00    9.3700E-
    3.2000E 01  2.2149E 04  -----+                                             5.3492E 01    2.4160E 00    1.0662E-
    3.4000E 01  2.2556E 04  ------+                                            5.5555E 01    2.4307E 00    1.2037E-
    3.6000E 01  2.2981E 04  -------+                                           5.7743E 01    2.4440E 00    1.3495E-
    3.8000E 01  2.3423E 04  --------+                                          6.0055E 01    2.4560E 00    1.5037E-
    4.0000E 01  2.3881E 04  --------+                                          6.2493E 01    2.4667E 00    1.6662E-
    4.2000E 01  2.4355E 04  ---------+                                         6.5056E 01    2.4760E 00    1.8371E-
    4.4000E 01  2.4844E 04  ----------+                                        6.7744E 01    2.4840E 00    2.0163E-
    4.6000E 01  2.5347E 04  -----------+                                       7.0557E 01    2.4907E 00    2.2038E-
    4.8000E 01  2.5863E 04  ------------+                                      7.3495E 01    2.4960E 00    2.3997E-
    5.0000E 01  2.6391E 04  -------------+                                     7.6558E 01    2.5000E 00    2.6039E-
    5.2000E 01  2.6932E 04  -------------+                                     7.9746E 01    2.5027E 00    2.8164E-
    5.4000E 01  2.7484E 04  --------------+                                    8.3060E 01    2.5040E 00    3.0373E-
    5.6000E 01  2.8047E 04  ---------------+                                   8.6498E 01    2.5040E 00    3.2665E-
    5.8000E 01  2.8621E 04  ----------------+                                  9.0061E 01    2.5027E 00    3.5041E-
    6.0000E 01  2.9205E 04  ----------------+                                  9.3750E 01    2.5000E 00    3.7500E-
    6.2000E 01  2.9700E 04  -----------------+                                 9.7063E 01    2.5120E 00    3.9708E-
    6.4000E 01  3.0194E 04  ------------------+                                1.0044E 02    2.5247E 00    4.1959E-
    6.6000E 01  3.0688E 04  -------------------+                               1.0388E 02    2.5380E 00    4.4250E-
    6.8000E 01  3.1181E 04  -------------------+                               1.0738E 02    2.5520E 00    4.6584E-
    7.0000E 01  3.1674E 04  --------------------+                              1.1094E 02    2.5667E 00    4.8959E-
    7.2000E 01  3.2166E 04  ---------------------+                             1.1456E 02    2.5820E 00    5.1376E-
    7.4000E 01  3.2657E 04  ---------------------+                             1.1825E 02    2.5980E 00    5.3834E-
    7.6000E 01  3.3147E 04  ----------------------+                            1.2200E 02    2.6147E 00    5.6334E-
    7.8000E 01  3.3636E 04  -----------------------+                           1.2581E 02    2.6320E 00    5.8876E-
    8.0000E 01  3.4124E 04  -----------------------+                           1.2969E 02    2.6500E 00    6.1459E-
    8.2000E 01  3.4612E 04  ------------------------+                          1.3363E 02    2.6687E 00    6.4084E-
    8.4000E 01  3.5098E 04  -------------------------+                         1.3763E 02    2.6880E 00    6.6750E-
    8.6000E 01  3.5584E 04  --------------------------+                        1.4169E 02    2.7080E 00    6.9459E-
    8.8000E 01  3.6068E 04  --------------------------+                        1.4581E 02    2.7287E 00    7.2208E-
    9.0000E 01  3.6552E 04  ---------------------------+                       1.5000E 02    2.7500E 00    7.5000E-
    9.2000E 01  3.6982E 04  ---------------------------+                       1.5375E 02    2.7667E 00    7.7500E-
    9.4000E 01  3.7406E 04  ----------------------------+                      1.5750E 02    2.7833E 00    8.0000E-
    9.6000E 01  3.7825E 04  ----------------------------+                      1.6125E 02    2.8000E 00    8.2500E-
    9.8000E 01  3.8238E 04  -----------------------------+                     1.6500E 02    2.8167E 00    8.5000E-
    1.0000E 02  3.8647E 04  -----------------------------+                     1.6875E 02    2.8333E 00    8.7500E-
```

Fig. 2.60 Printer-plot output for contact stress.

REFERENCES

1. *System/360 Continuous System Modeling Program User's Manual* GH20-0367-4, Program Number 360A-CX-16X, IBM Data Processing Division, White Plains, N.Y.

2. *System/360 Continuous System Modeling Program, Operations Guide*, Program Number 360A-CX-16X, IBM Data Processing Division, White Plains, N.Y.

3. *Continuous System Modeling Program III (CSMP III) Operations Guide*, Program Number 5734-XS9, SH19-7002-1, IBM Data Processing Division, White Plains, N.Y.

4. McCracken, Daniel D., *A Guide To Fortran IV Programming* 2nd ed. New York: John Wiley and Sons, Inc.

PROBLEMS

1 Write a CSMP program for solving the following nonlinear differential equation:

$$\frac{d^2x}{dt^2} + 25(1.0 + 0.1x^2)x = 0$$

subject to the initial conditions,

$$x(0) = 3.0$$
$$\dot{x}(0) = 0.0$$

Answer:†

$$x(t) = 2.9531 \text{ at TIME} = 1.0$$

2 A high-temperature oven is heated by an electrical element supplying 12,000 BTU/hr. The heat loss from the oven by convection and radiation is calculated to be the following.

$$q_{\text{loss}} = 2 \times 10^{-8}(T_o^4 - T_a^4) + 8.0(T_o - T_a) \text{ BTU/hr}$$

If the initial temperature is 530°R (70°F), determine the oven temperature at 0.05 hr, 0.1 hr, 0.3 hr, and 0.5 hr. The differential equation for this transient heat transfer problem is

$$C_t \dot{T}_o = q_{\text{supplied}} - q_{\text{loss}}$$

$$C_t = \text{thermal capacitance, 24 BTU/°R}$$

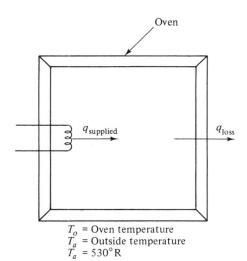

T_o = Oven temperature
T_a = Outside temperature
T_a = 530°R   **Fig. P2.2**

Answer:

$$\text{At TIME} = 0.5 \text{ hr, } T_o = 723.6°R$$

†Answers are given to the problems at particular discrete times. If the programs are written correctly then the solutions should agree at the designated value of TIME.

3 Given the system shown in Fig. P2.3, if the input is a unit-step (STEP(0.0)), write a program to find the printer-plot of the output. Use a FINTIM of 1.3 and an OUT-DEL of 0.026.

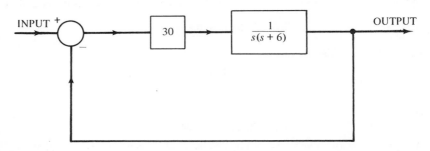

Fig. P2.3

Answer:

$$\text{Output} = 0.81395 \text{ at TIME} = 0.364$$

4 The period of vibration of a pendulum for small angles is

$$2\pi\sqrt{\frac{L}{g}}.$$

Using the capability of a PARAMETER card to make sequential runs, compare the periods of oscillation of a pendulum which has initial angles of 5°, 20°, 45°, 90°, and 135°. The equation of motion for the pendulum is

$$L\ddot{\theta} + g \sin(\theta) = 0$$

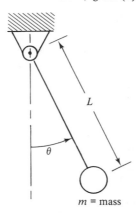

m = mass   **Fig. P2.4**

Answer:

$$5° \text{ period} = 6.286\sqrt{L/g}$$
$$20° \text{ period} = 6.331\sqrt{L/g}$$
$$45° \text{ period} = 6.534\sqrt{L/g}$$
$$90° \text{ period} = 7.416\sqrt{L/g}$$
$$135° \text{ period} = 9.600\sqrt{L/g}$$

5 A system is represented by the vector differential equation,

$$\dot{x} = Ax + bu$$

in which

$$x = \begin{bmatrix} x_1(t) \\ x_2(t) \\ x_3(t) \end{bmatrix} \qquad A = \begin{bmatrix} 0 & 1 & 0 \\ 0 & -10 & 1 \\ -16 & 0 & -2 \end{bmatrix} \qquad b = \begin{bmatrix} 0 \\ 0 \\ 16 \end{bmatrix} \qquad u = \text{unit-step}$$

Use CSMP to find $x_1(t)$, $x_2(t)$, and $x_3(t)$.

Answer:

$$\text{At TIME} = 1.9: x_1(t) = 0.793$$

6 The differential equation for the motion of the base of an unbalanced electric motor is shown below.

$$M\ddot{x} + c\dot{x} + kx = me\omega^2 \sin(\omega t)$$

$M = 0.1$ lb-sec²/in

$c = 1.2$ lb-sec/in

$k = 800$ lb/in

$me = 0.009$ lb-sec²

Assuming the system is initially at rest, find the time-history of the motion of the base. The motor speed increases according to the following formula.

$$\omega = 120\pi(1 - e^{-t/4})$$

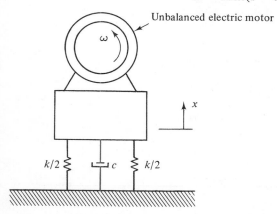

Unbalanced electric motor

Fig. P2.6

Answer:

When $\omega = 90$ rad/sec, $x = 0.067$ in.

7 The system diagram of a process having time delay is shown in Fig. P-2.7
Find the unit-step response of the system. Compare your results with those given in Fig. 2.14 which is the step response of the system above without delay. Use FINTIM = 8.0, and OUTDEL = 0.16.

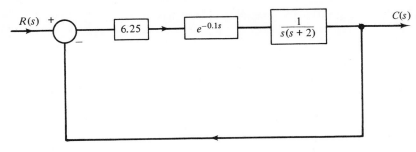

Fig. P2.7

Answer:
    Using $N = 20$ in the DELAY Function gives $c(t) = 1.4116$ at TIME $= 1.44$

8   Determine the time-history for the water level in each of the three tanks. Initially the water level in all tanks is 10ft. The cross-sectional areas of the tanks are shown below.

$$A_1 = 4 \text{ ft}^2$$
$$A_2 = 8 \text{ ft}^2$$
$$A_3 = 12 \text{ ft}^2$$

The general equation describing the height of water is

$$A_i \dot{H}_i = q_{in} - q_{out}$$

The flow through an orifice is given below.

$$q = A_{orifice}\sqrt{2g\Delta H}$$

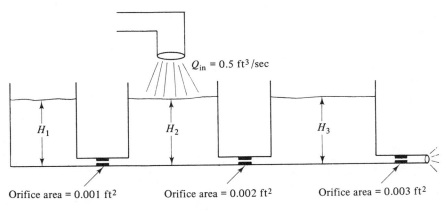

Fig. P2.8

Answer:

$$\text{At TIME} = 50 \text{ sec: } H_1 = 10.11 \text{ ft}$$
$$H_2 = 12.95 \text{ ft}$$
$$H_3 = 9.76 \text{ ft}$$

9   A space vehicle is reentering the earth's atmosphere. In the position shown, the vehicle is 200,000 ft from the surface of the earth and the speed is 25,000 mph. The aerodynamic drag is:

$$f_{total} = 0.5AC_d\rho V^2$$

where A = cross sectional area of vehicle = 80 ft^2

$\quad C_d$ = drag coefficient = 0.55

$\quad \rho = 0.00238\ e^{-h/24000}$ (slug/ft^3)

$\quad h$ = altitude in feet

The equations of motion in cylindrical coordinates are shown below.

$$m(\ddot{r} - r\dot{\theta}^2) = f_{drag\ radial} - mg$$
$$m(r\ddot{\theta} + 2\dot{r}\dot{\theta}) = -f_{drag\ tangential}$$

(a) Using this information, how fast does the vehicle strike the earth? (b) What is the total distance traveled by the vehicle? *Hint:* $s = \int_o^t v\ dt$ (c) How much energy is dissipated in aerodynamic drag? *Hint:* Energy $= \int_o^t \text{force}\ \dfrac{ds}{dt}\ dt$

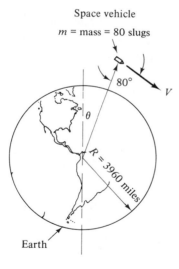

Space vehicle

m = mass = 80 slugs

$80°$

V

θ

$R = 3960\ miles$

Earth

Fig. P2.9

Answer:

(a) 224 ft/sec (b) 779,000 ft (c) 5.43 × 10^{10} ft-lb

10 The block diagram of a system containing amplifier saturation is given in Fig. P2.10. Use the function generator Y = LIMIT(P$_1$,P$_2$,X) given in Appendix I to find the response of the system if $r(t)$ = R(s) = 3.5u(t) where u(t) is a unit-step starting at $t = 0$. Select FINTIM = 4.0, OUTDEL = 0.08, and obtain the printer-plot output of $c(t)$.

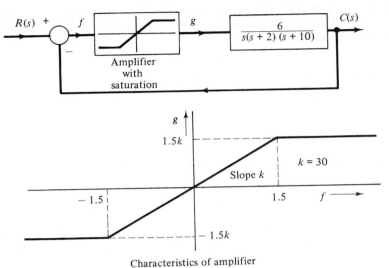

Characteristics of amplifier

Fig. P2.10

Answer:
$$\text{At TIME} = 1.28 \ c(t) = 5.2106$$

11 To test the impact characteristics of neoprene, a 36.6 lb weight is dropped on a doughnut-shaped specimen. The velocity at impact is 52.7 in/sec. The acceleration of the weight was measured to be the following:

Time (sec)	Acceleration (g's)	
0	0	$1 \ g = 386 \ \text{in/sec}^2$
0.002	−13.0	
0.003	−23.0	
0.004	−32.0	
0.005	−35.0	
0.006	−34.0	
0.007	−33.0	
0.008	−24.0	
0.01	−11.0	
0.012	0	
0.015	1.0	

Using the AFGEN function to represent data taken from this test, determine the time-history of the velocity and displacement of the top surface of the neoprene. Also determine the energy absorbed by the neoprene.

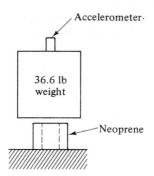

Fig. P2.11

Answer:

at TIME = 0.0075 secs

displacement = 0.208 in.

velocity = −15.0 in/sec

12 An automobile passes over a triangular bump at a speed of 20 ft/sec. Determine the maximum pitch and vertical motion of the center of gravity. Neglect the deflection of the tires and the effects of the unsprung mass. The automobile has the following parameters.

Weight = 4200 lb

Spring constants = front: 120 lb/in., rear: 180 lb/in.

Damping coefficients = front: 10 lb-sec/in., rear: 12.0 lb-sec/in.

Pitch moment of inertia = 40,000 lb-sec²/in.

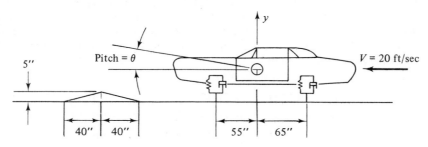

Fig. P2.12

Hint: $I_{\ddot{\theta}} = \Sigma$ torques due to springs and shock absorbers about center of gravity

$m\ddot{y} = \Sigma$ forces due to spring and shock absorbers.

Answer:

$y_{max} = 2.09$ in.; $\theta_{max} = 1.84°$

13 Same as the above problem, except the shock absorbers are more realistically represented by having different linear damping coefficients in the compression and rebound directions.

Damping coefficients = front: compression 6.0 lb-sec/in.; rebound
14.0 lb-sec/in.
rear: compression 8.0 lb-sec/in.; rebound
18.0 lb-sec/in.

Answer:

$$y_{max} = 1.588 \text{ in.}, \ \theta_{max} = 1.395°$$

14 A block diagram model of a position control system with tachometer feedback is given in Fig. P2.14.

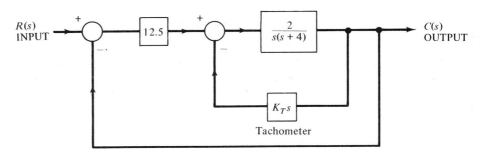

Fig. P2.14

Let $R(s)$ correspond to a unit-step [i.e., INPUT = STEP(0.0)] and find $c(t)$ for values of $K_T = 0.1, 1.5, 4.0$. Use the PARAMETER card for running multiple values of K_T. Specify FINTIM = 5.0, and OUTDEL = 0.1 for the printer-plot of the output. Also specify that ERROR and INPUT are listed on the same page which gives the printer-plot of the output. Use a LABEL card with your choice of wording for the title of the printer-plot output.

Answer:

$$c(t) = 1.2329 \quad \text{at TIME} = 0.7 \quad K_T = 1.0$$
$$c(t) = 1.0468 \quad \text{at TIME} = 0.9 \quad K_T = 1.5$$
$$c(t) = 0.9919 \quad \text{at TIME} = 1.9 \quad K_T = 4.0$$

15 The following Blasius equation is a transformed equation describing the laminar boundary layer on a flat plate.

$$\frac{d^3 f}{d\eta^3} + \frac{d^2 f}{d\eta^2} f = 0$$

Based on previous experience, the following boundary conditions will yield a satisfactory solution.

$$f(\eta = 0) = \frac{df}{d\eta}(\eta = 0) = 0 \qquad \frac{df}{d\eta}(\eta = 3) = 2$$

Using a trial and error solution determine the value of

$$\frac{d^2 f}{d\eta^2}(\eta = 0)$$

Answer:

$$\frac{d^2 f}{d\eta^2}(\eta = 0) = 1.328$$

16 Solve the following set of differential equations.

$$3\ddot{x}_1 + \ddot{x}_2 + 5\dot{x}_1^3 + 10x_1 = 10 + e^{-t}$$

$$\ddot{x}_1 + 4\ddot{x}_2 + 25\dot{x}_2 + 4x_2^3 = 0$$

where

$$x_1(0) = x_2(0) = 0$$

$$\dot{x}_1(0) = 2 \qquad \dot{x}_2(0) = 7$$

Answer:

$$\text{At TIME} = 1.0 \qquad x_1 = 1.276$$

17 The diagram below represents an experimentally measured water temperature increase in a steam generating boiler due to a step input of fuel. The fuel is natural

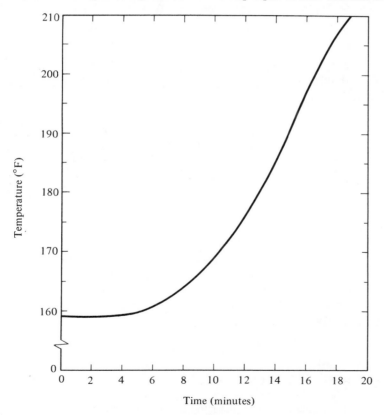

Fig. P2.17

gas and the step input level is 10,000 ft³/hr. The heat content of the gas is approximately 1000 Btu/ft³. A proposed model for matching the temperature increase (over the range $t = 0$ to $t = 19$ min) due to the step input of fuel is given by

$$\frac{T(s)}{U(s)} = \frac{K_2 \epsilon^{-\tau_d s}}{s(1 + \tau_1 s)(1 + \tau_2 s)}$$

where T has units of °F, K_2 has units of °F/ft³, U(input) is in ft³/min, and τ_d, τ_1, τ_2 have units of minutes. Use CSMP to assist you in determining appropriate values for K_2, τ_d, τ_1, and τ_2.

Answers:

$$K_2 = 0.04 \text{ °F/ft}^3, \quad \tau_d = 4 \text{ min}, \quad \tau_1 = 2 \text{ min}, \quad \tau_2 = 5 \text{ min}$$

18 A red hot cylindrical piece of iron is brought from a heat treating furnace and is allowed to cool in still air. Energy is removed from the piece by both convection and radiation. Assuming the iron remains at a uniform temperature, the following equation can be used to describe the transient temperature.

$$mc\frac{dT}{dt} = hA(T_a - T) + A\epsilon\sigma(T_a^4 - T^4)$$

where: $T_a = 545°R$, $\quad m = 161.7$ lbm, $\quad c = 0.13$ Btu-lbm/°R

$\quad\quad\quad A = 3.01$ ft², $\quad h = 0.18(T - T_a)^{0.33}$ Btu/ft²-hr $\quad$ (assuming turbulence flow over piece)

$\quad\quad\quad \epsilon = 0.95, \quad\quad \sigma = 0.1714 \times 10^{-8}$ Btu/hr-ft²-°R⁴

Use the FINISH card to determine how long it takes for the piece to cool from 1910°R to 1300°R.

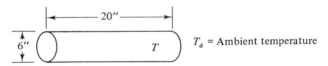

Fig. P2.18

Answer:

0.378 hours

19 The following equations describe the angular recoil of a hand gun. The assumption is made that the gun is loosely held so that no external moment is applied to the gun. The restraining force vector passes through the center of mass. The parameters for a 38 caliber pistol are shown below.

Bullet weight $= w = 0.0207$ pounds

Moment of inertia about center of mass $= I = 0.0105$ lb-in.-sec²

$d = 0.9$ inches

Cross-sectional area of bullet $= A = 0.1$ in.²

Find the angular recoil of the hand gun and muzzle velocity for the following description of barrel pressure.

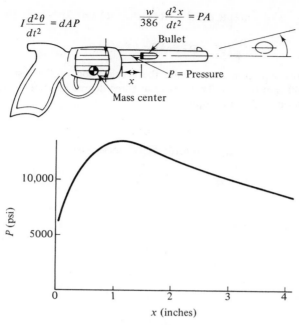

$$I\frac{d^2\theta}{dt^2} = dAP$$

$$\frac{w}{386}\frac{d^2x}{dt^2} = PA$$

Bullet

P = Pressure

x

Mass center

Fig. P2.19

Answer:

At x = 3.962 in., θ = 0.018212 radians,

Closed form solution:

$$x = 12600 \text{ in./sec, } \theta = xwd/(386I)$$

20 An automobile engine has a stroke of 3.70 in. and a connecting rod length of 5.84 in. as shown in Fig. P2.20. The following is an expression for the exact acceleration of the piston for a constant engine speed.

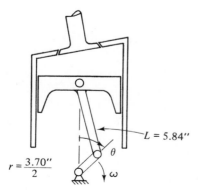

$L = 5.84''$

θ

$r = \dfrac{3.70''}{2}$

ω

Fig. P2.20

$$a = -\omega^2 \left[R \cos\theta + R^2 (L^2 - R^2 \sin^2\theta)^{-1/2} \cos 2\theta \right.$$

$$\left. + \frac{R^4}{4}(L^2 - R^2 \sin^2\theta)^{-3/2} \sin^2 2\theta \right]$$

Use the above equation to calculate the acceleration of the piston for an engine speed of 4800 rpm ($\omega = 160\,\pi$ rad/sec) at three degree increments during one complete revolution of the crankshaft.

Answer:

$$a_{max} = 3.355 \times 10^5 \text{ in/sec}^2 \text{ at } 141°$$

$$a_{max} = -6.155 \times 10^5 \text{ in/sec}^2 \text{ at } 0°$$

3

ADVANCED FEATURES
OF CSMP

The preceding chapter introduced the basic concepts and features that allow the reader to begin using CSMP. In this chapter, additional programming capabilities are described which permit considerable flexibility in the simulation of complex systems. The presentation is divided into five distinct areas: integration methods, data statements, translation control statements, data output, and subprograms. Each section is essentially a separate entity and consequently the material can be covered in any order.

Since integration is normally a pivotal point in most simulations, the first section concentrates on the various numerical integration routines available in CSMP.

Integration Methods

One of the most attractive features of CSMP is that the user seldom needs to become involved in specifying either the type or any of the details of the numerical integration method. For those who wish to specify the integration technique, there are five fixed-step and two variable-step methods available. CSMP III provides two additional techniques which are described in Chap. 5. If none of the seven methods are suitable, the user can supply his own integration subroutine. The following paragraphs describe the relative advantages and disadvantages of variable and fixed-step integration methods, error requirements, and give guidelines for choosing the best integration method.

Variable-Step Integration Methods

When the integration method is not specified, a variable-step Runge-Kutta method (CSMP name: RKS) is automatically used. This method is generally a good choice for most problems and is used in the example problems in Chap. 2. It is a sophisticated method that has the advantage of automatically adjusting the time increment of integration to meet the demands of the dynamic conditions of the simulation. Thus, the user is virtually assured a satisfactory solution.

The second variable-step integration technique is the fifth-order, predictor-corrector, Milne method (MILNE). This method is specified by including the following CSMP statement in the program.

METHOD MILNE

The MILNE method is similar to the RKS integration technique in that it uses rather sophisticated numerical algorithms and adjusts the step-size to meet changing conditions. MILNE integration has essentially the same advantages and disadvantages as the RKS technique. Generally, one will not know prior to the run which integration method will give the best simulation results. While the RKS method may give best results for one problem, the MILNE method may perform best for another type of problem.

Both methods have the disadvantage of sometimes using an extremely small integration step-size which results in excessive computer time. The step-size is controlled by algorithms which provide estimates of integration error.

Integration Error Requirements

In both variable-step methods, the absolute value of the estimated integration error (ABSERR) and the relative magnitude of the estimated error (RELERR) are compared with user-specified error-bounds. The step-size is then adjusted to meet the desired error criteria. The error limit of the absolute error dominates for large values of integrator output while the relative error is more important for small values of output. A detailed mathematical description of all integration methods and error criteria is contained in Appendix III.

If error-bounds are not specified, a value of 0.0001 is automatically used for both the absolute and relative errors for all RKS and MILNE integration. The user can specify either or both the relative and absolute errors for each integrator by using the RELERR and ABSERR statements. Examples of both statements are shown below.

RELERR X = 0.0002, Y = 5.0E-5, Z = 0.0005
ABSERR X = 0.0003, Y = 7.5E-5

The first statement sets the relative error for the X, Y, and Z integrators to the values shown. In the second statement, the absolute errors for only the X and Y integrators are changed as indicated.

The use of the RELERR and ABSERR cards enables the user to specify the allowable error for each separate integrator which, in effect, governs the size of the integration step. Computer time can be more efficiently used since step-size and computer time are directly related.

It is quite difficult to determine the exact quantitative effect that the relative and absolute errors have on step-size and solution accuracy. In order to show the relationship of error-bound to solution accuracy for one particular problem, the following second-order linear equation is solved using the variable-step Runge-Kutta method for values of relative and absolute error that ranged from 0.00005 to 0.1. The particular problem is the harmonic oscillator,

$$\ddot{y} + 4\pi^2 y = 0 \qquad\qquad (3.1)$$

with initial conditions of $y(0) = 1.0$, $\dot{y}(0) = 0$

A plot of the exact solution for $y(t)$ is given in Fig. 3.1. To illustrate the effect of ABSERR and RELERR error bounds, the exact solution is compared to the Runge-Kutta numerical solution.

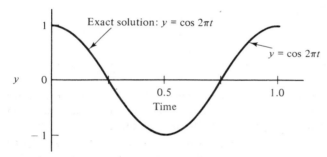

Fig. 3.1 Plot of the exact solution of $\ddot{y}(t) + 4^2 y(t) = 0$, with $y(0) = 1$, $\dot{y}(0) = 0$.

Table 3.1 contains a summary of the average absolute difference between the exact solution and numerical solution as a function of error-bound. The number of integration steps is also listed. The information in this table should not be directly applied to all problems but it does illustrate how error-bounds affect the accuracy of solution for one particular problem. Round-off errors in the numerical calculations limit the maximum accuracy of the solution. Consequently, the solution error cannot be expected to continue to become smaller indefinitely for increasingly smaller values of error-bound.

In some problems the error-bounds and the dynamics of the simulation do not control the integration step-size. Since the solution must be calculated at each output interval the integration step-size must be equal to or less than the magnitude of either PRDEL or OUTDEL. This means that for small output-intervals the step-size is determined by the print interval. For this condition the error-bound has no effect on the step size. In the numerical solution of Eq. (3.1), which generated the data contained in Table 3.1, the print interval was made sufficiently large to not affect the integration step-size.

Table 3.1

Solution Error of Eq. (3.1) Using RKS as a Function of ABSERR and RELERR.

Error-bound ABSERR and RELERR	Average Absolute Difference between Numerical and Exact Solutions	Number of Integration Steps
0.00005	5.48×10^{-6}	29
0.0001	7.06×10^{-5}	23
0.0002	7.95×10^{-5}	23
0.0005	1.01×10^{-4}	20
0.001	8.87×10^{-5}	20
0.002	5.88×10^{-4}	18
0.004	4.89×10^{-4}	17
0.005	4.89×10^{-4}	17
0.01	4.89×10^{-4}	17
0.02	1.69×10^{-3}	15
0.1	1.69×10^{-3}	15

As previously stated, it is not necessary to specify the integration step-size for either the RKS or MILNE integration methods. The first integration step is automatically set equal to $\frac{1}{16}$ of PRDEL or OUTDEL, whichever is smaller. If the user wishes to use a different value for the first step-size, this can be accomplished by specifying a value for DELT on the timer card. The definition of the DELT specification is as follows:

DELT. The value of DELT is the integration interval or step-size of the independent variable. If DELT is specified, it is automatically adjusted if necessary to be a submultiple of PRDEL or OUTDEL. If neither PRDEL or OUTDEL has been specified, DELT is adjusted to be a submultiple of FINTIM/100. When DELT is not specified, the first integration step is $\frac{1}{16}$ of the smaller value of PRDEL or OUTDEL. For either of the variable-step integration methods there is no need to specify a value for DELT unless the user feels that the first step-size ($\frac{1}{16}$ of smaller value of PRDEL or OUTDEL) is too large. When using a fixed-step integration method, the value of DELT should be carefully selected. Guidelines for choosing a value are discussed later.

For some problems, the error requirements and the dynamics of the solution may demand an extremely small step-size. A lower limit can be placed on the integration step-size by specifying a value for DELMIN on the timer card.

DELMIN. The value assigned to DELMIN on the timer card specifies the minimum allowable integration interval for the variable-step integration methods. If DELMIN is not specified, it is taken as FINTIM $\times 10^{-7}$. The following timer card shows a typical application where DELMIN is included.

TIMER FINTIM = 8.0,   PRDEL = 0.2,   DELMIN = 2.0E-8

If either of the variable-step integration methods attempts to use a value of DELT smaller than DELMIN, the run is terminated at that point with the appropriate message.

Fixed-Step Integration Methods

There are five fixed-interval integration techniques ranging from the sophisticated Runge-Kutta to an extremely simple rectangular integration. A listing of the five fixed-step methods in the order of decreasing complexity is given below.

Summary of Fixed-Step Integration Methods

CSMP Name	*Method*
RKSFX	Fourth-order Runge-Kutta with fixed interval
SIMP	Simpson's Rule integration
TRAPZ	Trapezoidal integration
ADAMS	Adams-Second Order
RECT	Rectangular integration

The integration technique is specified by the use of the METHOD card with the appropriate CSMP integration name. An example of specifying the trapezoidal method is shown below.

METHOD TRAPZ

Since these methods all use a fixed-integration interval, the value of DELT should be carefully chosen. Information for choosing a value for DELT is discussed in the following section.

Choosing the Integration Method

When specifying an integration method one must be concerned with obtaining sufficient accuracy without using excessive computing time. Generally, the selection of the best integration is an extremely complex decision. For problems that are not extremely complex or do not require a large amount of computer time, either the variable-step Runge-Kutta or Milne methods are probably the best choice. The error-bound can be specified to meet the desired accuracy and the step-size is then automatically adjusted to the changing dynamic conditions of the problem.

The solution of Eq. (3.1) by both the Milne and the variable-step Runge-Kutta methods were compared to the exact solution given in Fig. 3.1. The step-sizes on the first set of runs were entirely controlled by the default error-bound of 0.0001. In the second set of runs the error-bounds were not changed, but 100 lines of output were specified. From the results shown in Table 3.2, notice that the

Table 3.2

***Comparison of Numerical Solutions Using The Variable-
Step Milne and Runge-Kutta Methods***

	Integration Method	*Average Absolute Difference between Numerical and Exact Solutions of Eq. 3.1*	*Number of Integration Steps*
First set of runs, default error bound of 0.0001, 2 lines of output	Variable-step Runge-Kutta	7.06×10^{-5}	23
	Milne	1.08×10^{-5}	58
Second set of runs, default error bound of 0.0001, 100 lines of output	Variable-step Runge-Kutta	4.86×10^{-6}	104
	Milne	6.56×10^{-6}	112

smaller output interval of the second set of runs forced the integration step-size to be smaller and consequently reduced the error.

For certain types of problems, fixed-step integration methods may give better results. The following is a listing of three situations where fixed-step methods should be considered.

1. In some types of problems where sudden changes or discontinuities occur, the variable-step methods may demand an integration step which is smaller than the minimum allowed (DELMIN). If this occurs the run is terminated at that point. The user then has the choice of increasing the error requirements (RELERR and ABSERR), decreasing DELMIN, or using one of the fixed-step methods.

2. If the output interval is very small, the maximum step-size is constrained by the output and there is no need to use a sophisticated, time-consuming integration technique. For this situation a simple fixed-step method should be used.

3. When using elements such as IMPULS (Example 2.7) which involve critical time-sequencing relationships, it is good practice to use a fixed-step method that is a submultiple of the desired pulse interval.

The big problem in using a fixed-step method is choosing the proper integration step-size. DELT should be chosen such that it is sufficiently small to insure an accurate solution and not too small to result in excessive run time. There is no exact procedure for determining the proper step-size prior to making a simulation run. The dynamic response of the system determines the necessary step size.

The faster the response, the smaller should be the integration interval. It is advisable to make the first choice of DELT sufficiently small to ensure an accurate solution. The step-size is probably too large if a smaller integration interval results in a significantly different answer.

    In order to provide an estimate on how integration error is affected by step-size, the numerical solution of Eq. (3.1) was compared with the exact solution shown in Fig. 3.1. This was done for all five fixed-step integration methods using a step-size that ranged from 10 to 10,000 integration steps per cycle of the cosine function. The average absolute difference between the exact solution and the numerical solution is plotted in Fig. 3.2 as a function of step-size for the five integration methods.

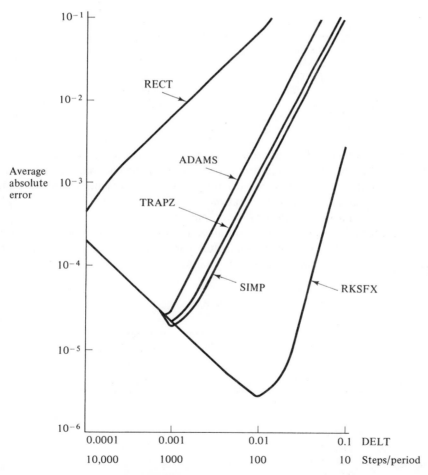

Fig. 3.2   Average absolute error as a function of step-size and integration method.

The results shown in Fig. 3.2 are only intended to show the relative accuracy of the various methods as a function of step-size. The curves also illustrate that a small integration interval does not necessarily give the greatest accuracy. If the frequency content of the solution can be estimated, Fig. 3.2 can be helpful in estimating the appropriate step-size.

As previously mentioned, the various integration methods are greatly different in complexity and consequently require different amounts of computing time. In order to provide an estimate on computing time for the five fixed-step methods, a program was run in which 10,000 integration steps were performed. The computer time required for only the numerical integration, using an IBM 360 Model 65 computer, is listed below.

Method	Time in Seconds
RKSFX	3.578
SIMP	2.513
TRAPZ	1.647
ADAMS	1.198
RECT	0.982

In many problems the most time-consuming portion of the simulation is not the result of the numerical integration, but is due to the calculations required at each integration step. The number of times that all statements in the Dynamic segment are executed for each integration step depends upon the method. Table 3.3 contains this information for all seven methods.

If the problem requires a large number of calculations for each integration step, a savings in computer time may result if a smaller step-size is used in conjunction with a less sophisticated integration method.

Table 3.3

Integration Method	Number of Times that all Statements in the Dynamic Segment are Executed for Each Integration Step
RKSFX & RKS	4
SIMP	3
MILNE	2
TRAPZ	2
ADAMS	1
RECT	1

The Variable KEEP

The CSMP variable KEEP is used to indicate when the numerical technique has reached the end of an integration step. KEEP is set equal to 1 when a valid integration step has been completed. For intermediate steps and for trial steps of the variable-step methods, KEEP equals 0. The KEEP variable was used in programs which calculated the average absolute error between the numerical and exact solution of Eq. (3.1).

An example of a program using Simpson's integration rule with a step-size of 0.001 seconds is shown in Fig. 3.3.

```
TITLE  PROGRAM TO CALCULATE THE AVERAGE ABSOLUTE ERROR OF EQUATION 3.1
INITIAL
CONSTANT   PI = 3.14159, SUM = 0.0, COUNT = 0.0
DYNAMIC

*   THE FOLLOWING 3 CARDS ARE USED TO SOLVE EQUATION 3.1, WHERE-
*   YDD = SECOND DERIVATIVE OF Y WITH RESPECT TO TIME
*   YD  = FIRST DERIVATIVE OF Y WITH RESPECT TO TIME

YDD = -4.0*PI*PI*Y
YD = INTGRL(0.0, YDD)
Y = INTGRL(1.0, YD)

*   THE FOLLOWING NOSORT SECTION IS REQUIRED TO USE THE "IF" STATEMENT

NOSORT

*   KEEP IS EQUAL TO 1 WHEN THE END OF A VALID INTEGRATION STEP IS REACHED

IF(KEEP. EQ .1)  GO TO 1
GO TO 2
1 SUM = SUM + ABS(Y - COS(2.0*PI*TIME))
COUNT = COUNT + 1.0
AAER = SUM/COUNT
2 CONTINUE
TERMINAL
TIMER  FINTIM = 1.0,   PRDEL = 0.1,   DELT = 0.001
METHOD  SIMP
PRINT  Y, SUM, COUNT, AAER
END
STOP
ENDJOB
```

Fig. 3.3 Program to calculate the average absolute error of Eq. 3.1.

In the program of Fig. 3.3, the variable KEEP was used to indicate when a valid integration step was reached. When KEEP equals 1, the variable SUM increases by the absolute difference between the numerical solution Y and the exact solution, $\cos(2.0*\pi*TIME)$. Also the number of valid integration steps is recorded by the variable COUNT. The quotient, SUM/COUNT is equal to the average absolute error AAER.

Table 3.3 shows that for Simpson's integration, statements in the Dynamic segment are executed three times at each integration step. This means that KEEP is equal to 1 every third time the program cycles through the Dynamic segment.

Changing Integration Technique and
Output Interval During a Run

CSMP offers the user the flexibility of changing integration methods, step-size, error requirements, and output-interval during a simulation run. The CSMP CONTINUE statement is used for this purpose. (In CSMP III, the END CONTINUE is used.) A brief description of the CONTINUE statement follows, while a detailed explanation is included in this chapter in the section on translation control statements.

A problem which illustrates the need to change step-size and output-interval is the simulation of the impact of two railroad freight cars. A small step-size is required during impact period because of the high forces and rapidly changing conditions. A much larger step-size and output-interval can be used for the post-impact motion.

Example 3.1

In this example it is desired to calculate the forces, accelerations, and deflections during the impact of two railroad cars. The motion after impact is also desired. Car number one is moving from left to right with a velocity of 8 ft/sec. It strikes and is coupled to a second car that is initially stationary. The impact first occurs at $t = 0$. The couplers are assumed to be connected to both cars by a linear-spring and damper. A constant rolling-resistance force opposing motion is also included. Figure 3.4 illustrates the simplified problem and gives the appropriate parameters.

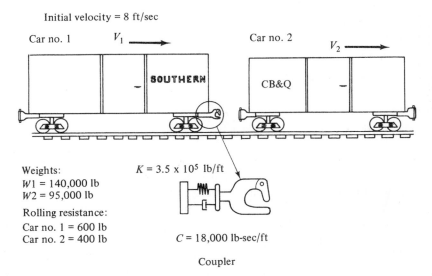

Fig. 3.4 Impacting freight cars.

Using Newton's second law, the equations describing the motion of the two freight cars are derived as follows.

$$\text{FORCE} = K*(X2 - X1)/2.0 + C*(V2 - V1)/2.0 \qquad (3.2)$$

$$\dot{V}1 = [\text{FORCE} - 600.0*\text{SIGN}(V1)]/W1/32.17 \qquad (3.3)$$

$$\dot{V}2 = [-\text{FORCE} - 400.0*\text{SIGN}(V2)]/W2/32.17 \qquad (3.4)$$

Symbols Used in CSMP
Program of Fig. 3.5

FORCE	Force between cars.
K	Spring constant of coupler spring, 3.5×10^5 lb/ft.
C	Damping constant of coupler, 18,000 lb-sec/ft.
ACC1 & ACC2	Accelerations of cars 1 & 2, respectively.
V1 & V2	Velocities of cars 1 & 2, respectively.
X1 & X2	Displacements of cars 1 & 2, respectively.
W1 & W2	Weights of cars 1 & 2, respectively.

Prior to running the program, it was estimated that the impact would occur in less than 1.0 sec. Therefore, FINTIM on the first TIMER card was set equal to 1.0 and the output-interval was chosen to be 0.02. For the post-impact motion FINTIM was set equal to 21.0 with an output-interval of 2.0. For illustration purposes, the fixed-step Runge-

```
TITLE   SIMULATION OF THE IMPACT OF TWO RAILROAD CARS.        THE OUTPUT
TITLE    INTERVAL AND INTEGRATION TECHNIQUE ARE CHANGED DURING THE RUN.
  CONSTANT   K = 3.5E5, C = 18000.0, W1 = 140000.0, W2 = 95000.0
  FORCE   = K*(X2 - X1)/2.0 + C*(V2 - V1)/2.0
  ACC1 = (FORCE - 600.0*SIGN(1.0,V1))/(W1/32.17)
  ACC2 = (-FORCE - 400*SIGN(1.0,V2))/(W2/32.17)
  V1  = INTGRL(8.0,ACC1)
  V2  = INTGRL(0.0,ACC2)
  X1  = INTGRL(0.0,V1)
  X2  = INTGRL(0.0,V2)
  DEFLCT = X1 - X2
  PRINT   FORCE, ACC1, ACC2, V1, V2, X1, X2, DEFLCT
  METHOD RKSFX
  TIMER   FINTIM = 1.0, PRDEL = 0.025, DELT = 0.0005

*  THE FOLLOWING CONTINUE CARD ALLOWS THE OUTPUT INTERVAL AND THE
*  INTEGRATION TECHNIQUE TO BE CHANGED DURING THE RUN.   THE CHANGES ARE MADE
*  AFTER 1.0 SECONDS, WHICH IS THE FINTIM TIME ON THE ABOVE TIMER CARD.

  CONTINUE
  METHOD MILNE
  RELERR  V1 = 0.0005, V2 = 0.0005, X1 = 0.0005, X2 = 0.0005
  ABSERR  V1 = 0.0005, V2 = 0.0005, X1 = 0.0005, X2 = 0.0005
  TIMER   FINTIM = 21.0, PRDEL = 2.0
END
STOP
ENDJOB
```

Fig. 3.5 Program for the simulation of the impact of two freight cars.

Kutta method (RKSFX) with a step-size of DELT = 0.0005 sec was used for the impact period and the variable-step Milne (MILNE) method was used for the motion after impact. Also for illustration purposes, the relative and absolute errors for all integrators of the Milne integration were set equal to 0.0005.

The program for simulating this system is shown in Fig. 3.5. One notes the use of the CONTINUE card to change the output-interval and integration technique. The CONTINUE card permits these changes without resetting the independent variable TIME.

The output for this simulation is contained in Figs. 3.6 and 3.7.

```
SIMULATION OF THE IMPACT OF TWO RAILROAD CARS.     THE OUTPUT       RKSFX    INTEGRATION
INTERVAL AND INTEGRATION TECHNIQUE ARE CHANGED DURING THE RUN.

TIME          FORCE        ACC1          ACC2          V1            V2            X1            X2            DEFLCT
0.0          -7.2000E 04  -1.6682E 01   2.4246E 01    8.0000E 00    0.0           0.0           0.0           0.0
2.5000E-02   -9.3825E 04  -2.1697E 01   3.1637E 01    7.5164E 00    7.0417E-01    1.9422E-01    8.4171E-03    1.8580E-01
5.0000E-02   -1.0758E 05  -2.4859E 01   3.6295E 01    6.9306E 00    1.5590E 00    3.7497E-01    3.6464E-02    3.3850E-01
7.5000E-02   -1.1344E 05  -2.6205E 01   3.8279E 01    6.2886E 00    2.4966E 00    5.4027E-01    8.7054E-02    4.5322E-01
1.0000E-01   -1.1201E 05  -2.5876E 01   3.7795E 01    5.6343E 00    3.4523E 00    6.8929E-01    1.6144E-01    5.2785E-01
1.2500E-01   -1.0425E 05  -2.4093E 01   3.5167E 01    5.0069E 00    4.3684E 00    8.2221E-01    2.5933E-01    5.6288E-01
1.5000E-01   -9.1367E 04  -2.1133E 01   3.0804E 01    4.4394E 00    5.1961E 00    9.4013E-01    3.7912E-01    5.6102E-01
1.7500E-01   -7.4724E 04  -1.7308E 01   2.5169E 01    3.9574E 00    5.8979E 00    1.0449E 00    5.1808E-01    5.2680E-01
2.0000E-01   -5.5740E 04  -1.2946E 01   1.8740E 01    3.5784E 00    6.4479E 00    1.1388E 00    6.7274E-01    4.6609E-01
2.2500E-01   -3.5798E 04  -8.3637E 00   1.1987E 01    3.3119E 00    6.8322E 00    1.2247E 00    8.3909E-01    3.8560E-01
2.5000E-01   -1.6171E 04  -3.8538E 00   5.3407E 00    3.1596E 00    7.0481E 00    1.3053E 00    1.0129E 00    2.9239E-01
2.7500E-01    2.0343E 03   3.2957E-01  -8.2431E-01    3.1164E 00    7.1032E 00    1.3835E 00    1.1901E 00    1.9341E-01
3.0000E-01    1.7936E 04   3.9837E 00  -6.2093E 00    3.1716E 00    7.0134E 00    1.4619E 00    1.3668E 00    9.5084E-02
3.2500E-01    3.0889E 04   6.9599E 00  -1.0595E 01    3.3099E 00    6.8010E 00    1.5428E 00    1.5397E 00    3.0394E-03
3.5000E-01    4.0493E 04   9.1668E 00  -1.3848E 01    3.5131E 00    6.4930E 00    1.6279E 00    1.7060E 00   -7.8133E-02
3.7500E-01    4.6593E 04   1.0569E 01  -1.5913E 01    3.7614E 00    6.1185E 00    1.7187E 00    1.8638E 00   -1.4503E-01
4.0000E-01    4.9257E 04   1.1181E 01  -1.6815E 01    4.0349E 00    5.7070E 00    1.8161E 00    2.0116E 00   -1.9547E-01
4.2500E-01    4.8741E 04   1.1062E 01  -1.6641E 01    4.3143E 00    5.2867E 00    1.9205E 00    2.1490E 00   -2.2851E-01
4.5000E-01    4.5458E 04   1.0308E 01  -1.5529E 01    4.5826E 00    4.8828E 00    2.0317E 00    2.2760E 00   -2.4432E-01
4.7500E-01    3.9929E 04   9.0372E 00  -1.3657E 01    4.8254E 00    4.5166E 00    2.1494E 00    2.3934E 00   -2.4405E-01
5.0000E-01    3.2746E 04   7.3867E 00  -1.1224E 01    5.0313E 00    4.2046E 00    2.2726E 00    2.5023E 00   -2.2964E-01
5.2500E-01    2.4525E 04   5.4976E 00  -8.4404E 00    5.1927E 00    3.9582E 00    2.4005E 00    2.6041E 00   -2.0363E-01
5.5000E-01    1.5867E 04   3.5080E 00  -5.5084E 00    5.3053E 00    3.7837E 00    2.5318E 00    2.7007E 00   -1.6892E-01
5.7500E-01    7.3274E 03   1.5459E 00  -2.6167E 00    5.3682E 00    3.6824E 00    2.6653E 00    2.7939E 00   -1.2857E-01
6.0000E-01   -6.1086E 02  -2.7824E-01   7.1402E-02    5.3837E 00    3.6512E 00    2.7998E 00    2.8854E 00   -8.5610E-02
6.2500E-01   -7.5608E 03  -1.8752E 00   2.4249E 00    5.3562E 00    3.6832E 00    2.9341E 00    2.9769E 00   -4.2837E-02
6.5000E-01   -1.3237E 04  -3.1796E 00   4.3471E 00    5.2923E 00    3.7688E 00    3.0672E 00    3.0699E 00   -2.7132E-03
6.7500E-01   -1.7463E 04  -4.1507E 00   5.7782E 00    5.1999E 00    3.8964E 00    3.1984E 00    3.1657E 00    3.2749E-02
7.0000E-01   -2.0169E 04  -4.7723E 00   6.6943E 00    5.0877E 00    4.0533E 00    3.3270E 00    3.2650E 00    6.2055E-02
7.2500E-01   -2.1378E 04  -5.0502E 00   7.1038E 00    4.9642E 00    4.2268E 00    3.4527E 00    3.3684E 00    8.4238E-02
7.5000E-01   -2.1201E 04  -5.0095E 00   7.0438E 00    4.8378E 00    4.4045E 00    3.5752E 00    3.4763E 00    9.8867E-02
7.7500E-01   -1.9813E 04  -4.6907E 00   6.5740E 00    4.7160E 00    4.5755E 00    3.6945E 00    3.5885E 00    1.0600E-01
8.0000E-01   -1.7442E 04  -4.1458E 00   5.7709E 00    4.6051E 00    4.7304E 00    3.8110E 00    3.7049E 00    1.0611E-01
8.2500E-01   -1.4342E 04  -3.4334E 00   4.7211E 00    4.5100E 00    4.8620E 00    3.9249E 00    3.8248E 00    1.0005E-01
8.5000E-01   -1.0781E 04  -2.6153E 00   3.5155E 00    4.4342E 00    4.9651E 00    4.0366E 00    3.9477E 00    8.8911E-02
8.7500E-01   -7.0223E 03  -1.7515E 00   2.2425E 00    4.3796E 00    5.0371E 00    4.1467E 00    4.0728E 00    7.3943E-02
9.0000E-01   -3.3066E 03  -8.9769E-01   9.8428E-01    4.3465E 00    5.0773E 00    4.2557E 00    4.1992E 00    5.6479E-02
9.2500E-01    1.5599E 02  -1.0203E-01  -1.8828E-01    4.3342E 00    5.0870E 00    4.3642E 00    4.3263E 00    3.7826E-02
9.5000E-01    3.1927E 03   5.9577E-01  -1.2166E 00    4.3406E 00    5.0691E 00    4.4725E 00    4.4533E 00    1.9222E-02
9.7500E-01    5.6803E 03   1.1674E 00  -2.0590E 00    4.3629E 00    5.0277E 00    4.5813E 00    4.5795E 00    1.7319E-03
1.0000E 00    7.5395E 03   1.5946E 00  -2.6886E 00    4.3977E 00    4.9679E 00    4.6907E 00    4.7045E 00   -1.3760E-02
```

Fig. 3.6 Motion of railroad cars during impact.

```
SIMULATION OF THE IMPACT OF TWO RAILROAD CARS.     THE OUTPUT       MILNE    INTEGRATION
INTERVAL AND INTEGRATION TECHNIQUE ARE CHANGED DURING THE RUN.

TIME         FORCE        ACC1          ACC2          V1            V2            X1            X2            DEFLCT
1.0000E 00   7.5395E 03   1.5946E 00  -2.6886E 00    4.3977E 00    4.9679E 00    4.6907E 00    4.7045E 00   -1.3760E-02
3.0000E 00   8.1178E 01  -1.1922E-01  -1.6294E-01    4.3517E 00    4.3583E 00    1.3679E 01    1.3679E 01   -1.2207E-04
5.0000E 00  -1.4982E 02  -1.7230E-01  -8.4720E-02    4.0830E 00    4.0768E 00    2.2113E 01    2.2113E 01    5.3406E-04
7.0000E 00  -2.8289E 02  -2.0288E-01  -3.9658E-02    3.8152E 00    3.7941E 00    2.9999E 01    2.9999E 01    5.3406E-04
9.0000E 00  -2.2676E 02  -1.8998E-01  -5.8663E-02    3.5452E 00    3.5146E 00    3.7338E 01    3.7338E 01   -2.7466E-04
1.1000E 01  -3.8365E 01  -1.4669E-01  -1.2246E-01    3.2848E 00    3.2209E 00    4.4127E 01    4.4130E 01   -3.0670E-03
1.3000E 01  -5.8007E 02  -2.7116E-01   6.0978E-02    2.9512E 00    3.0351E 00    5.0375E 01    5.0367E 01    7.6294E-03
1.5000E 01  -8.7971E 02  -3.4002E-01   1.6244E-01    2.7154E 00    2.7052E 00    5.6069E 01    5.6065E 01    4.5013E-03
1.7000E 01  -3.4298E 02  -2.1668E-01  -1.9308E-02    2.4556E 00    2.4107E 00    6.1215E 01    6.1215E 01   -3.5095E-04
1.9000E 01   4.4059E 02  -3.6630E-02  -2.8465E-01    2.1639E 00    2.1631E 00    6.5814E 01    6.5817E 01   -2.5635E-03
2.1000E 01  -1.9237E 02  -1.8208E-01  -7.0309E-02    1.8984E 00    1.8770E 00    6.9867E 01    6.9867E 01    0.0
```

Fig. 3.7 Post-impact motion of railroad cars.

In summary, choosing the best integration method is quite complicated, both from a theoretical and a practical point of view. No absolute set of guidelines can be written that will insure the optimum method is used for all problems. The user should be prepared to experiment in order to obtain the best method with regard to run-time and accuracy. *It should be emphasized that either of the variable-step integration methods using the default error-bounds will give satisfactory solutions for the vast majority of problems.* The additional computing time that may result when using this approach is usually considerably less expensive than the time the user would spend in choosing an optimum method. However, for complicated problems that will require a large amount of computing time, the user can justify spending time in deciding which method will give the best results.

User-Supplied Integration Method

In the event that none of these seven integration methods satisfies the user's requirements, an additional centralized integration method can be added to CSMP. The complete integration routine is entered into the program as a FORTRAN subroutine named CENTRL. The method is specified using the statement,

METHOD CENTRL

Example 3.2

To illustrate a program that employs a user-supplied integration technique, the following linear, first-order differential equation will be solved.

$$\frac{d(\text{CASH})}{dt} = \text{INTRST}*\text{CASH} \tag{3.5}$$

Initial condition: CASH($t = 0$) = \$1000.

Equation (3.5) governs the present value of a bank account where the money is continuously compounded. This method of compounding is used by many banks and savings and loan associations. The initial amount in the account is \$1,000.00.

$$\text{CASH} = \text{present value of bank account}$$
$$\text{INTRST} = \text{yearly interest rate} = 0.065\ (6.5\%)$$
$$t = \text{time in years}$$

The user supplied integration method is the CSMP Simpson method. Note that the only reason for using the CENTRL integration feature in this problem is to demonstrate its use.

Figure 3.8 shows a listing of the entire program. The comment cards are supplied to assist the reader to interpret the program. The IBM System Manual[1] as well as the section on subprograms should be used as a reference for this procedure. Most of the information contained in Fig. 3.8 was taken from the IBM System Manual.

Figure 3.9 gives the output for the program for a time period of 50 years. The accuracy of the numerical integration can be compared with Eq. (3.6) which is the analytical solution to Eq. (3.5).

$$\text{CASH} = 1000e^{0.065t} \tag{3.6}$$

```
*         PROGRAM TO ILLUSTRATE THE METHOD OF PROVIDING A USER-SUPPLIED
*         INTEGRATION TECHNIQUE TO THE PROGRAM.    IN THIS EXAMPLE, THE CSMP
*         SIMPSON METHOD IS USED.
LABEL HISTORY OF $1000 BANK ACCOUNT, 6.5 CONTINUOUSLY COMPUND INTEREST
      RENAME   TIME = YEARS
      CONSTANT  INTRST  = 0.065
      CASH  =  INTGRL(1000.0, INTRST*CASH )
      TIMER  FINTIM = 50.0,   OUTDEL = 1.0,   DELT = 0.05
      PRTPLT CASH
      METHOD   CENTRL
       END
       STOP
C     THE FOLLOWING IS THE FORTRAN SUBROUTINE "CENTRL"
      SUBROUTINE CENTRL
C     THE FOLLOWING 2 COMMON AND 5 EQUIVALENCE CARDS ARE REQUIRED FOR ALL
C     INTEGRATION METHODS.
      COMMON DDUM1(64),C(8000),NALARM,KPOINT,DDUM2(16),RANGE(400),H,KEEP
      COMMON DDUM3(1214),IFLAG(50),FAM(50),NOINTG,NOSYMB,SYMB(1)
      EQUIVALENCE (IFLAG(16),ISTEP),(IFLAG(19),DTIME),(IFLAG(40),ISTORE)
      EQUIVALENCE (IFLAG(18),ISTART),(IFLAG(10),IFIRST)
      EQUIVALENCE (C(1),TIME),(C(2),DELT),(C(3),DELMIN)
      EQUIVALENCE (C(4),FINTIM),(C(5),PRDEL),(C(6),OUTDEL)
      EQUIVALENCE (DDUM3(492),TNEXT),(DDUM2(10),KREL),(DDUM2(11),KABS)
C     NOINTG IS THE NUMBER OF INTEGRATORS
C     C  IS THE DYNAMIC STORAGE ARRAY FOR THE EXECUTION PHASE
C     H IS THE HIGHEST FREQUENCY OUTPUT SAMPLE RATE
C     DELT HAS BEEN MODIFIED TO BE A SUBMULTIPLE OF H
C     TNEXT IS THE TIME OF THE NEXT OUTPUT POINT
C
C     START OF PROGRAM
C     IF CENTRL IS AN ERROR TESTING INTEGRATION METHOD, SET NALARM TO
C     ONE WHEN ERROR TEST FAILS IN ORDER TO STOP RUN.
C     ISTART IS SET TO ZERO BY AN END OR CONTIN CARD TO START THE CASE
C     ISTEP IS SET TO ZERO TO INITIALIZE STATUS
      ISTEP=0
C     ISTORE IS THE NUMBER OF LOCATIONS USED IN THE C ARRAY
      ISTORE=KPOINT+3*NOINTG
      GO TO 3000
C
 2000 CONTINUE
C     FOR SPECIAL INTEGRATION, USER MUST INSERT HIS CODING HERE
C     AT THIS POINT VALUES ARE AVAILABLE FOR TIME ZERO.(Y,YDOT,YIC)
C     INTEGRATOR OUTPUTS ARE LOCATED IN C(N+6) FOR N=1 TO NOINTG
C     INTEGRATOR INPUTS ARE LOCATED IN C(N+6+NOINTG) FOR N=1 TO NOINTG
C     INTEGRATOR INITIAL CONDITIONS IN C(N+6+2NOINTG) FOR N=1 TO NOINTG
C
C     HISTORY OF OUTPUTS MUST BE STORED IN USER DIMENSIONED ARRAY OR
C     UPPER LOCATIONS OF C ARRAY
C
C     EXAMPLE FOR K HISTORY POINTS (REPLACE ARGUMENTS WITH CONSTANTS)
C     DIMENSION CHYS(K,1)
C     EQUIVALENCE(C(8000-300*K+1) ,CHYS(1))
C     IN THE SIMPSON INTEGRATION METHOD, THERE ARE 3 HISTORY POINTS,   K = 3
C     EQUIVALENCE (C(7101), CHYS(1))
C     DIMENSION CHYS(3,1),CY(1)
C     THE FOLLOWING IS THE CSMP SIMPSON INTEGRATION METHOD.
      EQUIVALENCE (C(7),CY(1))
      EQUIVALENCE (DDUM2(13),TLAST)
C
C     HISTORY IS STORED AS FOLLOWS
C
C     CHYS(J,I)=PAST VALUES AS REQUIRED
C     WHERE   J VARIES FROM 1 TO K HISTORY POINTS AND
C     WHERE I VARIES FROM 301-NOINTG TO 300
C
```

Fig. 3.8 Program illustrating the use of CENTRL.

```
      DO 2010 II=1,NOINTG
      I=301-II
      J=NOINTG+II
      CHYS(1,I)=CY(II)
      CHYS(2,I)=CY(J)
 2010 CY(II)=.5*DELT*CY(J)+CY(II)
      DTIME=DTIME+1.
      TIME=DTIME*.5*DELT+TLAST
      ISTART=1
      KEEP=0
      CALL UPDATE
      DO 2020 II=1,NOINTG
      I=301-II
      J=NOINTG+II
      CHYS(3,I)=CY(J)
 2020 CY(II)=.5*DELT*CY(J)+CY(II)
      DTIME=DTIME+1.
      TIME=DTIME*.5*DELT+TLAST
      CALL UPDATE
      DO 2030 II=1,NOINTG
      I=301-II
      J=NOINTG+II
 2030 CY(II)=CHYS(1,I)+.5*DELT*(CHYS(2,I)+4.*CHYS(3,I)+CY(J))/3.
 2040 CONTINUE
C     KEEP IS SET TO IDENTIFY POINT TO STORE
      KEEP = 1
      CALL UPDATE
 3000 CONTINUE
C     STATUS CHECKS FOR VARIABLE RANGES AND OUTPUT TIMES
      CALL STATUS
C     IFIRST IS SET TO 4 TO INDICATE THE END OF A CASE
      IF(IFIRST-4)2000,4000,4000
 4000 RETURN
      END
ENDJOB
```

Fig. 3.8   (Continued)

Data Statements

Three of the most useful and commonly used data statements are CON-
STANT, PARAMETER, and INCON. These three statements are completely
equivalent in CSMP and they were previously defined and used in Chap. 2. It
should be pointed out that these data statements cannot be used for assigning
values to subscripted variables. As in FORTRAN, subscripted variables require
special consideration. Following are descriptions of how data statements using
subscripted variables are used in CSMP programs.

Subscripted Variables

Subscripted variables provide a procedure for handling large arrays of related
data as are frequently found in systems represented by simultaneous equations.
Any valid symbol, except those which start with the letters I, J, K, L, M, or N,

		MINIMUM 1.0000E 03	CASH VERSUS YEARS	MAXIMUM 2.5786E 04
YEARS	CASH	I		I
0.0	1.0000E 03	+		
1.0000E 00	1.0672E 03	+		
2.0000E 00	1.1388E 03	+		
3.0000E 00	1.2153E 03	+		
4.0000E 00	1.2969E 03	+		
5.0000E 00	1.3840E 03	+		
6.0000E 00	1.4770E 03	+		
7.0000E 00	1.5762E 03	-+		
8.0000E 00	1.6820E 03	-+		
9.0000E 00	1.7950E 03	-+		
1.0000E 01	1.9155E 03	-+		
1.1000E 01	2.0441E 03	--+		
1.2000E 01	2.1814E 03	--+		
1.3000E 01	2.3279E 03	--+		
1.4000E 01	2.4843E 03	--+		
1.5000E 01	2.6511E 03	---+		
1.6000E 01	2.8291E 03	---+		
1.7000E 01	3.0191E 03	----+		
1.8000E 01	3.2219E 03	----+		
1.9000E 01	3.4383E 03	----+		
2.0000E 01	3.6692E 03	-----+		
2.1000E 01	3.9156E 03	-----+		
2.2000E 01	4.1785E 03	------+		
2.3000E 01	4.4591E 03	------+		
2.4000E 01	4.7586E 03	-------+		
2.5000E 01	5.0781E 03	--------+		
2.6000E 01	5.4191E 03	--------+		
2.7000E 01	5.7830E 03	---------+		
2.8000E 01	6.1714E 03	----------+		
2.9000E 01	6.5858E 03	-----------+		
3.0000E 01	7.0280E 03	------------+		
3.1000E 01	7.5000E 03	-------------+		
3.2000E 01	8.0036E 03	--------------+		
3.3000E 01	8.5411E 03	---------------+		
3.4000E 01	9.1147E 03	----------------+		
3.5000E 01	9.7268E 03	-----------------+		
3.6000E 01	1.0380E 04	-------------------+		
3.7000E 01	1.1077E 04	--------------------+		
3.8000E 01	1.1821E 04	---------------------+		
3.9000E 01	1.2615E 04	-----------------------+		
4.0000E 01	1.3462E 04	------------------------+		
4.1000E 01	1.4366E 04	--------------------------+		
4.2000E 01	1.5331E 04	---------------------------+		
4.3000E 01	1.6360E 04	-----------------------------+		
4.4000E 01	1.7459E 04	------------------------------+		
4.5000E 01	1.8632E 04	--------------------------------+		
4.6000E 01	1.9883E 04	----------------------------------+		
4.7000E 01	2.1218E 04	------------------------------------+		
4.8000E 01	2.2643E 04	--------------------------------------+		
4.9000E 01	2.4164E 04	--+		
5.0000E 01	2.5786E 04	--+		

Fig. 3.9 Output for program of Fig. 3.8.

may be subscripted. Examples of both valid and invalid subscripted variables are:

Valid	*Invalid*	
A(3)	JP(8)	(Invalid first symbol)
COUNT(5,7,6)	ABCDEFG(4,2)	(Excessive characters)
ZP123(7,2,8)	4X(9)	(First character cannot be a number)

As in FORTRAN, it is necessary to specify the following information before using subscripted variables.

1. What variables are subscripted.

2. How many subscripts there are for each subscripted variable.

3. The maximum size for each subscript.

This information can be specified by using either the DIMENSION or STORAGE statements.

DIMENSION Statement

Subscripted variables larger than one dimensional arrays [e.g. X(3, 5) Y(3, 7, 5)] must be listed on a DIMENSION card. The DIMENSION statement is handled exactly as in FORTRAN with the exception being that a virgule (/) must appear in the first column. The virgule indicates that the DIMENSION instruction is a FORTRAN specification statement (discussed below). An example of a valid statement is shown below.

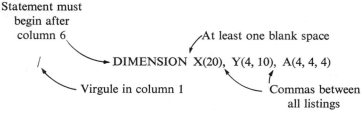

The compiler will assign 20 spaces to the one-dimensional array named X, 40(4 × 10) spaces to the two-dimensional array Y, and 64 (4 × 4 × 4) spaces to the three-dimensional array A. The user should not attempt to use a subscript larger than the maximum size as specified in the DIMENSION statement. Also, subscripts must never be smaller than 1. This means that negative numbers and zero cannot be used as subscripts. A seven-dimensional array, which is the limit of FORTRAN IV (Level G), is the largest that can be used in a CSMP program. The DIMENSION statement should be placed at the beginning of the program since it must appear before any subscripted variables are used.

FORTRAN Specification Cards

Cards containing a virgule (/) in column 1 are called FORTRAN specification cards. These cards are treated exactly as FORTRAN statements and consequently must follow a somewhat different format than usual CSMP statements. The following rules must be followed when using FORTRAN specification cards.

1. All FORTRAN specification cards must contain a virgule (/) in column 1.

2. Statements must be contained in columns 7–72.

3. A maximum of 10 FORTRAN specification cards are permitted in any one program.

4. Continuation cards are permitted and must contain a nonzero character in column 6 and a virgule in column 1.

The following DIMENSION statement illustrates the continuation of a FORTRAN specification instruction.

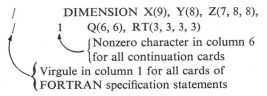

```
/        DIMENSION X(9), Y(8), Z(7, 8, 8),
/      1    Q(6, 6), RT(3, 3, 3, 3)
```
Nonzero character in column 6 for all continuation cards
Virgule in column 1 for all cards of FORTRAN specification statements

STORAGE Statement

One-dimensional arrays [e.g. X(5), Y(78)] can be handled by using the CSMP STORAGE statement. This card is used similarly to the DIMENSION statement, the exceptions being that it can only be used for one-dimensional arrays and that it is not a FORTRAN specification card. An example of a valid statement is

At least one blank space is necessary
STORAGE Z(90), Y(75), X(5)

The symbols following the STORAGE label represent subscripted variables with the appropriate number of storage locations contained within parentheses. A STORAGE statement may be continued to additional cards by using either three consecutive periods or multiple STORAGE cards. All STORAGE statements should be placed at the beginning of the program since they must appear before any subscripted variables are used. DIMENSION and STORAGE statements may be used in the same program.

Subscripted variables are subject to certain restrictions in CSMP programs. Special care must be taken to observe the following rules.

1 All subscripted variables must be declared in either a DIMENSION or STORAGE statement.

2 Subscripted variables cannot appear to the left of an equal sign except in a nosort section or within a PROCEDURE function. The use and definition of a PROCEDURE function is covered later in this chapter. Examples of statements that are *only* allowed in nosort sections or in PROCEDURE functions are:

$$Y(5) = SIN(3.0*TIME)$$
$$Y(1) = 3.14159*Q(3)$$

Examples of statements allowed in *all* sections of the program are:

$$Z = 4.0*X(4) + COS(Y(2))$$
$$Q = AFGEN(ABC, X(6))$$
$$P = INTGRL(8.7, Y(3))$$

3 The output of an INTGRL function cannot be a subscripted variable. Thus, the following statement type is *not allowed* anywhere in the program.

$$X(2) = INTGRL(2.3, Y)$$

4 Subscripted variables cannot be used in PRINT or PRTPLT statements. (This is not the case in CSMP III). In order to obtain a PRINT or PRTPLT output of subscripted variables, it is first necessary to set the dimensioned variables equal to nonsubscripted variables. This is illustrated in following examples.

Assigning Values to Subscripted Variables

There are three convenient methods for assigning values to subscripted variables.

1 The CSMP "TABLE" statement.

2 The FORTRAN "DATA" statement.

3 The FORTRAN "READ" statement.

A description of the use of these three statements along with a method of declaring variables as integers follows.

TABLE *Statement:* Subscripted variables that are listed on a STORAGE card may be assigned values using the TABLE statement. This is a CSMP data statement which allows values to be assigned to *one-dimensional* subscripted variables. The use of the TABLE card is similar to the other data statements (CONSTANT, PARAMETER, and INCON) the exception being that the TABLE statement can only be used for one-dimensional subscripted variables. In addition to the usual way of assigning values, the $K*n$ form can be used to assign K consecutive variables the value n. The following is an example of a valid statement.

TABLE X(1) = 22.56, X(9) = −0.891, X(3 − 6) = 4*6.74, Y(1) = 3.14

In the above example X(1), X(9), and Y(1) are assigned the values shown. X(3), X(4), X(5), and X(6) are all set equal to 6.74. All variables contained in a TABLE statement must be listed on a STORAGE card. TABLE statements may be continued to additional cards by using either three consecutive periods or multiple TABLE statements may be used.

DATA *Statement:* The FORTRAN DATA statement provides an additional means for assigning values to subscripted variables. The DATA card is a FORTRAN specification statement and must contain a virgule in column 1. Only subscripted variables should appear on a DATA card and these variables *must be listed* in a DIMENSION statement. Subscripted variables that are listed in a FORTRAN EQUIVALENCE statement must not appear on a DATA card. An example of a DATA and appropriate DIMENSION statement is

/   DIMENSION Y(3, 2)

/   DATA Y/6*3.14/

The above DATA statement assigns a value of 3.14 to all six variables of the two-dimensional array.

The order in which numerical values are assigned to a two-dimensional array is illustrated by the following statements.

/   DIMENSION Y(3, 2)

/   DATA Y/7.45, 2*5.89, −4.32, 2*3.33/

These statements will assign the numerical values of

Y(1, 1) = 7.45   Y(1, 2) = −4.32

Y(2, 1) = 5.89   Y(2, 2) =   3.33

Y(3, 1) = 5.89   Y(3, 2) =   3.33

A single DATA statement can also be used to assign values to more than one subscripted variable. For example,

/   DIMENSION X(3, 2, 2), Y(2)

/   DATA X/4*4.27, −7.16, 3*12.81, 4*0.71/, Y/2*87.3/

The above card assigns the values,

X(1, 1, 1) = 4.27	X(1, 2, 1) = 4.27	X(1, 1, 2) = 12.81	X(1, 2, 2) = 0.71
X(2, 1, 1) = 4.27	X(2, 2, 1) = −7.16	X(2, 1, 2) = 12.81	X(2, 2, 2) = 0.71
X(3, 1, 1) = 4.27	X(3, 2, 1) = 12.81	X(3, 1, 2) = 0.71	X(3, 2, 2) = 0.71
Y(1) = 87.3	Y(2) = 87.3		

A more comprehensive description of DATA statements can be found in many books on FORTRAN.[2-5]

FIXED *Statement:* When using subscripted variables, it is often necessary to use variables as integers for such purposes as indexing. Variables are specified as integers by the use of the FIXED statement. This instruction allows the user to declare selected variables as integers (fixed-point numbers). As previously stated, all variables in CSMP are automatically treated as real numbers (floating-point) unless otherwise specified. Consequently, any valid symbol that is to be treated as an integer must be included on a FIXED card. The following is an example of a valid FIXED statement.

FIXED I, A, Y6, J
↖At least one blank space

Continuation cards (. . .) or multiple FIXED cards may be used if necessary.

It should be remembered that integer variables cannot be used in either PRINT or PRTPLT statements. If the output of an integer variable is desired, it should first be set equal to a real variable. The following two statements could be used to print the value of the integer variable I.

Z = I   (This statement *must* be in the Dynamic segment.)
PRINT Z

READ *Statement:* The convenience and flexibility of the FORTRAN statement, READ (5, XYZ) can be used in CSMP programs to assign values to variables. The READ card must appear in a nosort section. It is usually placed in the Initial segment since this segment is only executed once at the beginning of the simulation. However, the READ statement can also be included in a nosort section of the Dynamic segment. The user should be aware that all statements in the Dynamic segment are executed from one to four times for each integration step. The exact number of executions depends upon the integration method and is given in Table 3.3. When using a READ instruction for nonsubscripted variables, the variable must appear somewhere else in the program before it can be included in a PRINT or PRTPLT statement.

A FORTRAN FORMAT statement must be used in conjunction with a READ card. The FORMAT card specifies the form in which numerical values are listed on data cards. The FORMAT card must be contained in a nosort section and any continuation cards must have a $ in card column 6. A detailed explanation on the use of READ and FORMAT statements can be found in books on FORTRAN.[2-5]

Numerical values used with the READ statements are contained on data cards. These cards must be placed between the labels DATA and ENDDATA. The DATA card must immediately follow the END card. The ENDDATA label follows the last data card and it must be punched in card columns 1–7. In CSMP III, the labels INPUT and ENDINPUT are used in place of DATA and END-DATA. It should be noted that this DATA card is a CSMP statement and is different from the FORTRAN DATA statement that was previously described. Portions of three programs which use READ and FIXED statements are shown below.

```
INITIAL
STORAGE X(4)
FIXED J
NOSORT
100 FORMAT(2F10.0)
READ(5, 100) (X(J), J = 1, 4)
DYNAMIC
   .  ⎫
   .  ⎬  Statements defining the run.
   .  ⎭
END
DATA
19.              24.
9.               47.
ENDDATA
STOP
ENDJOB
```

```
INITIAL
/        DIMENSION Y(2, 3)
FIXED K, J
NOSORT
50 FORMAT(3F14.2)
READ(5, 50)((Y(K, J), J = 1,3), K = 1, 2)
DYNAMIC
   .  ⎫
   .  ⎬  Statements defining the run.
   .  ⎭
END
DATA
13.42               3.56               −67.80
0.78                32.56              3.34
ENDDATA
STOP
ENDJOB
```

```
INITIAL
NOSORT
READ(5,999) P, Q, R, S
999 FORMAT(4E12.3)
DYNAMIC
   .  ⎫
   .  ⎬  Statements defining the run.
   .  ⎭
END
DATA
0.543E4          0.400E7          0.643E-4          0.484E3
ENDDATA
STOP
ENDJOB
```

Specification Form of INTGRL Statement

To facilitate the integration of subscripted variables, the *specification* or array form of the INTGRL statement can be used. This form of the INTGRL statement allows the use of subscripted variables as inputs, outputs, and initial conditions. The general form of the statement is

$$ZE = INTGRL(XOE, XE, N)$$

where   ZE = output of the integration

XOE = initial condition

XE = integrand

N = number of elements in the integrator array, N must be a literal integer constant

In order to use the above specification form of the INTGRL statement for an array of 20 in S/360 CSMP, the program must begin with DIMENSION and EQUIVALENCE cards.

```
/          DIMENSION  X(20),XO(20),Z(20)
/          EQUIVALENCE  (XE,X(1)), (XOE,XO(1)), (ZE,Z(1))
```

The above cards are not required when using CSMP III. See Chap. 5 for a detailed explanation and examples.

It is not permissible to use a STORAGE card to declare the subscripted variables that appear in the array form of INTGRL statements.

In conjunction with the above DIMENSION and EQUIVALENCE statements, the expression

$$ZE = INTGRL(XOE,XE,20)$$

will integrate 20 variables, namely

$$Z(1) = \int_0^t X(1)\, dt + XO(1)$$

$$Z(2) = \int_0^t X(2)\, dt + XO(2)$$

$$\vdots$$

$$Z(20) = \int_0^t X(20)\, dt + XO(20)$$

An example illustrating the use of subscripted variables with the specification form of the INTGRL statement is the following transient heat transfer problem.

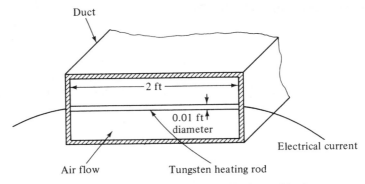

Fig. 3.10 Cross-section of tungsten heating rod in duct.

Example 3.3

Consider air flowing in a duct across a tungsten heating element. As electric current supplies energy at a uniform rate per unit-volume to the wire, energy is being removed from the circular heating element by radiation, convection, and conduction. The problem is to find the transient temperature distribution along the rod after the current is turned on. The temperature distribution of the rod is described by a partial differential equation having time and distance as the independent variables. Since partial differential equations cannot be directly solved using CSMP, a common method of solution is to divide the tungsten element into a number of small equal-sized elements. Each element is assumed to

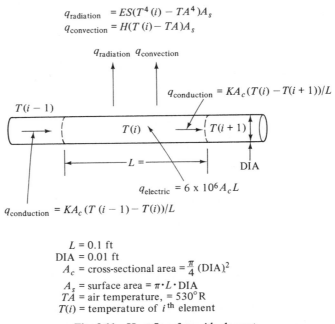

$$q_{radiation} = ES(T^4(i) - TA^4)A_s$$
$$q_{convection} = H(T(i) - TA)A_s$$

$$q_{conduction} = KA_c(T(i) - T(i+1))/L$$

$$q_{electric} = 6 \times 10^6 A_c L$$

$$q_{conduction} = KA_c(T(i-1) - T(i))/L$$

$L = 0.1$ ft
$DIA = 0.01$ ft
A_c = cross-sectional area $= \frac{\pi}{4}(DIA)^2$
A_s = surface area $= \pi \cdot L \cdot DIA$
TA = air temperature, $= 530°R$
$T(i)$ = temperature of ith element

Fig. 3.11 Heat flow from ith element.

have a uniform temperature. This method transforms a continuous system into a *lumped system* which is represented by a set of ordinary differential equations. The accuracy of the simulation improves as the element size decreases.

In this example, the heating element is divided into 20 equal-sized elements, each 0.1 ft long and 0.01 ft in diameter. The elements are sequentially numbered left to right.

Figure 3.11 shows the heat flow for the ith element. An energy balance on this element yields the following equation.

$$\frac{dT(i)}{dt} = [K \cdot AC(T(i-1) - 2.0T(i) + T(i+1))/L$$
$$- ES \cdot AS(T(i)^4 - TA^4) - H \cdot AS(T(i) - TA)$$
$$+ QELEC \cdot AC \cdot L]/(RO \cdot C \cdot AC \cdot L) \qquad (3.7)$$

where

$T(i)$ = temperature of ith element, °R

K = thermal conductivity, 94.0 Btu/hr-ft-°R

AC = cross-sectional area of tungsten rod, ft^2

ES = Stefan-Boltzmann constant times surface emissivity, 8.5×10^{-10} Btu/hr-ft^2-°R^4

TA = air temperatures in duct, 530°R

H = heat transfer coefficient for forced convection, 25.0 Btu/hr-ft^2-°R

AS = surface area of element, ft^2

RO = density of tungsten, 1208 lbm/ft^3

C = specific heat of tungsten, 0.032 Btu/lbm°R

$QELEC$ = electrical energy supplied per unit-volume, 6.0×10^6(Btu/hr-ft^3)

In the first equation of the array ($i = 1$), the $T(i-1)$ term must be set equal to the temperature of the left wall. In the same manner, the $T(i+1)$ term of the last equation ($i = 20$) must be set equal to the temperature of the right side wall.

The program for simulating this system is shown in Fig. 3.12.

There are several points which should be discussed regarding the program.

1. Note that it is necessary to use the DIMENSION and EQUIVALENCE statements to specify the subscripted variables used with the specification form of the INTGRL statement. These relationships are given below.

Subscripted variables	*EQUIVALENCE variables*
$T(i)$ = temperature of ith element	TE
$TI(i)$ = initial temperature of ith element	TIE
$TD(i) = \dfrac{dT(i)}{dt}$	TDE

```
TITLE    THE USE OF SUBSCRIPTED VARIABLES TO SOLVE A HEAT TRANSFER
TITLE    PROBLEM
INITIAL
/      DIMENSION  T(20),  TI(20),  TD(20)
/      EQUIVALENCE  ( TF,T(1) ), (TIE,TI(1) ), (TDE, TD(1) )
FIXED  I
CONSTANT  RO = 1208.0, C = 0.032, K = 94.0, H = 25.0, ES = 8.5E-10...
, DIA = 0.01, L = 0.1, TA = 530.0, TWALL = 530.0, QELEC = 6.0E6
AC = 3.14159*DIA*DIA/4.0
AS = 3.14159*DIA*L
NOSORT
DO 2 I = 1,20
  2 TI(I) = 530.0
DYNAMIC
NOSORT
TD(1) = (K*AC*(TWALL - 2.0*T(1) + T(2))/L - ES*AS*(T(1)**4   -     ...
TA**4) - H*AS*(T(1) - TA) + QELEC*AC*L)/(RO*C*AC*L)
TD(20) = (K*AC*(T(19) - 2.0*T(20) + TWALL)/L - ES*AS*(T(20)**4 -   ...
TA**4) - H*AS*(T(20) - TA) + QELEC*AC*L)/(RO*C*AC*L)
DO 1  I = 2,19
  1 TD(I) = (K*AC*(T(I-1) - 2.0*T(I) + T(I+1))/L - ES*AS*(T(I)**4 -   ...
TA**4) - H*AS*(T(I) - TA) + QELEC*AC*L)/(RO*C*AC*L)
SORT

*    THE FOLLOWING IS THE SPECIFICATION FORM OF THE INTGRL STATEMENT.  IT IS
*    USED TO INTEGRATE ARRAYS.

TE = INTGRL(TIE,TDE,20)

*    SUBSCRIPTED VARIABLES CAN NOT BE INCLUDED IN PRINT OR PRTPLT STATEMENTS.
*    CONSEQUENTLY, IT IS NECESSARY TO INTRODUCE THE FOLLOWING FOUR VARIABLES
*    FOR OUTPUT PURPOSES.

T1 = T(1)
T5 = T(5)
T10 = T(10)
T17 = T(17)
PRINT T1, T5, T10, T17
TIMER  FINTIM = 0.018,   PRDEL = 0.0006
END
STOP
ENDJOB
```

Fig. 3.12 Program to simulate temperature of tungsten rod.

2 The last portion of the Initial segment is changed to a nosort section to allow a DO loop to be used to set all values of initial temperatures TI(i) equal to 530°R. All calculations for initial conditions that are used in INTGRL statements must be performed in the Initial segment. Because TI(i) must be declared in a DIMENSION statement and not in a STORAGE statement, a TABLE card cannot be used to initialize the values of TI(i). Also, since TI(i) is contained in an EQUIVALENCE statement, a DATA card cannot be used to specify the initial values for TI(i).

3 The first portion of the Dynamic segment is changed to a nosort section to allow subscripted variables to appear on the left-hand side of the equal sign and also to permit the use of the DO loop.

4 If the INTGRL statement were written as

$$TE = INTGRL(530.0,TDE,20)$$

only the first term of the array T(1) would have the initial value of 530.0. All of the remaining terms would have initial value of zero. For this reason it is necessary to include the array TI(i) for nonzero initial conditions.

5 Since subscripted variables cannot be used in S/360 CSMP PRINT or PRTPLT statements, it is necessary to set all subscripted variables desired as output equal to nonsubscripted variables. In this example, the subscripted variables T(1), T(5), T(10), and T(17) are set equal to new variables for output purposes.

The resulting output for this simulation is shown in Fig. 3.13.

```
THE  USE  OF  SUBSCRIPTED  VARIABLES  TO  SOLVE  A  HEAT  TRANSFER            RKS        INTEGRATION
PROBLEM
```

TIME	T1	T5	T10	T17
0.0	5.3000E 02	5.3000E 02	5.3000E 02	5.3000E 02
6.0000E-04	6.1053E 02	6.1610E 02	6.1610E 02	6.1610E 02
1.2000E-03	6.7098E 02	6.8947E 02	6.8947E 02	6.8946E 02
1.8000E-03	7.1705E 02	7.5185E 02	7.5185E 02	7.5182E 02
2.4000E-03	7.5261F 02	8.0478E 02	8.0479E 02	8.0469E 02
3.0000F-03	7.8035E 02	8.4960E 02	3.4963E 02	8.4935E 02
3.6000E-03	8.0216E 02	8.8747E 02	8.8752E 02	8.8709E 02
4.2000E-03	8.1944F 02	9.1939E 02	9.1950E 02	9.1879E 02
4.8000F-03	8.3320E 02	9.4626E 02	9.4643E 02	9.4538E 02
5.4000E-03	8.4421E 02	9.6882E 02	9.6908E 02	9.6765E 02
6.0000E-03	8.5305E 02	9.8775E 02	9.8811E 02	9.8624E 02
6.6000E-03	8.6018E 02	1.0036E 03	1.0041E 03	1.0018E 03
7.2000E-03	8.6593E 02	1.0168E 03	1.0175E 03	1.0147E 03
7.8000F-03	8.7059E 02	1.0279E 03	1.0287E 03	1.0254E 03
8.4000E-03	8.7437F 02	1.0371E 03	1.0380E 03	1.0343E 03
9.0000F-03	8.7745E 02	1.0448E 03	1.0459E 03	1.0417E 03
9.6000E-03	8.7994E 02	1.0512E 03	1.0524E 03	1.0479E 03
1.0200E-02	8.8198E 02	1.0566E 03	1.0579E 03	1.0530E 03
1.0800E-02	8.8364E 02	1.0610E 03	1.0625E 03	1.0572E 03
1.1400E-02	8.8500E 02	1.0647E 03	1.0663E 03	1.0607E 03
1.2000F-02	8.8611E 02	1.0678E 03	1.0695E 03	1.0636E 03
1.2600E-02	8.8701E 02	1.0703E 03	1.0721E 03	1.0660E 03
1.3200E-02	8.8775E 02	1.0724E 03	1.0743E 03	1.0680E 03
1.3800E-02	8.8836E 02	1.0742E 03	1.0762E 03	1.0696E 03
1.4400E-02	8.8886E 02	1.0756E 03	1.0777E 03	1.0710E 03
1.5000F-02	8.8927E 02	1.0768E 03	1.0790E 03	1.0721E 03
1.5600E-02	8.8960E 02	1.0778E 03	1.0801E 03	1.0730E 03
1.6200E-02	8.8988E 02	1.0787E 03	1.0810E 03	1.0738E 03
1.6800E-02	8.9010E 02	1.0794E 03	1.0817E 03	1.0744E 03
1.7400F-02	8.9029E 02	1.0799E 03	1.0824E 03	1.0750E 03
1.8000E-02	8.9044E 02	1.0804E 03	1.0829E 03	1.0754E 03

Fig. 3.13 Output of program of Fig. 3.12.

The specification form of the INTGRL statement cannot be used for direct double-integration of arrays. For example, an acceleration array which is integrated into a velocity array cannot be directly integrated into a displacement array. The following example illustrates the correct procedure for performing double-integration of arrays.

Example 3.4

Consider the system of Fig. 3.14 which shows five uncoupled spring-mass units. The entire mass assembly is moving to the right with a common velocity of 50 ft/sec. The base strikes and then sticks to an immovable rigid surface. The problem is to find the resulting motion of all five masses after impact.

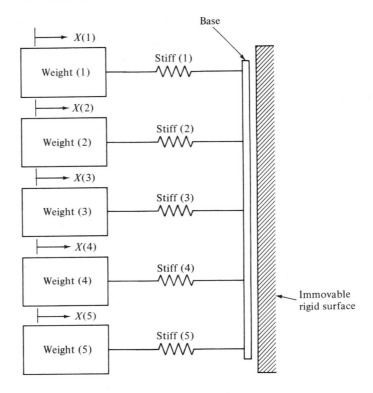

Initial velocity of all masses = 50 ft/sec

Fig. 3.14 Spring-mass system.

Using Newton's second law, the acceleration of the ith mass can be written as

$$A(i) = \frac{-\text{STIFF}(i)*X(i)}{\text{WEIGHT}(i)/32.17} \tag{3.8}$$

where STIFF(i) is the stiffness of the spring connected to the ith mass.

The program for calculating the system dynamics is shown in Fig. 3.15.

In the program, WEIGHT(i) and STIFF(i) are declared as subscripted variables by a STORAGE statement. This allows a TABLE card to be used to specify the values for WEIGHT(i) and STIFF(i).

Note that the entire Dynamic segment is changed to a nosort section. This assures that all statements will be executed in the exact order in which they appear in the program. The acceleration array A(J) is calculated and then integrated to obtain the velocity array V(J). It is not permissible to directly integrate this velocity array to obtain the displacement array. An intermediate array must be set equal to the velocity array. This is accomplished by the program statements

$$\text{DO } 2 \text{ J} = 1,5$$
$$2 \text{ VI(J)} = \text{V(J)}$$

```
TITLE PROGRAM TO CALCULATE THE DYNAMICS OF A MECHANICAL SYSTEM USING
TITLE SUBSCRIPTED VARIABLES.
INITIAL
/      DIMENSION X(5), V(5), A(5), V1(5), VELI(5)
/      EQUIVALENCE ( XE, X(1) ), (VE, V(1) ), (AE, A(1) )
/   1 , (V1E, V1(1) ), (VELIE, VELI(1) )
  STORAGE  WEIGHT(5),  STIFF(5)
  TABLE WEIGHT(1-2) = 2*13.6, WEIGHT(3) = 18.9, WEIGHT(4-5) = 2*11.5,...
  STIFF(1) = 135.0, STIFF(2-3) = 2*180.0, STIFF(4-5) = 2*320.0
  FIXED  J
NOSORT
  DO 3 J = 1,5
  3 VELI(J) = 50.0
DYNAMIC
NOSORT
  DO 1 J = 1,5
  1 A(J) = -STIFF(J)*X(J)/(WEIGHT(J)/32.17)
  VE = INTGRL(VELIE,AE,5)

*    CSMP DOES NOT PERMIT DIRECT DOUBLE INTEGRATION OF ARRAYS WHEN USING THE
*    SPECIFICATION FORM OF THE INTGRL STATEMENT.  CONSEQUENTLY, THE FOLLOWING
*    TWO STATEMENTS ARE USED TO GENERATE THE ARRAY V1(J) WHICH CAN BE
*    INTEGRATED TO FIND X(J).

  DO 2 J = 1,5
  2 V1(J) = V(J)
  XE = INTGRL(0.0,V1E,5)
  X1 = X(1)
  X2 = X(2)
  X3 = X(3)
  X4 = X(4)
  X5 = X(5)
TERMINAL
  TIMER FINTIM = 0.6, PRDEL = 0.02
  PRINT  X1, X2, X3, X4, X5
END
STOP
ENDJOB
```

Fig. 3.15 Program to simulate the dynamics of five masses.

The VI(J) array can now be integrated to obtain the displacement array. Since the VI(i) array is used in the specification form of an INTGRL statement, it must be included in the EQUIVALENCE statement. The output for this program is given in Fig. 3.16.

OVERLAY Statement

The FUNCTION statement which was defined in Example 2.4 is considered a data statement. It is used for specifying pairs of x–y coordinates for the function generating elements AFGEN and NLFGEN. When making more than one run, the OVERLAY statement can be used to change previously specified x–y values contained in FUNCTION statements. An example of an OVERLAY card is

OVERLAY TORQUE $= 0.0,4.5$, $3.0,7.8$, $6.0,11.9$, $10.0,19.0$

In this example, the FUNCTION defined as TORQUE will take on the new x–y values for the second run as given in the OVERLAY statement.

The format for an OVERLAY statement is the same as for a FUNCTION card with the following exceptions.

PROGRAM TO CALCULATE THE DYNAMICS OF A MECHANICAL SYSTEM USING   RKS   INTEGRATION
SUBSCRIPTED VARIABLES.

TIME	X1	X2	X3	X4	X5
0.0	0.0	0.0	0.0	0.0	0.0
2.0000E-02	9.7884E-01	9.7185E-01	9.7970E-01	9.4138E-01	9.4138E-01
4.0000E-02	1.8340E 00	1.7805E 00	1.8405E 00	1.5556E 00	1.5556E 00
6.0000E-02	2.4573E 00	2.2902E 00	2.4781E 00	1.6293E 00	1.6293E 00
8.0000E-02	2.7701E 00	2.4154E 00	2.8151E 00	1.1368E 00	1.1368E 00
1.0000E-01	2.7329E 00	2.1350E 00	2.8106E 00	2.4938E-01	2.4938E-01
1.2000E-01	2.3502E 00	1.4961E 00	2.4651E 00	-7.2470E-01	-7.2470E-01
1.4000E-01	1.6705E 00	6.0608E-01	1.8207E 00	-1.4469E 00	-1.4469E 00
1.6000E-01	7.7975E-01	-3.8576E-01	9.5533F-01	-1.6664E 00	-1.6664E 00
1.8000E-01	-2.0958E-01	-1.3128E 00	-2.5886E-02	-1.3069E 00	-1.3069E 00
2.0000E-01	-1.1724E 00	-2.0194E 00	-1.0040E 00	-4.9320E-01	-4.9320E-01
2.2000E-01	-1.9871E 00	-2.3870E 00	-1.8602E 00	4.9181E-01	4.9181E-01
2.4000E-01	-2.5506E 00	-2.3538E 00	-2.4909E 00	1.3059E 00	1.3059E 00
2.6000E-01	-2.7918E 00	-1.9253E 00	-2.8194E 00	1.6663E 00	1.6663E 00
2.8000E-01	-2.6802E 00	-1.1736E 00	-2.8026E 00	1.4476E 00	1.4476E 00
3.0000E-01	-2.2299E 00	-2.2487E-01	-2.4520E 00	7.2596E-01	7.2596E-01
3.2000E-01	-1.4977E 00	7.6164E-01	-1.8006E 00	-2.4791E-01	-2.4791E-01
3.4000E-01	-5.7630E-01	1.6203E 00	-9.3091E-01	-1.1356E 00	-1.1356E 00
3.6000E-01	4.1796E-01	2.2068E 00	5.1754E-02	-1.6288E 00	-1.6288E 00
3.8000E-01	1.3594E 00	2.4229E 00	1.0281E 00	-1.5559E 00	-1.5559E 00
4.0000E-01	2.1290E 00	2.2321E 00	1.8798E 00	-9.4239E-01	-9.4239E-01
4.2000E-01	2.6296E 00	1.6665E 00	2.5034E 00	-1.4542E-03	-1.4542E-03
4.4000E-01	2.7978E 00	8.2117E-01	2.8234E 00	9.3996E-01	9.3996E-01
4.6000E-01	2.6125E 00	-1.6207E-01	2.8008E 00	1.5548E 00	1.5548E 00
4.8000E-01	2.0970E 00	-1.1181E 00	2.4386E 00	1.6293E 00	1.6293E 00
5.0000E-01	1.3165E 00	-1.8864E 00	1.7805E 00	1.1377E 00	1.1377E 00
5.2000E-01	3.6959E-01	-2.3380E 00	9.0639E-01	2.5082E-01	2.5082E-01
5.4000E-01	-6.2400E-01	-2.3970E 00	-7.7623E-02	-7.2322E-01	-7.2322E-01
5.6000E-01	-1.5387E 00	-2.0535E 00	-1.0522E 00	-1.4459E 00	-1.4459E 00
5.8000E-01	-2.2590E 00	-1.3653E 00	-1.8992E 00	-1.6663E 00	-1.6663E 00
6.0000E-01	-2.6938E 00	-4.4777E-01	-2.5158E 00	-1.3075E 00	-1.3075E 00

Fig. 3.16   Output from program of Fig. 3.15.

1 The label OVERLAY is used in place of the label FUNCTION.

2 The number of *x–y* data pairs in an OVERLAY statement must not be more than the number of *x–y* pairs in the original FUNCTION statement.

3 Parentheses must not be used in OVERLAY statements to separate pairs of data.

A simple example which illustrates the use of an OVERLAY statement is the problem of calculating the volumetric flow of blood pumped by the human heart for two different pressure profiles.

Example 3.5

Figure 3.17 shows two time-histories of ventricular pressure in the human heart. Profile No. 1 represents a strong heart beat while profile No. 2 corresponds to a more average heart. An expression which approximates the flow rate of blood pumped by the heart is

$$\text{FLOW} = \frac{P}{575.0} \tag{3.9}$$

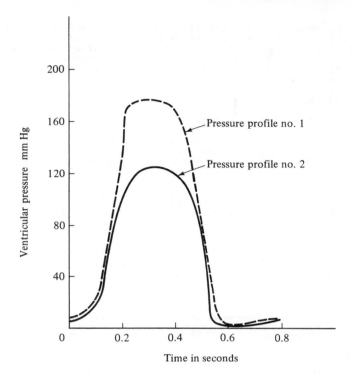

Fig. 3.17 Time-history of pressure in pumping chamber of the human heart.

where FLOW = volumetric flow rate in l/sec

P = ventricular pressure in mm-Hg

A single integration of FLOW with respect to time yields the total amount of blood that has flowed from the heart during one beat.

$$\text{VOLUME} = \int_0^t \text{FLOW } dt \tag{3.10}$$

Figure 3.18 shows the program for solving for the total flow for both pressure profiles. Data representing the first time-history of pressure is entered on the FUNCTION card with the label PRESS. The OVERLAY statement contains the data for the second pressure profile used for the function PRESS in the second run. Figure 3.19 contains the PRTPLT output resulting from the first pressure profile and Fig. 3.20 shows the output for the second time-history of pressure. Notice in the program of Fig. 3.18 that the OVERLAY statement appears immediately after the first END card. The END card permits the simulation to accept new data and control statements for another run. Note also that the END card sets the independent variable (TIME) to zero and resets all initial conditions. A detailed discussion on the use of the END statement is contained in the following section on translation control statements.

```
LABEL BLOOD FLOW PROBLEM TO ILLUSTRATE THE USE OF AN OVERLAY STATEMENT.
  FUNCTION  PRESS = (0.0,8.0),(0.05,12.0),(0.1,21.0),(0.15,80.0),        ...
   (0.2,140.0),(0.25,174.0),(0.28,178.0),(0.3,177.0),(0.32,176.0),       ...
   (0.35,175.0),(0.4,165.0),(0.45,138.0),(0.5,86.0),(0.55,19.0),         ...
   (0.6,3.0),(0.62,3.5),(0.64,4.0),(0.67,5.0),(0.7,6.0),(0.8,8.0)
  P = NLFGEN(PRESS,TIME)
  FLOW = P/575.0
  VOLUME = INTGRL(0.0,FLOW)
  PRTPLT VOLUME (P,FLOW)
  TIMER  FINTIM = 0.8,   OUTDEL = 0.04

*    THE FOLLOWING END CARD RESETS THE INITIAL CONDITIONS, RESETS TIME TO ZERO,
*  PREVIOUSLY DEFINED FUNCTION TO THE VALUES SHOWN.
*  ANOTHER RUN.   THE OVERLAY STATEMENT CHANGES THE DATA CONTAINED IN THE

  END
  OVERLAY PRESS = 0.0,5.0, 0.05,8.0, 0.1,18.0, 0.15,63.0, 0.2,100.0, ...
   0.25,120.0, 0.3,124.0, 0.35,124.0, 0.4,120.0, 0.5,65.0, 0.54,6.0, ...
   0.56,4.0, 0.58,2.0, 0.59,2.0, 0.61,2.2, 0.65,2.5, 0.7,3.0, 0.8,5.0
  END
  STOP
  ENDJOB
```

Fig. 3.18 Program to calculate blood flow and to illustrate the use of an OVERLAY statement.

```
BLOOD FLOW PROBLEM TO ILLUSTRATE THE USE OF AN OVERLAY STATEMENT.   PAGE   1

                           MINIMUM           VOLUME VERSUS TIME            MAXIMUM
                           0.0                                            1.0574E-01
 TIME          VOLUME       I                                                  I        P              FLOW
0.0            0.0          +                                                          8.0000E 00   1.3913E-02
4.0000E-02     6.6782E-04   +                                                          1.1200E 01   1.9478E-02
8.0000E-02     1.6116E-03   +                                                          1.6800E 01   2.9217E-02
1.2000E-01     3.2800E-03   -+                                                         3.8600E 01   6.7130E-02
1.6000E-01     7.8047E-03   ---+                                                       9.1920E 01   1.5986E-01
2.0000E-01     1.5868E-02   -------+                                                   1.4000E 02   2.4348E-01
2.4000E-01     2.6722E-02   ------------+                                              1.6928E 02   2.9440E-01
2.8000E-01     3.8947E-02   ------------------+                                        1.7800E 02   3.0957E-01
3.2000E-01     5.1268E-02   ------------------------+                                  1.7600E 02   3.0609E-01
3.6000E-01     6.3458E-02   ------------------------------+                            1.7383E 02   3.0232E-01
4.0000E-01     7.5282E-02   -----------------------------------+                       1.6500E 02   2.8696E-01
4.4000E-01     8.6120E-02   ----------------------------------------+                  1.4476E 02   2.5176E-01
4.8000E-01     9.5085E-02   --------------------------------------------+              1.0980E 02   1.9096E-01
5.2000E-01     1.0107E-01   -----------------------------------------------+           6.1000E 01   1.0609E-01
5.6000E-01     1.0342E-01   ------------------------------------------------+          1.1720E 01   2.0383E-02
6.0000E-01     1.0374E-01   -------------------------------------------------+         3.0000E 00   5.2173E-03
6.4000E-01     1.0398E-01   -------------------------------------------------+         4.0000E 00   6.9565E-03
6.8000E-01     1.0430E-01   -------------------------------------------------+         5.3333E 00   9.2754E-03
7.2000E-01     1.0471E-01   -------------------------------------------------+         6.5641E 00   1.1416E-02
7.6000E-01     1.0520E-01   -------------------------------------------------+         7.4461E 00   1.2950E-02
8.0000E-01     1.0574E-01   -------------------------------------------------+         8.0000E 00   1.3913E-02
```

Fig. 3.19 Blood flow for pressure-pulse.

Translation Control Statements

Translation control statements specify how structure statements are to be treated. They also are used for run control purposes. For example, the INITIAL, DYNAMIC, and TERMINAL statements, which were defined in Chap. 2, are translation control statements. They specify how groups of structure statements are to be handled. The END and CONTINUE statements, which have previously

```
BLOOD FLOW PROBLEM TO ILLUSTRATE THE USE OF AN OVERLAY STATEMENT.   PAGE   1

                      MINIMUM            VOLUME VERSUS TIME          MAXIMUM
                        0.0                                          7.4928E-02
   TIME       VOLUME     I                                            I         P            FLOW
  0.0        0.0         +                                                   5.0000E 00   8.6957E-03
  4.0000E-02 4.3130E-04  +                                                   7.4000E 00   1.2870E-02
  8.0000E-02 1.1042E-03  +                                                   1.3160E 01   2.2887E-02
  1.2000E-01 2.4927E-03  -+                                                  3.1800E 01   5.5304E-02
  1.6000E-01 6.0895E-03  ----+                                               7.1040E 01   1.2355E-01
  2.0000E-01 1.2068E-02  --------+                                           1.0000E 02   1.7391E-01
  2.4000E-01 1.9692E-02  -------------+                                      1.1736E 02   2.0410E-01
  2.8000E-01 2.8155E-02  -------------------+                                1.2432E 02   2.1621E-01
  3.2000E-01 3.6804E-02  -------------------------+                          1.2448E 02   2.1649E-01
  3.6000E-01 4.5445E-02  ----------------------------+                       1.2352E 02   2.1482E-01
  4.0000E-01 5.3930E-02  ----------------------------------+                 1.2000E 02   2.0870E-01
  4.4000E-01 6.1833E-02  ---------------------------------------+            1.0552E 02   1.8351E-01
  4.8000E-01 6.8379E-02  ---------------------------------------------+      8.1013E 01   1.4089E-01
  5.2000E-01 7.2734E-02  ------------------------------------------------+   3.8143E 01   6.6336E-02
  5.5000E-01 7.3638E-02  ------------------------------------------------+   4.0000E 00   6.9564E-03
  6.0000E-01 7.3812E-02  ------------------------------------------------+   2.0667E 00   3.5942E-03
  6.4000E-01 7.3971E-02  ------------------------------------------------+   2.4375E 00   4.2391E-03
  6.8000E-01 7.4151E-02  ------------------------------------------------+   2.7833E 00   4.8406E-03
  7.2000E-01 7.4361E-02  ------------------------------------------------+   3.2933E 00   5.7275E-03
  7.6000E-01 7.4615E-02  ------------------------------------------------+   4.0400E 00   7.0261E-03
  8.0000E-01 7.4928E-02  ------------------------------------------------+   5.0000E 00   8.6956E-03
```

Fig. 3.20   Blood flow for pressure-pulse.

been used and briefly described, can be used for run-control purposes and consequently are also translation control statements.

This section is concerned with the following six statements and how they can be used for changing data and output statements for sequential runs, resetting integration methods and error requirements, and other special features pertaining to run control techniques.

> END
> RESET
> CONTINUE
> STOP
> ENDJOB
> ENDJOB STACK

END

The END card must appear at least once in every program. For programs where control and data statements are not changed and only one run is desired, the END card simply appears immediately after the last structural statement. All examples in Chap. 2, with the exception of Example 2.10, illustrated this use of the END statement.

The END card can be used to permit the simulation to accept new data and control statements for sequential runs as illustrated in this chapter in the program of Fig. 3.18. When the END card is not followed immediately by a STOP card, another run is automatically initiated incorporating the changes that follow the END card. At the start of the new run, the initial conditions are automatically reset and the independent variable (TIME) is set to zero. The following simple example program illustrates how this is accomplished.

$$\text{INCON} \quad \text{YO} = 2.0$$

.

.

.

$$\text{Y} = \text{INTGRL(YO, V)}$$

.

.

.

```
END
INCON   YO = 7.5
END
STOP
ENDJOB
```

In the above example, the initial value for Y in the first run is 2.0. For the second run the initial value of Y is 7.5.

Example 3.6

Several important points can be illustrated by incorporating an extra END card in the program in Fig. 3.3. Figure 3.21 shows this program where an END card has been added to provide one additional run. The first run of the program is identical to the original program in Fig. 3.3. In the second run, the integration method, output, and timer variables are changed as specified by the cards following the first END statement.

Note that SUM and COUNT are set equal to zero for the second run. This is necessary even though these variables were originally set equal to zero in the Initial segment. If they are not reset to zero, the final values of SUM and COUNT in the first run will be used as the starting values in the second run.

For the variables that appear on the TIMER card, it is only necessary to specify changes. In this problem, DELT is the only timer variable that changes for the second run.

The two pages of output from the program in Fig. 3.21 is shown in Fig. 3.22.

All changes that were made for the second run, in the previous example, nullified the corresponding instruction specified in the first run. Not all statements, however, will cancel previous instructions. The following is a list of the statements that when used for an additional run will completely nullify the corresponding instructions of previous runs.

```
FINISH
METHOD
PREPARE
PRINT
TITLE
```

The following is a list of five additive instructions. These statements do not nullify previous ones, but simply provide additional instructions.

```
ABSERR & RELERR
LABEL
PRTPLT
RANGE
```

```
TITLE   PROGRAM TO CALCULATE THE AVERAGE ABSOLUTE ERROR OF EQUATION 3.1
INITIAL
CONSTANT   PI = 3.14159, SUM = 0.0, COUNT = 0.0
DYNAMIC

*    THE FOLLOWING 3 CARDS ARE USED TO SOLVE EQUATION 3.1, WHERE-
*    YDD = SECOND DERIVATIVE OF Y WITH RESPECT TO TIME
*    YD  = FIRST DERIVATIVE OF Y WITH RESPECT TO TIME

YDD = -4.0*PI*PI*Y
YD = INTGRL(0.0, YDD)
Y = INTGRL(1.0, YD)

*   THE FOLLOWING NOSORT SECTION IS REQUIRED TO USE THE "IF" STATEMENT

NOSORT

*   KEEP IS EQUAL TO 1 WHEN THE END OF A VALID INTEGRATION STEP IS REACHED

IF(KEEP. EQ .1)  GO TO 1
GO TO 2
1 SUM = SUM + ABS(Y - COS(2.0*PI*TIME))
COUNT = COUNT + 1.0
AAER = SUM/COUNT
2 CONTINUE
TERMINAL
TIMER   FINTIM = 1.0,   PRDEL = 0.1,   DELT = 0.001
METHOD  SIMP
PRINT   Y, SUM, COUNT, AAER

*   THE FOLLOWING "END" CARD RESETS THE INITIAL CONDITIONS AND SETS THE
*   INDEPENDENT VARIABLE (TIME) TO ZERO AND INITIATES ANOTHER RUN USING THE
*   CHANGES THAT FOLLOW THE END CARD.

END
    PRINT   Y, YD, YDD, SUM, COUNT, AAER
    TIMER   DELT = 0.004
    CONSTANT   SUM = 0.0, COUNT = 0.0
    METHOD  TRAPZ
END
STOP
ENDJOB
```

Fig. 3.21 Program to illustrate the use of END statement.

```
PROGRAM TO CALCULATE THE AVERAGE ABSOLUTE ERROR OF EQUATION 3.1  SIMP       INTEGRATION

  TIME          Y            SUM          COUNT        AAER
0.0           1.0000E 00   0.0          1.0000E 00   0.0
1.0000E-01    8.0902E-01   1.3816E-04   1.0100E 02   1.3680E-06
2.0000F-01    3.0902E-01   4.0048E-04   2.0100E 02   1.9925E-06
3.0000E-01   -3.0901E-01   5.8232E-04   3.0100E 02   1.9346E-06
4.0000E-01   -8.0900E-01   1.7282E-03   4.0100E 02   4.3098E-06
5.0000E-01   -9.9997E-01   4.1801E-03   5.0100E 02   8.3434E-06
6.0000E-01   -8.0899E-01   7.4814E-03   6.0100E 02   1.2448E-05
7.0000E-01   -3.0900E-01   1.0343E-02   7.0100E 02   1.4755E-05
8.0000E-01    3.0900E-01   1.1399E-02   8.0100E 02   1.4231E-05
9.0000F-01    8.0897E-01   1.3523E-02   9.0100E 02   1.5009E-05
1.0000E 00    9.9994E-01   1.8603E-02   1.0010E 03   1.8585E-05
```

Fig. 3.22 Output of program of Fig. 3.21.

PROGRAM TO CALCULATE THE AVERAGE ABSOLUTE ERROR OF EQUATION 3.1   TRAPZ   INTEGRATION

TIME	Y	YD	YDD	SUM	COUNT	AAER
0.0	1.0000E 00	0.0	-3.9478E 01	0.0	1.0000E 00	0.0
1.0000E-01	8.0898E-01	-3.6935E 00	-3.1937E 01	3.4213E-04	2.6000E 01	1.3159E-05
2.0000E-01	3.0889E-01	-5.9759E 00	-1.2195E 01	2.3875E-03	5.1000E 01	4.6813E-05
3.0000E-01	-3.0920E-01	-5.9753E 00	1.2207E 01	6.4546E-03	7.6000E 01	8.4929E-05
4.0000E-01	-8.0917E-01	-3.6918E 00	3.1945E 01	1.0940E-02	1.0100E 02	1.0832E-04
5.0000E-01	-1.0000E 00	2.0339E-03	3.9478E 01	1.2990E-02	1.2600E 02	1.0309E-04
6.0000E-01	-8.0879E-01	3.6951E 00	3.1930E 01	1.5985E-02	1.5100E 02	1.0586E-04
7.0000E-01	-3.0859E-01	5.9765E 00	1.2182E 01	2.4672E-02	1.7600E 02	1.4018E-04
8.0000E-01	3.0951E-01	5.9746E 00	-1.2219E 01	3.6836E-02	2.0100E 02	1.8326E-04
9.0000E-01	8.0936E-01	3.6902E 00	-3.1952E 01	4.7795E-02	2.2600E 02	2.1148E-04
1.0000E 00	1.0000E 00	-4.0638E-03	-3.9478E 01	5.2229E-02	2.5100E 02	2.0808E-04

Fig. 3.22 (Continued)

As an example, consider a program having the following structure.

⋮

Main body of program

⋮

Control statements for first run
{
TIMER FINTIM = 4.0, OUTDEL = 0.1, PRDEL = 0.2
PRINT A, B, C
PRTPLT X
PRTPLT Y
RANGE X
ABSERR X = 0.001, Y = 0.003
LABEL OUTPUT OF "X" INTEGRATOR
LABEL OUTPUT OF "Y" INTEGRATOR
}
END

Control statements for second run
{
PRINT C
PRTPLT Z
RANGE Y, Z
ABSERR Z = 0.0004, Y = 0.005
LABEL OUTPUT OF "Z" INTEGRATOR
}
END
STOP
ENDJOB

For the first run of the above example program, the following occurs:

1 A, B, and C are printed at the interval specified on the timer card.

2 A printer-plot of the variable X will be made with the heading: OUTPUT OF "X" INTEGRATOR.

3 A printer-plot of the variable Y will be made with the heading: OUTPUT OF "Y" INTEGRATOR.

4 The maximum and minimum values of X are listed by the RANGE statement.

5 Absolute errors of 0.001 and 0.003 will be used for the X and Y integrators, respectively.

For the second run, the following occurs:

1 The variable C is printed at the interval specified on the only timer card.

2 As in the first run, printer-plots will be made for X and Y with the same headings.

3 An additional printer-plot of the variable Z will be made with the heading: OUTPUT OF "Z" INTEGRATOR.

4 The maximum and minimum values of X, Y, and Z are listed by the RANGE statements.

5 Absolute errors of 0.001, 0.005, and 0.0004 will be used for the X, Y, and Z integrators, respectively.

In this example PRTPLT, LABEL, RANGE, and ABSERR are the additive statements.

It may be necessary in some simulations to eliminate certain control statements in going from one run to another. For this purpose the RESET card should be used.

RESET

This statement allows the user to nullify certain control instructions used in previous runs.

The RESET card can be used for the following statements.

> ABSERR & RELERR
> FINISH
> LABEL
> PREPARE
> PRINT
> PRTPLT
> RANGE

The RESET card should be placed immediately after the END or CONTINUE card to insure its proper use. RESET PRTPLT will nullify all previous PRTPLT and LABEL instructions. RESET LABEL nullifies only LABEL statements. A card containing only RESET on it will nullify all previous LABEL, PRTPLT, PREPARE, PRINT, and RANGE instructions.

A TITLE statement cannot be nullified by using a RESET card. The use of a TITLE instruction after an END or CONTINUE card will nullify the previous TITLE card.

The following example illustrates the use of the END and RESET statements along with the multiple run capability of a PARAMETER card. This example also shows the use of a CSMP random number generator.

Example 3.7

Random number signal sources can be very useful in simulating problems associated with many fields, including Industrial Engineering. An example of a typical problem involves determining the random distribution of the stiffness of a coil spring that is manufactured in high volume. The wire diameter and inside diameter of the spring are random variables that can best be described by a Gaussian distribution having a mean and standard deviation. The stiffness of a coil spring which is shown in Fig. 3.23 depends on the wire size and inside diameter of the coil as expressed by

$$K = \frac{G * DW^4}{8(DI + DW)^3 N} \tag{3.11}$$

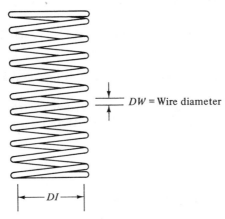

DW = Wire diameter

DI = inside diameter of spring

Fig. 3.23   Coil spring.

where G = modulus of elasticity for steel in shear = 11,500,000 psi

N = number of active coils = 10

DW = diameter of wire

DI = inside diameter of spring coil

In addition, the following program parameter definitions are used.

MEANI = mean inside diameter of spring = 1.000 in.

SIGMAI = standard deviation of inside diameter of spring: 0.001, 0.002, and 0.005 in.

MEANW = mean wire diameter = 0.100 in.

SIGMAW = standard deviation of wire diameter: 0.00033, 0.0006, 0.0015 in.

Since there are three different standard deviations for both the wire size and inside spring diameter, a total of nine runs will be required to simulate all combinations.

CSMP has signal sources which provide random numbers from two distributions. These are the Gaussian and the uniform distributions. For the Gaussian distribution, the statement

X = GAUSS(N, M, S)

gives a signal having a normal distribution

> where N = any *odd* integer (used as a *seed* value); in CSMP III use any integer
>
> M = mean value
>
> S = standard deviation

Only at each valid integration step (when KEEP = 1) does the GAUSS signal source provide a number. It is not activated during trial or immediate integration steps.

A uniform random number distribution between 0 and 1 can be provided by the following RNDGEN function.

$$X = RNDGEN(N)$$

where N is any *odd* integer which is also used as a seed value. When using two or more random-number signal sources, each statement should use a different seed value.

Using the GAUSS signal source, the following CSMP statements can be used to describe the random numbers representing the wire and inside spring diameters.

$$DW = GAUSS(1, MEANW, SIGMAW)$$
$$DI = GAUSS(3, MEANI, SIGMAI)$$

The standard deviation of the stiffness can be expressed by the following formula.

$$SIGMAK = \sqrt{\frac{n \sum_{i=1}^{n} K_i^2 - (\sum_{i=1}^{n} K_i)^2}{n(n-1)}} \tag{3.12}$$

> where K_i = stiffness of ith spring
>
> n = number of springs in sample

Using the above expressions for the random variables DW and DI, and Eq. (3.12) for the standard deviation of the spring stiffness, a program for simulating production runs for all nine combinations of standard deviations of wire and inside diameters is shown in Fig. 3.24. Note the method used in this program to reset COUNT, SUMK, and SUMK2 to zero. This procedure should be used when using the multiple-run capability of a PARAMETER card. The output for the very last run (SIGNAW = 0.0015, SIGMAI = 0.005), which involves a simulation of 3000 springs, is given in Fig. 3.25.

CONTINUE

The CONTINUE card can be used to change data or control statements without resetting initial conditions or the independent variable (TIME). It is used in place of the END card. When the program encounters a CONTINUE statement, it will accept the changes in control and data statements that follow this card. The program will then continue from the point where the previous run was ended. The CONTINUE statement provides, in effect, an interrupt point where changes in control and data statements can be made. The program in Fig. 3.5 illustrates the use of the CONTINUE card to change integration methods, integration error requirements, and timer variables during the run.

In CSMP III, the END CONTINUE label is used in place of CONTINUE. When using the CONTINUE card, the following rules must be followed.

```
TITLE  PROGRAM TO SIMULATE THE VARIANCE IN THE STIFFNESS IN THE HIGH
TITLE  VOLUME PRODUCTION OF A COIL SPRING.
  RENAME  TIME = NUMBER
  PARAMETER  SIGMAW  =  (0.00033, 0.0006, 0.0015)
  CONSTANT  COUNT = 0.0, SUMK = 0.0, SUMK2 = 0.0, MEANW = 0.1,      ...
    MEANI = 1.0, SIGMAI = 0.001, N = 10.0, G = 11.5E6
  DW   =  GAUSS(1, MEANW, SIGMAW)
  DI   =  GAUSS(1, MEANI, SIGMAI)
  K    =  G*(DW**4)/(8.0*((DW + DI)**3)*N)
NOSORT

*    THE FOLLOWING 5 CARDS ARE USED TO RESET COUNT, SUMK, AND SUMK2 TO ZERO
*    AT THE BEGINNING OF EACH RUN.

  IF(NUMBER.NE.0.0)  GO TO 2
  COUNT = 0.0
  SUMK = 0.0
  SUMK2 = 0.0
2 CONTINUE
  IF(KEEP.NE.1)  GO TO 1
  COUNT  =  COUNT  +  1.0
  SUMK   =  SUMK  +  K
  SUMK2  =  SUMK2  +  K*K

*    THE FOLLOWING "IF" STATEMENT IS USED TO AVOID DIVIDING BY ZERO
*    IN THE CALCULATION OF THE STANDARD DEVIATION (SIGMAK)

  IF(COUNT.LT. (PRDEL - 2.0) )  GO TO 1
  SIGMAK  =  SQRT((COUNT*SUMK2 - SUMK*SUMK)/(COUNT*(COUNT - 1.0)))
  MEANK   =  SUMK/COUNT
1 CONTINUE
  PRINT  SIGMAK,  MEANK
  RANGE  K,  DW,  DI,  SIGMAK,  MEANK
  TIMER  FINTIM = 5000.0,  DELT = 1.0,  PRDEL = 100.0
  METHOD  RECT
  END
  RESET  RANGE
  RANGE  K, DW, DI
  CONSTANT  SIGMAI  = 0.002
  TIMER  FINTIM = 3000.0,  PRDEL = 200.0
  PRINT  SIGMAK,  MEANK,  K,  DW,  DI
  END
  CONSTANT  SIGMAI  = 0.005
  END
  STOP
  ENDJOB
```

Fig. 3.24 Program to simulate the production of a coil spring.

1 The multiple run capability of the PARAMETER, CONSTANT, or INCON statements must not be used in a run where a CONTINUE card is used.

2 A program containing a TERMINAL segment should not be used with a CONTINUE statement.

3 A new value for FINTIM should be specified on a timer card for every use of a CONTINUE card. If a CONTINUE statement is initiated by a FINISH card, the output (PRDEL and OUTDEL) will be incremented from the time at which the FINISH condition was encountered.

4 The FORTRAN CONTINUE statement is distinguished from the CSMP CONTINUE card by a statement number. Figure 3.24 illustrates the use

PROGRAM TO SIMULATE THE VARIANCE IN THE STIFFNESS IN THE HIGH   INTGRL NOT USED   SIGMAW= 1.5000E-03
VOLUME PRODUCTION OF A COIL SPRING.

NUMBER	SIGMAK	MEANK	K	DW	DI
0.0	3.0245E-01	1.0796E 01	9.3799E 00	9.5150E-02	9.8383E-01
2.0000E 02	6.1168E-01	1.0836E 01	1.0007E 01	9.7835E-02	9.9805E-01
4.0000E 02	6.4029E-01	1.0804E 01	1.1095E 01	1.0075E-01	1.0003E 00
6.0000E 02	6.4603E-01	1.0787E 01	1.2347E 01	1.0324E-01	9.9448E-01
8.0000E 02	6.3315E-01	1.0804E 01	1.1378E 01	1.0165E-01	1.0033E 00
1.0000E 03	6.3632E-01	1.0797E 01	1.1053E 01	1.0133E-01	1.0095E 00
1.2000E 03	6.2593E-01	1.0797E 01	1.0644E 01	1.0010E-01	1.0061E 00
1.4000E 03	6.2863E-01	1.0797E 01	1.1533E 01	1.0183E-01	1.0007E 00
1.6000E 03	6.3020E-01	1.0798E 01	1.0704E 01	9.9853E-02	1.0013E 00
1.8000E 03	6.2993E-01	1.0804E 01	1.0589E 01	9.9512E-02	1.0006E 00
2.0000E 03	6.2875E-01	1.0804E 01	1.1122E 01	1.0015E-01	9.9135E-01
2.2000E 03	6.2451E-01	1.0811E 01	1.0604E 01	9.9617E-02	1.0015E 00
2.4000E 03	6.2795E-01	1.0810E 01	1.0789E 01	1.0025E-01	1.0038E 00
2.6000E 03	6.2814E-01	1.0805E 01	1.0583E 01	9.9897E-02	1.0061E 00
2.8000E 03	6.2734E-01	1.0805E 01	1.0528E 01	9.9399E-02	1.0011E 00
3.0000E 03	6.2547E-01	1.0804E 01	1.0589E 01	9.9600E-02	1.0018E 00

Range Output

PROBLEM DURATION 0.0   TO   3.0000E 03

VARIABLE	MINIMUM	NUMBER	MAXIMUM	NUMBER
K	8.6505E 00	3.8300E 02	1.2991E 01	2.4100E 03
DW	9.4510E-02	3.8300E 02	1.0530E-01	1.6500E 02
DI	9.7930E-01	2.0200E 02	1.0220E 00	5.6500E 02

Print Output

Fig. 3.25 Output of Fig. 3.24 for last run.

of the FORTRAN CONTINUE card. It can only be used in a nosort section or within a PROCEDURE function with a statement number.

An example of a complicated program which makes use of several END, RESET, and CSMP CONTINUE statements follows.

.
.
.

Main body of program

.
.
.

1st portion of 1st run
{
CONSTANT Q = −13.5, R = 9.0, S = 11.0
LABEL PRINTER-PLOT OF "X"
PRTPLT X
TIMER FINTIM = 1.0, OUTDEL = 0.02
}
CONTINUE

2nd portion of 1st run
{
RESET PRTPLT
TIMER FINTIM = 1.1, DELT = 0.0005, PRDEL = 0.001
METHOD ADAMS
CONSTANT Q = −17.0, R = 8.0
PRINT X, Y
}
CONTINUE

3rd portion of 1st run
{
TIMER FINTIM = 2.4, PRDEL = 0.02
METHOD RKS
PRTPLT Y
LABEL Y = DISTANCE
}
CONTINUE

4th portion of 1st run
{
RELERR Z = 0.00002
TIMER FINTIM = 3.8
FINISH Y = 100.0
}
END

2nd, 3rd, & 4th runs
{
RESET
PARAMETER S = (12.0, 12.3, 12.7)
PRTPLT X,Y
TIMER FINTIM = 2.0
FINISH Y = 126.0
END
STOP
}
ENDJOB

The above example program makes two complete runs. In the first run, the CONTINUE card is used three times to make changes in parameters, output, and integration method. The following is a summary of the first run.

1 The first portion of the first run, which ends at FINTIM = 1.0, uses the following parameters and timer variables.

$$Q = -13.5$$
$$R = 9.0$$
$$S = 11.0$$
$$OUTDEL = 0.02$$

A printer-plot of the variable X is made with the following heading: PRIN-TER-PLOT OF "X". Since the integration method is not specified, the variable-step Runge-Kutta method is used.

2 The second portion of the first run terminates at FINTIM = 1.1. The following parameters and timer variables are changed to the values shown.

$$Q = -17.0$$
$$R = 8.0$$
$$DELT = 0.0005$$
$$PRDEL = 0.001$$

Adams integration is used. The RESET PRTPLT card nullifies the previous PRTPLT and LABEL statements and the only output is given by the PRINT X,Y statement.

3 The third portion of the first run ends at FINTIM = 2.4. There are no changes in parameter values and the integration method is changed back to variable-step Runge-Kutta (RKS). The output includes the previous PRINT X,Y statement and the additional PRTPLT Y card with the following heading: Y = DISTANCE.

4 For the fourth and last portion of the first run, the run will terminate at a time of 3.8, or when the variable Y first reaches the value of 100.0. The relative error for the Z integrator is changed from the usual value of 0.0001 to 0.00002. All PRINT and PRTPLT outputs remain the same as in the third portion.

The first END card indicates the completion of the first run. All initial conditions are reset to the original values and the independent variable (TIME) is set to zero. The multiple run capability of a PARAMETER card is used to make three additional runs using the specified values of the parameter S. Note that because of the use of the multiple run PARAMETER card, a CONTINUE statement cannot be used in this portion of the program. The RESET card nullifies the previous PRTPLT, RELERR, LABEL, PRINT, and FINISH statements. Consequently, the only output will be printer-plots of the variables X and Y. The output interval will be the value of OUTDEL as specified in the first portion of the first run. The STOP and ENDJOB cards have been used without explanation in all programs. Their use is described as follows.

STOP

The STOP card must follow the last END card to signify the completion of the last run. It is also used to separate any user-supplied FORTRAN subprogram from the CSMP program. All subprograms must follow the STOP card.

ENDJOB

This card is used to signify the end of a job. In all previous programs, which have not included FORTRAN subroutines, the ENDJOB statement follows the STOP card. When FORTRAN subroutines are used, the ENDJOB must follow them. This is illustrated in the section in this chapter on subprograms. ENDJOB is one of five statements that must begin in card column 1. The other statements are: COMMON, COMMON MEN, ENDDATA, and ENDJOB STACK.

ENDJOB STACK

This statement, when used in place of the ENDJOB card, allows another CSMP job to directly follow. However, most computer installations prohibit the practice of "stacking runs". It is recommended that the user check with the computer installation before using the ENDJOB STACK statement. The ENDJOB label must begin in card column 1 and STACK must begin in column 9. A blank card should follow the ENDJOB STACK card.

Data Output

The majority of CSMP output is ordinarily handled by the five output statements: PRINT, PRTPLT, RANGE, TITLE, and LABEL. The user, however, has a wide range of other methods for printing data. These include the use of FORTRAN output statements, a DEBUG subroutine that provides output of all variables, a PREPARE statement that is used to allow plotting of variables on offline X–Y plotters, and a DECK statement that returns a punched deck of the program. CSMP III provides additional output capabilities that are covered in Chap. 5. This section describes the use of the four types of output that are available in both S/390 CSMP and CSMP III.

FORTRAN Output

The entire output capability of FORTRAN, with the exception of the PUNCH instruction, is available to the CSMP user. This includes the use of FORTRAN WRITE, PRINT, and FORMAT statements. Methods of using these output instructions are similar to those in FORTRAN. The primary restriction is that all FORTRAN output statements must be contained in nosort sections or in PROCEDURE functions. A summary of general rules follows later in this section. The reader should refer to specialized texts on FORTRAN, because of the generality and complexity of FORTRAN output statements.[2-5]

It is not obvious when it is desirable or necessary to use FORTRAN output capabilities. Some typical situations are listed as follows.

1 It may be necessary to print values from either the Initial or Terminal segments. These values would be printed only one time at the beginning or

end of the run. CSMP PRINT or PRTPLT statements cannot be used for this purpose. Consequently, the only choice is to use FORTRAN output statements.

2 Subscripted variables and integers cannot be used in PRINT and PRTPLT statements in S/360 CSMP. FORTRAN output statements can be used for these types of variables.

3 A maximum of fifty variables (including TIME) can be included in a S/360 CSMP PRINT statement. Essentially any number of variables can be incorporated in FORTRAN output statements.

4 The user has no control over output format when using PRINT and PRTPLT statements. Conversely, the format flexibility of FORTRAN offers considerable freedom in the style of output data.

5 CSMP PRINT and PRTPLT statements are executed at regular intervals as specified on the TIMER card. With the proper use of FORTRAN logical control and output statements, it is possible to have output at any point and at any interval during the run.

6 Since integration must be performed at each PRINT and PRTPLT interval, CSMP output statements usually affect the integration interval of variable-step methods. FORTRAN output instructions do not affect the integration step-size and as a result FORTRAN WRITE statements can be used to monitor each step in variable-step integration.

Statements in the Initial and Terminal segments are executed only once during a run. Consequently, FORTRAN output instructions can be utilized in these two segments with no special consideration other than being used in nosort sections or PROCEDURE functions. However, when FORTRAN output statements are used in the Dynamic segment, special care must be used. For example, depending upon the integration method, FORTRAN output instructions are executed up to four times for each integration step (see Table 3.3) and for each trial integration step if a variable step method is used. To provide FORTRAN output only at the end of valid integration steps, the following type of statement is recommended.

IF(KEEP. EQ. 1) WRITE(6,100) X,Y,Z

The above will print the values of X,Y, and Z using FORMAT number 100 at each valid integration step.

The following is a summary of some general rules for using FORTRAN output statements.

1 All FORTRAN output and FORMAT statements must be in nosort sections or PROCEDURE functions.

2 Continuation cards used with FORTRAN output and FORMAT statements *must* contain a $ in card column 6.

3 FORTRAN output instructions in the Initial and Terminal segments are executed only one time. Statements in the Dynamic segment are printed up to four times for each integration step. It is recommended that the following use of the IF statement be employed to provide output only at the end of valid integration steps.

IF(KEEP. EQ. 1) WRITE(6,100) X, Y, Z

4 The FORTRAN PRINT statement can be used in a CSMP program as part of a logical IF statement. For example, the following FORTRAN PRINT statement is valid.

IF(TIME. GT. 1.9) PRINT 100, X, Y, Z

The following PRINT instruction cannot be used.

PRINT 100, X, Y, Z

It should be pointed out that the FORTRAN PRINT statement offers no advantages over the FORTRAN WRITE instruction. For this reason, it is recommended that the FORTRAN PRINT statement not be used.

5 FORTRAN FORMAT instructions can be located in any nosort section or procedure function of the program. The most logical location is in a nosort section of the Initial segment.

6 The FORTRAN PUNCH instruction cannot be used in a CSMP program. If punched output on data cards is desired, the user should employ a FOR-TRAN WRITE statement. In a FORTRAN program, the following statement provides punch output.

WRITE(7,100) X, Y, Z

The above statement is not valid in a CSMP program. Unit number 7 is a work data set used by the CSMP system. The user should specify in JCL cards that another unit number be set in reserve for punched output. This unit number is then used in the WRITE instruction to provide punched output.

The following orbital mechanics problem is used to illustrate the use of FORTRAN output statements in all three segments.

Example 3.8

This problem involves the two-dimensional motion of a space vehicle in an orbit that passes over both the North and South Poles. In the initial condition, the vehicle is in a circular orbit at an altitude of 300,000 m. At the start of the simulation, rockets on the space vehicle are ignited for 120 sec which provide a thrust in the tangential direction. At the end of the 120 sec rocket burn, the only force acting on the vehicle is gravity.

To eliminate the problem of programming the effect of variable vehicle mass, it is assumed that the thrust of the rockets provides a constant acceleration of 4.905 m/sec^2 (0.5 g). A cylindrical coordinate system, as shown in Fig. 3.26, is ideally suited to describe the two-dimensional motion.

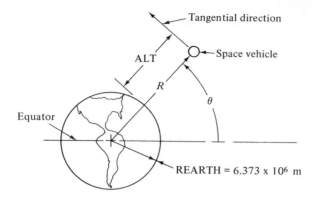

Fig. 3.26 Cylindrical coordinates of a space vehicle in a polar orbit.

The nonlinear equation set is

$$\ddot{r} = r\omega^2 - g \tag{3.13}$$

$$\dot{\omega} = \frac{(A - 2\dot{r}\omega)}{r} \tag{3.14}$$

$$\theta = \frac{180}{\pi} \int_0^t \omega \, dt \tag{3.15}$$

Program symbols

where r = radius from center of earth, m   R

θ = angular position, deg   ANGLE

ω = angular velocity, radians/sec   OMEGA

g = acceleration of gravity,   G
        $3.983 \times 10^{14}/r^2$, m/sec^2

A = acceleration resulting from rocket thrust, m/sec^2 A

$A = 4.905$ TIME $\lesssim$ 120.0 sec

$A = 0$ TIME $>$ 120.0 sec

The program, shown in Fig. 3.27, which simulates the motion of the space vehicle is divided into the three segments. In the Initial segment the initial radius RO and initial angular velocity OMEGAO are calculated. The angular velocity of a satellite in a circular orbit is given by Eq. (3.16).

$$\text{OMEGAO} = \sqrt{\frac{g}{r}} \quad \text{(rad/sec)} \tag{3.16}$$

Note that a FORTRAN WRITE statement is used with an appropriate FORMAT card to print the velocity in the circular orbit in mph. A WRITE instruction is also included in the Initial segment to provide a heading for output from the Dynamic segment. In the Dynamic segment, a logical IF statement is used in the first nosort section to assign the proper value of A. A sort section is then used to calculate $\ddot{r}$ (RDD) and $\dot{\omega}$

```
LABEL   PROGRAM TO CALCULATE THE ORBIT OF SATELLITE.
INITIAL
  CONSTANT  REARTH = 6.373E6, C = 3.983E14, ALTO = 300000.0
  RO = REARTH + ALTO
  G = C/(RO*RO)
  OMEGAO = SQRT(G/RO)
  VEL = OMEGAO*RO
  MPH = VEL/0.44704
NOSORT
  100 FORMAT(1H1,26HINITIAL VELOCITY IN MPH = ,F10.0)
  101 FORMAT(4E28.6)
  102 FORMAT(2F25.2)
  103 FORMAT(1H1,12X,21HFINAL VELOCITY IN MPH,7X,23HFINAL ALTITUDE IN MI
     $LES)
  104 FORMAT(/,16X,'ALTITUDE IN METERS',10X,'ANGLE IN DEGREES',12X,'VELO
     $CITY IN METERS/SEC',6X,'TIME IN SECONDS')
  WRITE(6,100) MPH
  WRITE(6,104)
DYNAMIC
NOSORT
  A = 0.0
  IF(TIME.LE.120.0)  A = 4.905
SORT
  G = C/(R*R)
  RDD = R*OMEGA*OMEGA - G
  OMEGAD = (A - 2.0*RD*OMEGA)/R
  RD = INTGRL(0.0,RDD)
  R = INTGRL(RO,RD)
  OMEGA = INTGRL(OMEGAO,OMEGAD)
  ANGLE = (180.0/3.14159)*INTGRL(0.0,OMEGA)
NOSORT
  ALT = R - REARTH
  VEL = SQRT(RD*RD + (R*OMEGA)**2)
  IF(TIME.GT.120.0)  GO TO 1
  IF(KEEP.EQ.1)  WRITE(6,101) ALT, ANGLE, VEL, TIME
  1 CONTINUE
TERMINAL
  TIMER  FINTIM = 9500.0, OUTDEL = 120.0
  ABSERR R = 1.0
  FINISH ANGLE = 360.0
  PRTPLT ALT  (ANGLE, VEL, G)
  MPH = VEL/0.44704
  MILES = ALT/1609.344
  WRITE(6,103)
  WRITE(6,102) MPH, MILES
END
STOP
ENDJOB
```

Fig. 3.27 Program to simulate the motion of a space vehicle and to demonstrate the use of FORTRAN output statements.

(OMEGAD) and integrations are performed to calculate $\dot{r}$ (RD), r, ω (OMEGA), and θ (THETA). The Dynamic section is changed back to a nosort section to allow the use of a FORTRAN WRITE statement. The WRITE statement is executed at each valid integration step during the first 120.0 sec of the simulation. This WRITE statement does not affect the step-size of the variable interval integration methods. Notice that the value of OUTDEL is equal to 120.0. This means an integration step must be performed at TIME = 120.0 sec which insures the program will simulate the total burn of the rockets. Also note that the allowable absolute error for the R integrator is increased to 1.0. The default absolute error of 0.0001 provides tighter control than necessary for very large output of the R integrator.

The output from the FORTRAN WRITE and FORMAT instructions is given in Fig. 3.28.

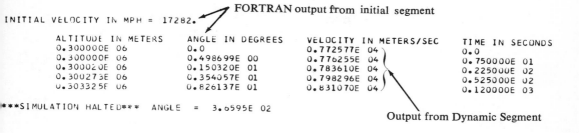

FINAL VELOCITY IN MPH   FINAL ALTITUDE IN MILES
      18650.59   187.03 ◄──── Output from Terminal Segment

Fig 3.28   Output from FORTRAN statements in Fig. 3.27.

Note that the first two lines of Fig. 3.28 were initiated in the Initial segment. The following five lines which describe the motion during the 120.0 sec rocket burn period were printed by the WRITE statement in the Dynamic segment. Since all integration steps are printed and the FORTRAN WRITE statement does not affect the integration step size of the variable-step Runge-Kutta method, the progress of the numerical integration can be monitored. Note that the first integration step is 7.5 sec which is $\frac{1}{16}$ of the output interval. In the second step the integration interval is increased to 15.0 sec, and for the third step the interval is 30.0 secs. For the fourth and last integration step during the rocket burn, a step size of 67.5 sec is used. The final velocity and altitude which was printed from the Terminal segment is shown in the last two lines of Fig. 3.28.

The time-history of the altitude is shown by the PRTPLT output of Fig. 3.29.

Note that the FORTRAN output statements had absolutely no effect on the PRTPLT output. This is not the case if both FORTRAN WRITE and PRINT statements are used in the Dynamic segment of the same program. FORTRAN and PRINT output statements are printed in the order in which they are generated. Consequently, FORTRAN and PRINT outputs are mixed together and can be confusing to read. Consider the example of using the following PRINT statement used with a PRDEL of 15.0 in the program of Fig. 3.27.

PRINT ALT, ANGLE, VEL, G

The portion of the mixed output is shown in Fig. 3.30.

DEBUG Subroutine

It is often useful when developing or checking out new programs to know the values of all variables. The DEBUG subroutine can be used for this purpose. It will print the current values of all nonsubscripted variables at each successive integration step. Any diagnostic messages that may appear are printed at the point where they occur. The subroutine is called by the following statement.

CALL DEBUG(N, T)

```
PROGRAM TO CALCULATE THE ORBIT OF SATELLITE.                    PAGE   1
                        MINIMUM              ALT    VERSUS TIME      MAXIMUM
                        3.0000E 05                                  2.9358E 06
    TIME          ALT          I                                         I      ANGLE          VEL            G
0.0           3.0000E 05       +                                               0.0          7.7258E 03     8.9447E 00
1.2000E 02    3.0533E 05       +                                               8.2614E 00   8.3107E 03     8.9357E 00
2.4000E 02    3.2387E 05       +                                               1.6823E 01   8.3132E 03     8.8810E 00
3.6000E 02    3.6495E 05       -+                                              2.5306E 01   8.2695E 03     8.7730E 00
4.8000E 02    4.2542E 05       --+                                             3.3662E 01   8.2057E 03     8.6177E 00
6.0000E 02    5.0366E 05       ---+                                            4.1849E 01   8.1240E 03     8.4227E 00
7.2000E 02    5.9773E 05       -----+                                          4.9834E 01   8.0272E 03     8.1969E 00
8.4000E 02    7.0538E 05       -------+                                        5.7590E 01   7.9182E 03     7.9495E 00
9.6000E 02    8.2429E 05       ---------+                                      6.5101E 01   7.7999E 03     7.6890E 00
1.0800E 03    9.5208E 05       -------------+                                  7.2360E 01   7.6752E 03     7.4230E 00
1.2000E 03    1.0864E 06       ---------------+                                7.9362E 01   7.5465E 03     7.1580E 00
1.3200E 03    1.2251E 06       -----------------+                              8.6113E 01   7.4162E 03     6.8991E 00
1.4400E 03    1.3662E 06       -------------------+                            9.2620E 01   7.2862E 03     6.6499E 00
1.5600E 03    1.5077E 06       ---------------------+                          9.8892E 01   7.1583E 03     6.4133E 00
1.6800E 03    1.6479E 06       -----------------------+                        1.0494E 02   7.0338E 03     6.1910E 00
1.8000E 03    1.7854E 06       -------------------------+                      1.1079E 02   6.9137E 03     5.9840E 00
1.9200E 03    1.9189E 06       ---------------------------+                    1.1645E 02   6.7991E 03     5.7929E 00
2.0400E 03    2.0472E 06       -----------------------------+                  1.2193E 02   6.6906E 03     5.6177E 00
2.1600E 03    2.1693E 06       ------------------------------+                 1.2724E 02   6.5888E 03     5.4583E 00
2.2800E 03    2.2843E 06       --------------------------------+               1.3242E 02   6.4941E 03     5.3142E 00
2.4000E 03    2.3916E 06       ----------------------------------+             1.3746E 02   6.4068E 03     5.1850E 00
2.5200E 03    2.4904E 06       -----------------------------------+            1.4238E 02   6.3272E 03     5.0700E 00
2.6400E 03    2.5802E 06       -------------------------------------+          1.4720E 02   6.2556E 03     4.9688E 00
2.7600E 03    2.6606E 06       --------------------------------------+         1.5193E 02   6.1919E 03     4.8807E 00
2.8800E 03    2.7313E 06       ---------------------------------------+        1.5658E 02   6.1364E 03     4.8053E 00
3.0000E 03    2.7918E 06       ----------------------------------------+       1.6117E 02   6.0892E 03     4.7420E 00
3.1200E 03    2.8419E 06       ----------------------------------------+       1.6570E 02   6.0502E 03     4.6905E 00
3.2400E 03    2.8816E 06       -----------------------------------------+      1.7018E 02   6.0195E 03     4.6504E 00
3.3600E 03    2.9105E 06       -----------------------------------------+      1.7463E 02   5.9972E 03     4.6215E 00
3.4800E 03    2.9285E 06       ------------------------------------------+     1.7906E 02   5.9833E 03     4.6036E 00
3.6000E 03    2.9358E 06       ------------------------------------------+     1.8348E 02   5.9777E 03     4.5964E 00
3.7200E 03    2.9320E 06       ------------------------------------------+     1.8790E 02   5.9806E 03     4.6001E 00
3.8400E 03    2.9174E 06       -----------------------------------------+      1.9232E 02   5.9918E 03     4.6146E 00
3.9600E 03    2.8920E 06       ----------------------------------------+       1.9677E 02   6.0114E 03     4.6400E 00
4.0800E 03    2.8558E 06       ---------------------------------------+        2.0124E 02   6.0394E 03     4.6764E 00
4.2000E 03    2.8090E 06       --------------------------------------+         2.0575E 02   6.0757E 03     4.7242E 00
4.3200E 03    2.7518E 06       -------------------------------------+          2.1032E 02   6.1203E 03     4.7837E 00
4.4400E 03    2.6844E 06       -----------------------------------+            2.1495E 02   6.1732E 03     4.8551E 00
4.5600E 03    2.6070E 06       ---------------------------------+              2.1965E 02   6.2342E 03     4.9391E 00
4.6800E 03    2.5202E 06       -------------------------------+                2.2444E 02   6.3033E 03     5.0361E 00
4.8000E 03    2.4242E 06       -----------------------------+                  2.2933E 02   6.3804E 03     5.1466E 00
4.9200E 03    2.3196E 06       --------------------------+                     2.3433E 02   6.4652E 03     5.2712E 00
5.0400E 03    2.2070E 06       ------------------------+                       2.3946E 02   6.5576E 03     5.4105E 00
5.1600E 03    2.0870E 06       ----------------------+                         2.4473E 02   6.6571E 03     5.5649E 00
5.2800E 03    1.9606E 06       --------------------+                           2.5016E 02   6.7636E 03     5.7350E 00
5.4000E 03    1.8287E 06       ------------------+                             2.5575E 02   6.8763E 03     5.9211E 00
5.5200E 03    1.6923E 06       ----------------+                               2.6154E 02   6.9947E 03     6.1230E 00
5.6400E 03    1.5528E 06       --------------+                                 2.6752E 02   7.1179E 03     6.3405E 00
5.7600E 03    1.4115E 06       ------------+                                   2.7372E 02   7.2449E 03     6.5727E 00
5.8800E 03    1.2701E 06       ----------+                                     2.8015E 02   7.3744E 03     6.8182E 00
6.0000E 03    1.1304E 06       --------+                                       2.8682E 02   7.5048E 03     7.0744E 00
6.1200E 03    9.9439E 05       ------+                                         2.9374E 02   7.6343E 03     7.3380E 00
6.2400E 03    8.6422E 05       ----+                                           3.0092E 02   7.7606E 03     7.6044E 00
6.3600E 03    7.4218E 05       ---+                                            3.0835E 02   7.8813E 03     7.8674E 00
6.4800E 03    6.3063E 05       --+                                             3.1603E 02   7.9936E 03     8.1201E 00
6.6000E 03    5.3193E 05       --+                                             3.2395E 02   8.0946E 03     8.3539E 00
6.7200E 03    4.4835E 05       -+                                              3.3207E 02   8.1815E 03     8.5598E 00
6.8400E 03    3.8196E 05       -+                                              3.4038E 02   8.2513E 03     8.7289E 00
6.9600E 03    3.3449E 05       +                                               3.4883E 02   8.3017E 03     8.8529E 00
7.0800E 03    3.0725E 05       +                                               3.5737E 02   8.3309E 03     8.9253E 00
7.2000E 03    3.0099E 05       +                                               3.6595E 02   8.3376E 03     8.9420E 00
```

Fig. 3.29   PRTPLT output for the program of Fig. 3.27.

N is equal to the number of outputs. An output, which includes a listing of the values of all variables, is printed for all trial and intermediate integration steps as well as for valid steps. N must be an integer constant. The output will begin at TIME = T or at the first integration step after TIME = T.

The DEBUG subroutine can only be called from nosort sections or PROCEDURE functions. It is recommended that DEBUG be called at the end of the Dynamic section. This insures that all computations of the dynamic simulation are completed before DEBUG is called. DEBUG may not be called from the Ter-

CSMP

TIME	ALT	ANGLE	VEL	G	RKS	INTEGRATION
0.0	3.0000E 05	0.0	7.7258E 03	8.9447E 00	0.773036E 04	0.937500E 00
	0.300000E 06		0.622078E-01		0.773956E 04	0.281250E 01
	0.300000E 06		0.186734E 00		0.775795E 04	0.656250E 01
	0.300005E 06		0.436232E 00		0.779932E 04	0.150000E 02
1.5000E 01	3.0001E 05	9.9977E-01	7.7993E 03	8.9446E 00	0.787284E 04	0.300000E 02
	0.300049E 06		0.209900E 01			
3.0000E 01	3.0005E 05	2.0090E 00	7.8728E 03	8.9445E 00	0.794628E 04	0.450000E 02
	0.300171E 06		0.302769E 01			
4.5000E 01	3.0017E 05	3.0277E 00	7.9463E 03	8.9442E 00	0.801958E 04	0.600000E 02
	0.300409E 06		0.405579E 01			
6.0000E 01	3.0041E 05	4.0558E 00	8.0196E 03	8.9436E 00	0.809271E 04	0.750000E 02
	0.300803E 06		0.509327E 01			
7.5000E 01	3.0080E 05	5.0933E 00	8.0927E 03	8.9425E 00	0.816564E 04	0.900000E 02
	0.301393E 06		0.614008E 01			
9.0000E 01	3.0139E 05	6.1401E 00	8.1656E 03	8.9409E 00	0.823830E 04	0.105000E 03
	0.302219E 06		0.719613E 01			
1.0500E 02	3.0222E 05	7.1961E 00	8.2383E 03	8.9387E 00	0.831068E 04	0.120000E 03
	0.303321E 06		0.826135E 01			
1.2000E 02	3.0332E 05	8.2614E 00	8.3107E 03	8.9358E 00		
1.3500E 02	3.0474E 05	9.3317E 00	8.3153E 03	8.9320E 00		
1.5000E 02	3.0647E 05	1.0402E 01	8.3134E 03	8.9273E 00		
1.6500E 02	3.0852E 05	1.1471E 01	8.3112E 03	8.9218E 00		
1.8000E 02	3.1089E 05	1.2539E 01	8.3087E 03	8.9155E 00		
1.9500E 02	3.1358E 05	1.3607E 01	8.3058E 03	8.9084E 00		
2.1000E 02	3.1658E 05	1.4674E 01	8.3026E 03	8.9004E 00		
2.2500E 02	3.1990E 05	1.5740E 01	8.2990E 03	8.8915E 00		
2.4000E 02	3.2353E 05	1.6804E 01	8.2951E 03	8.8819E 00		
2.5500E 02	3.2747E 05	1.7868E 01	8.2909E 03	8.8715E 00		
2.7000E 02	3.3172E 05	1.8930E 01	8.2863E 03	8.8602E 00		
2.8500E 02	3.3628E 05	1.9991E 01	8.2814E 03	8.8482E 00		

Output from FORTRAN Statements

Output from PRINT Statement

Fig. 3.30 Example of mixed FORTRAN and CSMP output.

131

minal segment since this segment is not executed until the end of the run. Logical control instructions can be used with DEBUG but they should not be used to "branch around" the CALL DEBUG statement at TIME = 0. The DEBUG subroutine can be used several times in one program.

Figure 3.31 shows a DEBUG statement added to the program of Example 3.8, which simulated the motion of a space vehicle. To simplify the program, all FORTRAN output statements are removed.

```
LABEL  PROGRAM TO CALCULATE THE ORBIT OF SATELLITE USING CALL DEBUG
INITIAL
   CONSTANT  REARTH = 6.373E6, C = 3.983E14, ALTO = 300000.0
   RO = REARTH + ALTO
   G = C/(RO*RO)
   OMEGAO = SQRT(G/RO)
DYNAMIC
NOSORT
   A = 0.0
   IF(TIME.LE.120.0)  A = 4.905
SORT
   G = C/(R*R)
   RDD = R*OMEGA*OMEGA - G
   OMEGAD = (A - 2.0*RD*OMEGA)/R
   RD = INTGRL(0.0,RDD)
   R = INTGRL(RO,RD)
   OMEGA = INTGRL(OMEGAO,OMEGAD)
   ANGLE = (180.0/3.14159)*INTGRL(0.0,OMEGA)
NOSORT
   ALT = R - REARTH
   VEL = SQRT(RD*RD + (R*OMEGA)**2)

*    THE FOLLOWING DEBUG STATEMENT WILL PRINT THE VALUE OF ALL VARIABLES USED
*    IN THE PROGRAM FOR 4 ITERATIONS.    THE OUTPUT WILL START AT TIME = 1600,
*    OR ON THE FIRST INTEGRATION STEP AFTER TIME = 1600.

   CALL DEBUG(4,1600.0)
TERMINAL
   TIMER  FINTIM = 9500.0, OUTDEL = 120.0
   ABSERR R = 1.0
   FINISH ANGLE = 360.0
   PRTPLT  ALT  (ANGLE, VEL, G)
END
STOP
ENDJOB
```

Fig. 3.31 Program to illustrate the use of CALL DEBUG.

The DEBUG output is shown in Fig. 3.32. Note that the first output occurred at TIME = 1620.0 which is the first integration step after TIME = 1600.0. Since KEEP was equal to 0 in all four DEBUG printings, none of the outputs represent a valid integration step. The six-digit symbols beginning with ZZ are generated by the CSMP program.

PREPARE

If the user's computer installation has offline plotting facilities, data generated by CSMP programs can be plotted on offline X–Y plotters. The PREPARE statement allows the user to specify up to forty-nine variables for offline plotting pur-

```
CSMP                                                    RKS      INTEGRATION

       DEBUG OUTPUT    KEEP= 0
TIME  = 1.6200E 03      DELT  = 1.2000E 02    DELMIN= 9.5000E-04    FINTIM= 9.4800E 03    PRDEL =  0.0
                        OUTDEL= 1.2000E 02    RD    = 1.1693E 03    R     = 7.9510E 06    OMEGA =  8.8021E-04
                        ZZ0005= 1.7793E 00    RDD   =-1.4011E-01    ZZ0008= 1.1727E 03    OMEGAD= -2.5890E-07
                        ZZ0009= 8.8810E-04    ZZ0002= 0.0          RO    = 6.6730E 06    OMEGA0=  1.1578E-03
                        ZZ0006= 0.0          REARTH= 6.3730E 06    C     = 3.9829E 14    ALTO  =  3.0000E 05
                        G     = 6.3003E 00    A     = 0.0          ANGLE = 1.0195E 02    ALT   =  1.5780E 06
                        VEL   = 7.0955E 03    ZZ0007= 0.0

       DEBUG OUTPUT    KEEP= 0
TIME  = 1.6200E 03      DELT  = 1.2000E 02    DELMIN= 9.5000E-04    FINTIM= 9.4800E 03    PRDEL =  0.0
                        OUTDEL= 1.2000E 02    RD    = 1.1693E 03    R     = 7.9510E 06    OMEGA =  8.8020E-04
                        ZZ0005= 1.7793E 00    RDD   =-1.4012E-01    ZZ0008= 1.1693E 03    OMEGAD= -2.5889E-07
                        ZZ0009= 8.8021E-04    ZZ0002= 0.0          RO    = 6.6730E 06    OMEGA0=  1.1578E-03
                        ZZ0006= 0.0          REARTH= 6.3730E 06    C     = 3.9829E 14    ALTO  =  3.0000E 05
                        G     = 6.3002E 00    A     = 0.0          ANGLE = 1.0195E 02    ALT   =  1.5780E 06
                        VEL   = 7.0955E 03    ZZ0007= 0.0

       DEBUG OUTPUT    KEEP= 0
TIME  = 1.6500E 03      DELT  = 1.2000E 02    DELMIN= 9.5000E-04    FINTIM= 9.4800E 03    PRDEL =  0.0
                        OUTDEL= 1.2000E 02    RD    = 1.1651E 03    R     = 7.9861E 06    OMEGA =  8.7244E-04
                        ZZ0005= 1.805TE 00    RDD   =-1.6642E-01    ZZ0008= 1.1693E 03    OMEGAD= -2.5456E-07
                        ZZ0009= 8.8020E-04    ZZ0002= 0.0          RO    = 6.6730E 06    OMEGA0=  1.1578E-03
                        ZZ0006= 0.0          REARTH= 6.3730E 06    C     = 3.9829E 14    ALTO  =  3.0000E 05
                        G     = 6.2450E 00    A     = 0.0          ANGLE = 1.0346E 02    ALT   =  1.6131E 06
                        VEL   = 7.0641E 03    ZZ0007= 0.0

       DEBUG OUTPUT    KEEP= 0
TIME  = 1.6500E 03      DELT  = 1.2000E 02    DELMIN= 9.5000E-04    FINTIM= 9.4800E 03    PRDEL =  0.0
                        OUTDEL= 1.2000E 02    RD    = 1.1643E 03    R     = 7.9860E 06    OMEGA =  8.7257E-04
                        ZZ0005= 1.8055E 00    RDD   =-1.6491E-01    ZZ0008= 1.1651E 03    OMEGAD= -2.5443E-07
                        ZZ0009= 8.7244E-04    ZZ0002= 0.0          RO    = 6.6730E 06    OMEGA0=  1.1578E-03
                        ZZ0006= 0.0          REARTH= 6.3730E 06    C     = 3.9829E 14    ALTO  =  3.0000E 05
                        G     = 6.2452E 00    A     = 0.0          ANGLE = 1.0344E 02    ALT   =  1.6130E 06
                        VEL   = 7.0649E 03    ZZ0007= 0.0

**SIMULATION HALTED***  ANGLE = 3.6595E 02
```

Fig. 3.32   DEBUG output form program of Fig. 3.31.

poses. Its use for preparing the variables X, Y, and Z for plotting is illustrated below.

PREPARE X, Y, Z

Using the PREPARE card is similar to using the PRINT statement; the similarities are listed below.

1 The CSMP continuation device (...) may be used for continuation of the PREPARE statement to additional cards.

2 Only one PREPARE statement should be used.

3 The independent variable TIME is automatically included.

4 The heading contained in TITLE statements is specified in the data set generated by the PREPARE card and is available to the X–Y plotting programs.

5 The output interval is specified by the timer variable OUTDEL. Note, the output interval for the PRINT statement is PRDEL.

The data from the PREPARE card is generated on the I/O device specified as number 15. Depending upon the computer operating system, this data will either be stored on a disk-storage device or on magnetic tape.

The simulation of the dynamic response of a pneumatic piston-cylinder system illustrates the use of the PREPARE statement.

Example 3.9

A common engineering problem involves simulating the response of a piston driven by pressurized air flowing into a cylinder. This is a reasonably complex problem since it involves the fluid- and thermodynamics of a compressible fluid. Figure 3.33 shows a drawing of the system where air is flowing at a subsonic velocity through a small orifice to the cylinder.

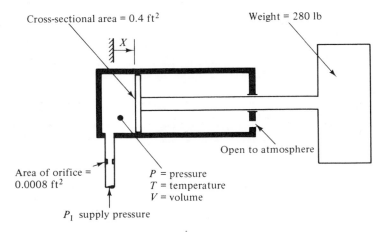

Fig. 3.33 Piston cylinder system.

The mass flow rate for subsonic velocity of an ideal gas, without losses, through an orifice is

$$\dot{m} = C\left[\left(\frac{P}{P_1}\right)^{2/\gamma} - \left(\frac{P}{P_1}\right)^{(\gamma+1)/\gamma}\right]^{1/2} \tag{3.17}$$

$$\text{where } C = A_1 P_1 \sqrt{\frac{2g_c\gamma}{(\gamma-1)RT_1}} \tag{3.18}$$

An energy balance on the air in the adiabatic cylinder yields the following expression for the time rate of change of cylinder air temperature.

$$\dot{T} = \frac{\left(-\dfrac{PA\dot{X}}{J} + \dot{m}c_p T_1 - \dot{m}c_v T\right)}{c_v m} \tag{3.19}$$

$$T(0) = T_o$$

The volume inside the cylinder is

$$V = V_o + A \cdot X \tag{3.20}$$

Assuming thermodynamic equilibrium, the perfect gas law can be used to solve for the cylinder pressure.

$$P = \frac{m \cdot R \cdot T}{V} \tag{3.21}$$

Newton's second law can be used to calculate the acceleration of the piston

$$\ddot{X} = \frac{A(P - P_o)}{\text{WEIGHT}/g_c} \tag{3.22}$$

Symbols Used in
Program of Fig. 3.34

where	A_1 = area of orifice, 0.0008 ft^2	A1
	P_1 = supply pressure, 3900.0 lbf/ft^2	P1
	T_1 = temperature of incoming air, 560°R	T1
	R = gas constant, 53.35 ft-lbf/lbm-°R	R
	X = movement of piston, ft	X
	γ = ratio of specific heats, 1.4	GAMMA
	A = area of piston, 0.4 ft^2	A
	c_p = specific heat of air at constant pressure, 0.24 Btu/lbm-°R	CP
	c_v = specific heat of air at constant volume, 0.171 Btu/lbm-°R	CV
	T_o = initial temperature of air in cylinder, 530°R	TO
	P_o = atmospheric pressure, 2116.8 lbf/ft^3	PO
	M_o = initial mass of air in cylinder	MO
	V_o = initial volume of cylinder, 0.06 ft^3	VO
	g_c = 32.17 lbm-ft/lbf-sec^2	GC
WEIGHT	= weight of piston and attached mass, 280.0 lbm	WEIGHT
	J = 778 ft-lbf/Btu	J

Equations (3.17), (3.19), and (3.22) are contained in the program of Fig. 3.34 and are integrated to determine the response.

The PREPARE card in the program of Fig. 3.34 is used to generate a data set for the offline plotting of the variables X, XD, and XDD.

Once the data set is generated, the user must supply a program to interface directly with the plotter. In this example the velocity XD and acceleration XDD are plotted as a function of displacement X. The interfacing program that was used at the University of Tennessee for a CALCOMP 763 plotter is shown in Fig. 3.35. It should be noted that each computer installation will require its own interfacing program.

Figure 3.36 shows the PRINT output from the program of Fig. 3.34. Figure 3.37 is the resulting offline X–Y plot.

DECK

In simulation studies it is often necessary to make several runs from the same program changing only parameters and output statements. Computer time can be

```
INITIAL
TITLE PROGRAM TO STUDY THE DYNAMICS OF A PNEUMATIC PISTON-CYLINDER
TITLE AND TO ILLUSTRATE THE USE OF THE PREPARE STATEMENT.
   CONSTANT  P1 = 3900.0, T1 = 560.0, A1 = 0.0008, P0 = 2116.8,      ...
   T0 = 530.0, V0 = 0.06, A = 0.4, WEIGHT = 280.0, CP = 0.24,        ...
   CV = 0.171, R = 53.35, J = 778.0, GAMMA = 1.4, GC = 32.17
   M0 = P0*V0/(R*T0)
   C = A1*P1*SQRT(2.0*GC*GAMMA/(R*T1*(GAMMA - 1.0)))
DYNAMIC
   MD = C*SQRT((P/P1)**(2.0/GAMMA) - (P/P1)**((GAMMA + 1.0)/GAMMA))
   V = V0 + A*X
   TD = (-P*A*XD/J + MD*CP*T1 - MD*CV*T)/(M*CV)
   P = M*R*T/V
   XDD = (P - P0)*A/(WEIGHT/GC)
      M = INTGRL(M0,MD)
      T = INTGRL(T0,TD)
      XD = INTGRL(0.0,XDD)
      X = INTGRL(C.C,XD)
   PRINT  X, XD, XDD, T, TD, M, MD, P
   FINISH X = 1.0
   TIMER  FINTIM = 5.0, PRDEL = 0.01, OUTDEL = 0.005

*      THE FOLLOWING PREPARE CARD IS USED TO GENERATE A DATA SET THAT
*   CAN BE USED FOR OFF LINE PLOTTING OF THE VARIABLES X, XD, AND XDD.

   PREPARE  X, XD, XDD

END
STOP
ENDJOB
```

Fig. 3.34 Program to simulate the response of a pneumatic piston cylinder system and to illustrate the use of the PREPARE statement.

saved in this situation by making a punched card deck of the translated CSMP program. An understanding of the inner workings of the CSMP program is obviously not necessary for successfully programming. However, the following brief description of the CSMP program will enable the reader to better understand the advantages of using a translated punched deck. For the reader who is interested in more detail, the CSMP System Manual[1] is recommended.

Once the user's program is read into the computer, the CSMP program begins to build the subroutine UPDATE. This is called the translation phase. In generating UPDATE, three primary tasks are accomplished.

1 The statements in the sort sections are placed in the proper order.

2 The proper transfer of control for the various segments and sections is established.

3 Common statements are established to make the proper variables available between UPDATE and the CSMP modules.

In the next step, the FORTRAN G compiler is invoked and the linkage editor then links UPDATE with the various CSMP execution modules. Control is then passed to MAIN which is the main or calling program. During the execution of the program, MAIN calls the appropriate subroutine to perform the necessary opera-

```
//   EXEC   FORTGCLG,GOREG=96K
//FORT.SYSIN   DD   *
      DIMENSION IBUF(664)
      DIMENSION TIME(1000),X(1000),Y(1000),Z(1000),XDOT(1000)
      DIMENSION ICODE(20),ICODE1(20)
      DO 5 I = 1,11
      READ (8,80) DUMMY
   5  CONTINUE
  80  FORMAT (A4)
      READ (5,60) ICODE
      READ (5,60) ICODE1
  60  FORMAT ( 20A4 )
      NOPTS = 110
      DO 10 I = 1,NOPTS
      READ (8,END=20) TIME(I),X(I),Y(I),Z(I)
  10  CONTINUE
  20  CALL PLOTS (IBUF,664,60.0)
      CALL ZIPOFF
      YAXIS = 9.0
      XAXIS = 10.0
      DO 100  J = 1,3
      NPTS = NOPTS
      NPTS = NPTS / J
      NST = NPTS * J +1
      NDEL = NPTS * J + J + 1
      CALL SCALE (TIME,XAXIS,NPTS,J)
      CALL SCALE (X,XAXIS,NPTS,J)
      CALL SCALE (Y,YAXIS,NPTS,J)
      CALL SCALE (Z,YAXIS,NPTS,J)
      CALL SYMBOL (7.0,9.03,0.14,10,0.0,-1)
      CALL SYMBOL (7.5,9.0,0.14,ICODE,0.0,80)
      CALL SYMBOL (7.0,8.7,0.14,01,0.0,-1)
      CALL SYMBOL (7.5,8.64,0.14,ICODE1,0.0,80)
      CALL AXIS (-0.75,0.0,3HXDD,+3,YAXIS,90.0,Z(NST),Z(NDEL))
      CALL AXIS (0.0,0.0,2HXD,+2,YAXIS,90.0,Y(NST),Y(NDEL))
      CALL AXIS (0.0,0.0,1HX,-1,XAXIS,0.0,X(NST),X(NDEL))
      CALL FLINE (X,Y,NPTS,J,+1,10)
      CALL FLINE (X,Z,NPTS,J,+1,01)
      CALL PLOT (15.0,0.0,-3)
 100  CONTINUE
      CALL PLOT (0.0,0.0,999)
      NSTOP = NOPTS
      WRITE (6,63) (TIME(L),X(L),Y(L),Z(L),L=1,NSTOP)
  63  FORMAT (1H ,4(E14.6,5X))
      STOP
      END
//GO.PLOTTAPE   DD   SYSCUT=P
//GO.FT08F001   DD   DSN=PLOTTER,DCB=(RECFM=VS,LRECL=255,BLKSIZE=2550),
//   UNIT=2400,DISP=(OLD,KEEP),LABEL=(,SL,,IN),VOL=SER=002330
//GO.SYSIN   DD   *
XD   (VELOCITY)
XDD  (ACCELERATION)
/*
```

Fig. 3.35 Program used at The University of Tennessee to interface between the data set generated by the PREPARE card and a CALCOMP 763 plotter.

tion; integration, printing, plotting, initialization, etc. This total operation of the CSMP program is quite complicated. A strong background in all areas of programming is required if the process is to be understood.

When the label DECK is included in a program with the appropriate JCL (job control language) cards, a punched card deck is made. The deck includes the CSMP subroutine UPDATE and all user-supplied subroutines, a listing of sym-

```
PROGRAM TO STUDY THE DYNAMICS OF A PNEUMATIC PISTON-CYLINDER          RKS        INTEGRATION
AND TO ILLUSTRATE THE USE OF THE PREPARE STATEMENT.

  TIME         X           XC          XDD          T           TD          M           MD          P
0.0          0.0         C.0        -1.0098E-04   5.3000E 02   3.9930E 03   4.4918E-03   7.0071E-02   2.1168E 03
1.5000E-02   1.2369E-03   2.4369E-01   3.1264E 01   5.7555E 02   2.0834E 03   5.5107E-03   6.4301E-02   2.7971E 03
3.0000E-02   9.2445E-03   8.6853E-01   4.9013E 01   5.9341E 02   3.7445E 02   6.4049E-03   5.5627E-02   3.1833E 03
4.5000E-02   2.7946E-02   1.6260E 00   4.9603E 01   5.9023E 02  -6.6606E 02   7.2246E-03   5.5247E-02   3.1961E 03
6.0000E-02   5.7628E-02   2.3086E 00   4.0511E 01   5.7696E 02  -1.0122E 03   8.0899E-03   6.0393E-02   2.9983E 03
7.5000E-02   9.6383E-02   2.8258E 00   2.8966E 01   5.6178E 02  -9.7191E 02   9.0332E-03   6.5095E-02   2.7471E 03
9.0000E-02   1.4167E-01   3.1876E 00   1.8363E 01   5.4847E 02  -7.9159E 02   1.0033E-02   6.7948E-02   2.5164E 03
1.0500E-01   1.9114E-01   3.3905E 00   9.6998E 00   5.3812E 02  -5.9082E 02   1.1065E-02   6.9372E-02   2.3279E 03
1.2000E-01   2.4282E-01   3.4834E 00   2.9908E 00   5.3063E 02  -4.1209E 02   1.2110E-02   6.9953E-02   2.1819E 03
1.3500E-01   2.9520E-01   3.4886E 00  -2.0441E 00   5.2559E 02  -2.6475E 02   1.3161E-02   7.0102E-02   2.0723E 03
1.5000E-01   3.4715E-01   3.4289E 00  -5.7183E 00   5.2254E 02  -1.4676E 02   1.4212E-02   7.0057E-02   1.9924E 03
1.6500E-01   3.9783E-01   3.3225E 00  -8.3020E 00   5.2107E 02  -5.3236E 01   1.5262E-02   6.9948E-02   1.9362E 03
1.8000E-01   4.4667E-01   3.1842E 00  -1.0010E 01   5.2084E 02   2.0588E 01   1.6311E-02   6.9841E-02   1.8990E 03
1.9500E-01   4.9326E-01   3.0257E 00  -1.1009E 01   5.2160E 02   7.8645E 01   1.7358E-02   6.9765E-02   1.8772E 03
2.1000E-01   5.3738E-01   2.8568E 00  -1.1430E 01   5.2314E 02   1.2396E 02   1.8404E-02   6.9731E-02   1.8681E 03
2.2500E-01   5.7895E-01   2.6852E 00  -1.1376E 01   5.2527E 02   1.5877E 02   1.9450E-02   6.9735E-02   1.8693E 03
2.4000E-01   6.1796E-01   2.5175E 00  -1.0930E 01   5.2785E 02   1.8475E 02   2.0496E-02   6.9772E-02   1.8790E 03
2.5500E-01   6.5452E-01   2.3585E 00  -1.0165E 01   5.3077E 02   2.0311E 02   2.1543E-02   6.9830E-02   1.8956E 03
2.7000E-01   6.8879E-01   2.2138E 00  -9.1415E 00   5.3391E 02   2.1479E 02   2.2591E-02   6.9899E-02   1.9179E 03
2.8500E-01   7.2102E-01   2.0857E 00  -7.9167E 00   5.3718E 02   2.2051E 02   2.3640E-02   6.9968E-02   1.9445E 03
3.0000E-01   7.5146E-01   1.9770E 00  -6.5430E 00   5.4050E 02   2.2086E 02   2.4690E-02   7.0029E-02   1.9744E 03
3.1500E-01   7.8043E-01   1.8899E 00  -5.0702E 00   5.4378E 02   2.1637E 02   2.5741E-02   7.0074E-02   2.0065E 03
3.3000E-01   8.0827E-01   1.8252E 00  -3.5463E 00   5.4697E 02   2.0755E 02   2.6792E-02   7.0099E-02   2.0396E 03
3.4500E-01   8.3530E-01   1.7835E 00  -2.0170E 00   5.4999E 02   1.9490E 02   2.7843E-02   7.0102E-02   2.0729E 03
3.6000E-01   8.6189E-01   1.7645E 00  -5.2546E-01   5.5280E 02   1.7898E 02   2.8895E-02   7.0083E-02   2.1054E 03
3.7500E-01   8.8835E-01   1.7673E 00   8.8864E-01   5.5534E 02   1.6039E 02   2.9946E-02   7.0045E-02   2.1361E 03
3.9000E-01   9.1501E-01   1.7906E 00   2.1899E 00   5.5760E 02   1.3975E 02   3.0996E-02   6.9993E-02   2.1645E 03
4.0500E-01   9.4216E-01   1.8323E 00   3.3486E 00   5.5953E 02   1.1772E 02   3.2046E-02   6.9933E-02   2.1897E 03
4.2000E-01   9.7006E-01   1.8902E 00   4.3415E 00   5.6112E 02   9.4977E 01   3.3094E-02   6.9872E-02   2.2113E 03
4.3500E-01   9.9893E-01   1.9616E 00   5.1519E 00   5.6237E 02   7.2167E 01   3.4142E-02   6.9814E-02   2.2289E 03

***SIMULATION HALTED***    X    =   1.0063E 00

4.3875E-01   1.0063E 00   1.9813E 00   5.3248E 00   5.6263E 02   6.6527E 01   3.4403E-02   6.9801E-02   2.2327E 03
```

Fig. 3.36 PRINT output from the simulation of the piston cylinder system.

bols used in the execution phase, and all data, execution, and output statements. When the translated deck is used for initiating a run, generating the subroutine UPDATE is not necessary and consequently, computer time is saved.

Any of the data, execution, and output control statements can be changed in the translated deck. Minor modifications can even be made in the structure statements of the translated punched deck. A complete discussion of the use of translated decks is included in the CSMP System Manual.[1]

Figure 3.38 shows the deck statement inserted in the program of Example 3.9. A listing of the resulting punched card deck is shown in Fig. 3.39. Included in this listing are the JCL cards that are required at The University of Tennessee's computing center to make additional runs.

Additional runs can be made by changing the data, execution, and output control statements that appear at the end of the punched deck. It is suggested that the user contact the computer installation to determine the required JCL cards.

User Defined Functions and Subroutines

Functions most often encountered in digital simulations have been programmed as a fundamental part of CSMP. Several of these functional blocks were

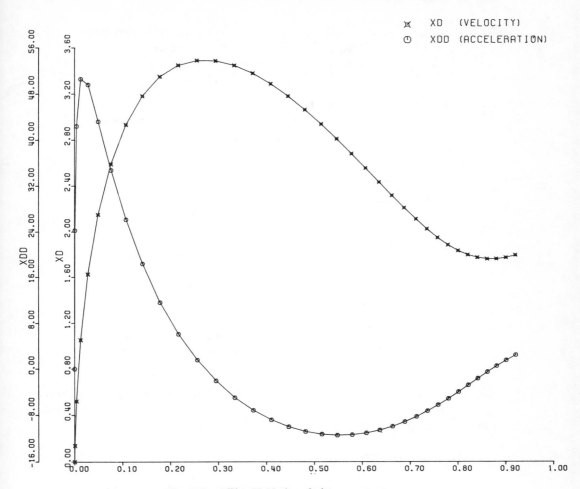

Fig. 3.37   Offline X–Y plot of piston response.

presented in Chap. 2 and the remaining ones are given in Appendix I. Such functions are typically of the form $Y = REALPL(IC, P, X)$. It is impossible to foresee the various types of relationships that might be needed in every situation and thus an all-encompassing set of expressions cannot be pre-programmed. Also, the storage requirements for including such flexibility would soon become completely unreasonable. Having recognized this, CSMP was written so that users could easily develop their own functions relative to a particular problem or even to a given discipline. Such capability is available through the use of MACRO, PRO-CEDURE, and standard FORTRAN subprograms.

The purpose of this section is to present basic guidelines for developing and using extended functional capabilities. The MACRO type function will be intro-

```
TITLE   PROGRAM TO STUDY THE DYNAMICS OF A PNEUMATIC PISTON-CYLINDER
TITLE   AND TO ILLUSTRATE THE USE OF THE DECK STATEMENT
INITIAL
   CONSTANT  P1 = 3900.0,  T1 = 560.0,  A1 = 0.0008,  PO = 2116.8,      ...
   TO = 530.0,  VO = 0.06,  A = 0.4,  WEIGHT = 280.0,  CP = 0.24,       ...
   CV = 0.171,  R = 53.35,  J = 778.0,  GAMMA = 1.4,  GC = 32.17
   MO = PO*VO/(R*TO)
   C = A1*P1*SQRT(2.0*GC*GAMMA/(R*T1*(GAMMA - 1.0)))
DYNAMIC
   MD = C*SQRT((P/P1)**(2.0/GAMMA) - (P/P1)**((GAMMA + 1.0)/GAMMA))
   V = VO + A*X
   TD = (-P*A*XD/J + MD*CP*T1 - MD*CV*T)/(M*CV)
   P = M*R*T/V
   XDD = (P - PO)*A/(WEIGHT/GC)
      M = INTGRL(MO,MD)
      T = INTGRL(TO,TD)
      XD = INTGRL(0.0,XDD)
      X = INTGRL(0.0,XD)
   PRINT  X, XD, XDD, T, TD, M, MD, P
   FINISH X = 1.0
   TIMER  FINTIM = 5.0,  PRDEL = 0.015

*     THE FOLLOWING DECK STATEMENT WHEN USED WITH THE PROPER JCL CARDS
*     WILL RETURN A PUNCHED CARD DECK OF THE TRANSLATED PROGRAM.

   DECK

END
STOP
ENDJOB
```

Fig. 3.38 Program to simulate the response of a pneumatic piston cylinder system and to illustrate the use of the DECK statement.

duced first and will be followed by the PROCEDURE and FORTRAN subprograms. As usual, example programs are given which illustrate the use of these functions in a complete simulation.

MACRO FUNCTIONS

In many respects a macro function can be compared to a FORTRAN subroutine although they are definitely not the same. As a simple illustration suppose the following set of equations is used within the body of a more comprehensive simulation.

$$Ax(t) + By(t) = f_1(t) \tag{3.23}$$

$$Cx(t) + Dy(t) = f_2(t) \tag{3.24}$$

We assume that A, B, C, and D are fixed parameters while $f_1(t)$ and $f_2(t)$ are time varying functions defined elsewhere in the simulation. The problem is to solve for $x(t)$ and $y(t)$. A matrix form of solution is

$$\begin{bmatrix} x(t) \\ y(t) \end{bmatrix} = \begin{bmatrix} A & B \\ C & D \end{bmatrix}^{-1} \begin{bmatrix} f_1(t) \\ f_2(t) \end{bmatrix} \tag{3.25}$$

```
      SUBROUTINE UPDATE
      COMMON   ZZ9901(5),IZ9901,ZZ9902,IZ9902,ZZ9903,IZ9903,ZZ9991(54)
      COMMON TIME
     1,DELT    ,DELMIN,FINTIM,PRDEL  ,OUTDEL,M      ,T      ,XD     ,X
     1,MD      ,TD     ,XDD    ,ZZ0007,MO    ,TO     ,ZZ0004,ZZ0006,P1
     1,T1      ,A1     ,PO     ,VO     ,A     ,WEIGHT,CP     ,CV     ,R
     1,J       ,GAMMA ,GC     ,C      ,V     ,P
      COMMON ZZ9992(7966),NALARM,IZ9993,ZZ9994(417),KEEP,ZZ9995(489)
     $,IZ0000,ZZ9996(824),IZ9997,IZ9998,ZZ9999( 81)
      REAL       M
     1,MO      ,J
      GO TO(39995,39996,39997,39998),IZ0000
C     SYSTEM SEGMENT OF MODEL
39995 CONTINUE
      IZ9993=   35
      IZ9997=    4
      IZ9998=   34
      READ(5,39990)(ZZ9999(IZ9999),IZ9999=1,  81)
39990 FORMAT(18A4)
      IZ9901=  310019
      IZ9902=  340032
      IZ9903=      81
      GO TO 39999
C     INITIAL SEGMENT OF MODEL
39996 CONTINUE
      MO=PO*VO/(R*TO)
      C=A1*P1*SQRT(2.0*GC*GAMMA/(R*T1*(GAMMA-1.0)))
      GO TO 39999
C     DYNAMIC SEGMENT OF MODEL
39997 CONTINUE
      V=VO+A*X
      P=M*R*T/V
      MD=C*SQRT((P/P1)**(2.0/GAMMA)-(P/P1)**((GAMMA+1.0)/GAMMA))
C     M        =INTGRL   (MO         ,MD         )
      TD=(-P*A*XD/J+MD*CP*T1-MD*CV*T)/(M*CV)
C     T        =INTGRL   (TO         ,TD         )
      XDD=(P-PO)*A/(WEIGHT/GC)
C     XD       =INTGRL   (ZZ0004     ,XDD        )
C     X        =INTGRL   (ZZ0006     ,XD         )
      ZZ0007=XD
      GO TO 39999
C     TERMINAL SEGMENT OF MODEL
39998 CONTINUE
39999 CONTINUE
      RETURN
      END
//LKED.SYSLIB DD
// DD
// DSN=UTCC.APPLIB,DISP=SHR
//LKED.SYSIN DD DSN=UTCC.PAFMLIB(LKEDCSMP),DISP=SHR
//GO.FT13F001 DD UNIT=WORK,SPACE=(CYL,(1,1)),
// DCB=(RECFM=VBS,LRECL=200,BLKSIZE=1004)
//GO.FT14F001 DD UNIT=WORK,SPACE=(CYL,(1,1)),
// DCB=(RECFM=VBS,LRECL=200,BLKSIZE=1004)
//GO.SYSIN DD *
```

JCL cards added to the punched deck

```
TIME    DELT    DELMINFINTIMPRDEL  OUTDELM       T      XD      X       MD      TD      SYMB   1
XDD     ZZ0007MO      TO      ZZ0004ZZ0006P1      T1      A1      PO      VO      A       SYMB   2
WEIGHTCP       CV      R       J       GAMMA GC      C       V       P               SYMB   3
                                                                                SYMB   4
                                                                                SYMB   5
```

These data, execution, and output control statements can be changed.

```
TITLE PROGRAM TO STUDY THE DYNAMICS OF A PNEUMATIC PISTON-CYLINDER
TITLE AND TO ILLUSTRATE THE USE OF THE DECK STATEMENT
   CONSTANT  P1 = 3900.0, T1 = 560.0, A1 = 0.0008, PO = 2116.8,
   TO = 530.0, VO = 0.06, A = 0.4, WEIGHT = 280.0, CP = 0.24,
   CV = 0.171, R = 53.35, J = 778.0, GAMMA = 1.4, GC = 32.17
   PRINT  X, XD, XDD, T, TD, M, MD, P
   FINISH X = 1.0
   TIMER  FINTIM = 5.0, PRDEL = 0.01
END
STOP
```

Fig. 3.39 Listing of cards returned by DECK statement from program in Fig. 3.38.

A macro giving the solution for $x(t)$ and $y(t)$ can be formulated as follows

```
MACRO  ANS1, ANS2 = INVERT(A, B, C, D, FUN1, FUN2)
       DETERM = A*D − C*B
       A11 = D/DETERM
       A12 = −B/DETERM
       A21 = −C/DETERM
       A22 = A/DETERM
       ANS1 = A11*FUN1 + A12*FUN2
       ANS2 = A21*FUN1 + A22*FUN2
ENDMAC
```

The calculations as specified in the macro could be invoked from the program by the following statement.

$$P, Q = INVERT(6.0, 7.1, −2.3, 18.0, N1, N2)$$

The above instruction will execute the statements of the INVERT macro and consequently solve for P and Q from the following equation set.

$$6P + 7.1Q = N1 \tag{3.26}$$

$$−2.3P + 18P = N2 \tag{3.27}$$

At another point within the main program the macro might again be invoked by

$$X, Y = INVERT(PAR1,PAR2**2, F-M,RING,IN1,IN2)$$

More will be said later about the function outputs and the arguments of the macro name.

A macro is a parallel-type function. This means that the statements written between MACRO and ENDMAC will be arranged by the CSMP sort algorithm to insure that calculations are performed in the correct order. In addition, the function defining the macro (INVERT in the above example) will be sorted within other statements of the overall program.

Fundamental rules which must be observed when using macros are

1 Macro definition cards must be placed at the beginning of the program and precede any structural statements even though the structural statements are in an INITIAL segment. Generally, the only cards placed before a macro are those which relate to FIXED, STORAGE, HISTORY, MEMORY, and RENAME.

2 The very first card of the macro is a translational control statement containing the word MACRO and is followed, on the same card, by the expression representing the function. At least one blank space must follow MACRO and the first output variable. The last card of the macro must contain the letters ENDMAC. For a single output variable one might have

$$\text{MACRO FORCE} = \text{SHARP(MASS, TORQUE, CONST1)}$$

.
.
.

Macro statements

.
.
.

ENDMAC

3 If necessary, the function defining the macro may be continued up to a total of four cards using the customary . . . for continuation. As an illustration one might have,

$$\text{MACRO OUT} = \text{CALCU(M21,M22, . . .}$$
$$\text{M31, M32, IN1, IN2)}$$

Statements within the body of the macro cannot be continued from one card to another using the . . . feature. Therefore, continuation cards cannot appear in a macro other than the first statement defining the function.

4 There are no restrictions on the number of output variables, input variables, and parameters used in defining a macro other than that the total function statement cannot exceed four cards. The macro outputs and macro arguments must be separated by commas.

5 All variables and parameters associated with the macro definition are "dummy" names with the exception of the unique name assigned to the function definition. Thus,

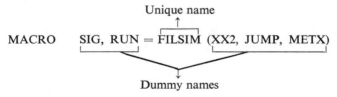

When using this macro, the label FILSIM must be retained; however, SIG, RUN, XX2, JUMP, and METX can be replaced by other names. For example, the above macro could also be invoked by the statement

$$\text{TRIG, DECAY} = \text{FILSIM(6.4, CCX, (4.0 − B)/A)}$$

in which case the following replacements are made by the macro

	Replaces	
TRIG	⟶	SIG
DECAY	⟶	RUN
6.4	⟶	XX2
CCX	⟶	JUMP
(4.0 − B)/A	⟶	METX

6 Once a macro function is defined it can be used several times in one program. For example,

$$AA, \; BX = FILSIM(IC1, \; DUM, \; RY)$$
$$KY, \; NY = FILSIM(NUM, \; DEM, \; RELX)$$
$$RED, \; WWW = FILSIM(3.2, \; A–M, \; LLP)$$

can all be used to invoke the previously defined macro FILSIM (see 5 above) in a single simulation.

7 The arguments used in a macro statement can be literals, variables, subscripted variables, and standard FORTRAN expressions. However, system macros (CMPXPL, MODINT, LEDLAG, REALPL and INTGRL) cannot be used as arguments. Consider the macro definition given earlier

$$MACRO \quad SIG, \; RUN = FILSIM(XX2, \; JUMP, \; METX)$$

The following two statements are valid.

$$X, \; Y = FILSIM(Q**2, \; M(2), \; SQRT(B**2 - 4*A*C))$$
$$RX, \; MY = FILSIM(SINE(0.1, \; 400.0, \; 0.0), \; STEP(0.0), \; AA)$$

Statements which cannot be used include

$$X, \; Y = FILSIM(9.3, \; \underline{INTGRL(0.0, \; Z)}, \; AA)$$

$$\uparrow$$
not allowed as arguments
$$\downarrow$$

$$X, \; Y = FILSIM(9.3, \; \overline{REALPL(0.0, \; 0.2, \; W)}, \; AA)$$

8 Neither a user defined macro function nor a system macro can be used as part of an expression in a structure statement. As an example, the following is an invalid instruction.

$$Q = SHARP(3.3, \; T, \; -4.71)/3.75$$

9 Variables defined within the structure statements of a macro are not available for program output through the use of PRINT, PRTPLT, PREPARE, and RANGE. For a variable to be available as output data it must be defined as one of the output names of the macro.

10 A macro may be invoked within another macro. For example suppose the following macro is defined.

$$MACRO \; Y = STATE(A11, \; A12, \; B1, \; R, \; IC1)$$
$$X1DOT = A11*X1 + B1*R + A12$$
$$X1 = INTGRL(IC1, \; X1DOT)$$
$$Y = X1$$
$$ENDMAC$$

The macro STATE may be used in another macro called SIMULA as illustrated by

```
MACRO OUTX = SIMULA(INPUT, DIST, S11, S12, SB1, SIC1)
    XX = STATE(S11, S12, SB1, GAIN*INPUT, SIC1)
    QQ = INTGRL(0.0, DIST)
    OUTX = XX + QQ
ENDMAC
```

11 As noted previously, macro structure statements may appear in any order since the sorting algorithm will place them in a correct sequence for computation. Macro functions can be used either in the structure statements of a procedure function or in a nosort section. A macro so used *will not be sorted* and furthermore, *a macro cannot invoke another macro when used in this way*.

12 Neither CSMP data nor control statements may appear between MACRO and ENDMAC. Also, FORTRAN control and input/output statements cannot appear unless they are embedded within a procedure function. Input and output statements can reference FORMAT statements but the FORMAT statements cannot appear in a macro. With the above exception, any CSMP or FORTRAN structure statements, including a subroutine call, may appear within a macro.

Perhaps the best way to understand the rules and guidelines for formulating macros is to see how they are applied in several examples.

A good policy to follow when developing a macro is to test the operation of the function before incorporating it into a more comprehensive simulation. The following example presents the development of a rather simple, but useful, macro and a test for asserting the correct operation of the function so generated.

Example 3.10

The desired operating characteristics of an on-off device are shown in Fig. 3.40. The actual device might represent a relay with deadspace, although here the purpose is to make the characteristics as general as possible. Once developed, the macro can be used as a subcomponent in a more extensive system.

The macro may be generated by

```
MACRO  PLAY = RELAY(YP1, YP2, XP1, XP2, FLASH)
           IF(FLASH.GT.XP2.AND.FLASH.LT.XP1) GO TO 99
           IF(FLASH.GE.XP1) GO TO 98
           IF(FLASH.LE.XP2) GO TO 97
    99     PLAY = 0.0
           GO TO 96
    98     PLAY = YP1
           GO TO 96
    97     PLAY = YP2
    96     CONTINUE
ENDMAC
```

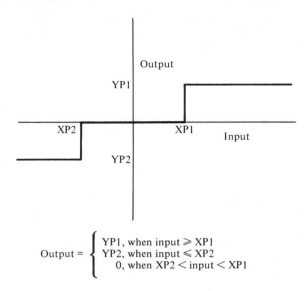

$$\text{Output} = \begin{cases} \text{YP1, when input} \geqslant \text{XP1} \\ \text{YP2, when input} \leqslant \text{XP2} \\ \quad 0, \text{ when XP2} < \text{input} < \text{XP1} \end{cases}$$

Fig. 3.40   Desired characteristics of Relay-type device.

The output, **PLAY**, and arguments YP1, YP2, XP1, and XP2, and FLASH are dummy symbols and can be replaced by other symbols when invoked from the main program. Notice that FORTRAN branching and logic statements have been used. Since the structure statements of a macro are sorted, the user would encounter difficulty using the program unless steps are taken to insure the statements are not sorted. One way to avoid sorting is to place a card with the letters PROCEDURAL immediately following the macro definition card. This will be explained more fully in the next section dealing with procedure functions. Another way to avoid sorting is to place a NOSORT card before the program statement which invokes the macro. A program which uses this approach and also checks the performance of the macro is given in Fig. 3.41. The macro function is invoked by the statement

OUTPUT = RELAY(LIMIT1, LIMIT2, BOUND1, BOUND2, INPUT)

A sine wave with frequency of 6.28 radians/sec is used as the device input. The values of LIMIT1, LIMIT2, BOUND1, and BOUND2 have been specified on a Parameter card. The output from the program is given in Fig. 3.42, where one will note that the function is operating correctly.

If the NOSORT card were omitted from the program in Fig. 3.41 the program would not run and an appropriate error message would appear.

The major emphasis in developing CSMP was placed upon the creation of a simulation language for handling problems which involve differentiation and integration. As shown in Example 3.11, however, one can write CSMP programs which do not solve differential equations. While such programs may certainly be written in FORTRAN or other higher-level languages, CSMP quite often gives a simpler formulation without using DO loops and subscripted variables. The following will present one such use of CSMP.

```
MACRO  PLAY = RELAY(YP1, YP2, XP1, XP2, FLASH)
       IF(FLASH.GT.XP2.AND.FLASH.LT.XP1) GO TO 99
       IF(FLASH.GE.XP1) GO TO 98
       IF(FLASH.LE.XP2) GO TO 97
   99  PLAY = 0.0
       GO TO 96
   98  PLAY = YP1
       GO TO 96
   97  PLAY = YP2
   96  CONTINUE
ENDMAC
PARAM LIMIT1 = 1.2, LIMIT2 = -1.8, BOUND1 = 1.0, BOUND2 = -1.0
*NOSORT
INPUT = 3.0*SINE(0.0,6.28,0.0)
OUTPUT = RELAY(LIMIT1, LIMIT2, BOUND1, BOUND2, INPUT)
TIMER FINTIM = 1.0, OUTDEL = 0.02
PRTPLT OUTPUT(INPUT)
LABEL  PERFORMANCE OF RELAY MACRO - EXAMPLE 3-10
END
STOP
ENDJOB
```

*Placing the Macro in a Nosort section permits FORTAN logic and branching within the Macro.

Fig. 3.41   Program test for relay macro (Example 3.10).

Example 3.11

A mortgage of \$32,000 is to be financed through a bank on a 25 year 9% interest note. We would like to write a program which will calculate and print the following: (1) the current month; (2) the monthly payment; (3) the interest paid each month on the loan; (4) the running total interest; (5) the amount applied each month toward reducing the original mortgage; (6) the total equity; and (7) the balance left to pay on the loan.

A standard equation for determining the fixed monthly payments on a fixed-period loan is given by

$$\text{Monthly payment} = \frac{(P*IRM*S)}{(S-1)} \tag{3.28}$$

$$\text{where } S = (1 + IRM)^{\text{PERIOD}} \tag{3.29}$$

and

P = original mortgage value

IRM = monthly interest rate

This calculation is performed only one time and can therefore be placed in the INITIAL segment of the program.

We make the following definitions for programming convenience.

MONINT = monthly interest

BALANC = balance remaining on the loan

MPAY = amount paid each month toward reducing the loan

PAY = fixed monthly payments

EQUITY = total amount paid off on the loan

TOTINT = running total interest paid

```
                          MINIMUM              OUTPUT VERSUS TIME          MAXIMUM
                          -1.8000E 00                                     1.2000E 00
   TIME        OUTPUT      I                                              I        INP
0.0            0.0         ------------------------------+                         0.0
2.0000E-02     0.0         ------------------------------+                         3.75
4.0000E-02     0.0         ------------------------------+                         7.45
6.0000E-02     1.2000E 00  -------------------------------------------------+      1.10
8.0000E-02     1.2000E 00  -------------------------------------------------+      1.44
1.0000E-01     1.2000E 00  -------------------------------------------------+      1.76
1.2000E-01     1.2000E 00  -------------------------------------------------+      2.05
1.4000E-01     1.2000E 00  -------------------------------------------------+      2.31
1.6000E-01     1.2000E 00  -------------------------------------------------+      2.51
1.8000E-01     1.2000E 00  -------------------------------------------------+      2.71
2.0000E-01     1.2000E 00  -------------------------------------------------+      2.85
2.2000E-01     1.2000E 00  -------------------------------------------------+      2.94
2.4000E-01     1.2000E 00  -------------------------------------------------+      2.99
2.6000E-01     1.2000E 00  -------------------------------------------------+      2.99
2.8000E-01     1.2000E 00  -------------------------------------------------+      2.94
3.0000E-01     1.2000E 00  -------------------------------------------------+      2.85
3.2000E-01     1.2000E 00  -------------------------------------------------+      2.71
3.4000E-01     1.2000E 00  -------------------------------------------------+      2.5
3.6000E-01     1.2000E 00  -------------------------------------------------+      2.31
3.8000E-01     1.2000E 00  -------------------------------------------------+      2.05
4.0000E-01     1.2000E 00  -------------------------------------------------+      1.76
4.2000E-01     1.2000E 00  -------------------------------------------------+      1.44
4.4000E-01     1.2000E 00  -------------------------------------------------+      1.10
4.6000E-01     0.0         ------------------------------+                         7.50
4.8000E-01     0.0         ------------------------------+                         3.80
5.0000E-01     0.0         ------------------------------+                         4.70
5.2000E-01     0.0         ------------------------------+                        -3.71
5.4000E-01     0.0         ------------------------------+                        -7.41
5.6000E-01    -1.8000E 00  +                                                      -1.00
5.8000E-01    -1.8000E 00  +                                                      -1.44
6.0000E-01    -1.8000E 00  +                                                      -1.75
6.2000E-01    -1.8000E 00  +                                                      -2.04
6.4000E-01    -1.8000E 00  +                                                      -2.30
6.6000E-01    -1.8000E 00  +                                                      -2.52
6.8000E-01    -1.8000E 00  +                                                      -2.71
7.0000E-01    -1.8000E 00  +                                                      -2.85
7.2000E-01    -1.8000E 00  +                                                      -2.94
7.4000E-01    -1.8000E 00  +                                                      -2.99
7.6000E-01    -1.8000E 00  +                                                      -2.99
7.8000E-01    -1.8000E 00  +                                                      -2.94
8.0000E-01    -1.8000E 00  +                                                      -2.85
8.2000E-01    -1.8000E 00  +                                                      -2.71
8.4000E-01    -1.8000E 00  +                                                      -2.53
8.6000E-01    -1.8000E 00  +                                                      -2.31
8.8000E-01    -1.8000E 00  +                                                      -2.05
9.0000E-01    -1.8000E 00  +                                                      -1.77
9.2000E-01    -1.8000E 00  +                                                      -1.45
9.4000E-01    -1.8000E 00  +                                                      -1.11
9.6000E-01     0.0         ------------------------------+                        -7.54
9.8000E-01     0.0         ------------------------------+                        -3.85
1.0000E 00     0.0         ------------------------------+                        -9.56
```

Fig. 3.42 CSMP output for program given in Fig. 3.41.

The basic calculations required are

MONINT = IRM*BALANC
MPAY = PAY − MONINT
EQUITY = MPAY + PAST EQUITY
BALANC = P − EQUITY
TOTINT = MONINT + PAST TOTAL INTEREST

These equations must be solved 300 times (twenty five years with twelve payments each year). Each time the set of equations is solved, certain quantities must be updated similarly to a difference equation.

There are several ways these equations can be programmed in CSMP but a macro approach will be given since this is the major topic of discussion.

We first write the macro

MACRO   MONINT,MPAY=CALCU1(IRM,BALANC,PAY)
    MONINT = IRM*BALANC
    MPAY = PAY − MONINT
ENDMAC

where IRM and PAY are fixed parameters and BALANC is an input variable. MONINT and MPAY are output variables of the macro. The statement selected to call the macro is

MONINT,MPAY = CALCU1(IRM,BALANC,PAY)

As explained earlier, the user is only required to retain CALCU1 in the call statement and all other terms can be replaced by perhaps more meaningful terms. In this case it is convenient not to change the output and argument symbols.

Another macro is now written for the remaining calculations.

MACRO   EQUITY,BALANC,TOTINT=CALCU2(MPAY,PASPAY, . . .
            P,MONINT,PSTINT)
    EQUITY = MPAY + PASPAY
    BALANC = P − EQUITY
    TOTINT = MONINT + PSTINT
    PASPAY = EQUITY
    PSTINT = TOTINT
ENDMAC

The call statement to the macro will be

EQUITY,BALANC,TOTINT=CALCU2(MPAY,PASPAY,P,MONINT,PSTINT)

The comments made regarding symbols used in the previous macro also apply here. Note in this last macro that PASPAY and PSTINT are used for updating the equity and total interest. At this point a word of caution is in order.

In problems which do not involve integration, calculations in the program are made at intervals of DELT. If DELT is not specified, a value of FINTIM/100 is automatically used. In this problem, all calculations must be performed once for each month. Since the unit of time is months, it is essential that DELT be specified equal to one.

A program for performing the calculations of this example is given in Fig. 3.43. The first five printed output lines are as shown in Fig. 3.44. As a matter of convenience we

```
*********          EXAMPLE  3-11          $$$$$$$$$$
*   BALANC  = BALANCE TO PAY ON THE LOAN
*   MCNINT  = INTEREST FOR THE CURRENT MONTH
*   TOTINT  = TOTAL INTEREST PAID TO DATE
*   MPAY    = AMOUNT PAID ON LOAN FOR THE CURRENT MONTH
*   EQUITY  = CURRENT AMCUNT PAID-OFF ON THE LOAN
*   IRM     = THE MONTHLY INTEREST RATE (IN DECIMAL FORM)
*   PAY     = THE MCNTHLY PAYMENT (A CONSTANT)

MACRO  MONINT, MPAY = CALCU1(IRM,BALANC,PAY)
    MONINT = IRM*BALANC
    MPAY    = PAY - MONINT
ENDMAC


MACRO  EQUITY, BALANC, TOTINT = CALCU2(MPAY, PASPAY,...
        P, MONINT, PSTINT)
    EQUITY = MPAY + PASPAY
    BALANC = P - EQUITY
    TOTINT = MONINT + PSTINT
    PASPAY = EQUITY
    PSTINT = TOTINT
ENDMAC


RENAME TIME = MONTH


INITIAL
PARAMETER P = 32000.0, I = 9.0, PERIOD = 300.0
INCON       PSTINT = 0.0, TOTINT = 0.0, MONINT = 0.0,...
            EQUITY = 0.0, PASPAY = 0.0
BALANC = P
IRM = I/1200.0
S = (1.0 + IRM)**PERIOD
PAY = (P*IRM*S)/(S - 1.0)


DYNAMIC
NOSORT
      IF(MONTH.EQ.0.0) GO TO 7
MCNINT, MPAY = CALCU1(IRM,BALANC,PAY)
EQUITY,BALANC,TOTINT = CALCU2(MPAY,PASPAY,P,MONINT,PSTINT)
    7  CONTINUE
TIMER FINTIM =300.0, PRDEL = 1.0, DELT = 1.0, OUTDEL = 10.0
PRINT PAY, MONINT, TOTINT,MPAY,EQUITY,BALANC
TITLE  OUTPUT FOR EXAMPLE 3-11
PRTPLT EQUITY(TOTINT)
END
STOP
ENDJOB
```

Fig. 3.43 Program for calculating the amortization of a mortgage.

change TIME to MONTH by the rename statement, as shown in the program. Note that the two macros are placed at the beginning of the program. Also, PSTINT,TOTINT, MONINT,EQUITY, and PASPAY are initialized to zero and BALANC set equal to P($32,000). After the NOSORT card a FORTRAN branching statement,

IF(MONTH.EQ.O.O) GO TO 7

is placed. This obviously causes the program to by-pass calculations, as desired, at

OUTPUT FOR EXAMPLE 3-11   INTGRL NOT USED

MONTH	PAY	MONINT	TOTINT	MPAY	EQUITY	BALANC
0.0	2.6855E 02	0.0	0.0	0.0	0.0	3.2000E 04
1.0000E 00	2.6855E 02	2.4000E 02	2.4000E 02	2.8546E 01	2.8546E 01	3.1971E 04
2.0000E 00	2.6855E 02	2.3979E 02	4.7979E 02	2.8760E 01	5.7305E 01	3.1943E 04
3.0000E 00	2.6855E 02	2.3957E 02	7.1936E 02	2.8975E 01	8.6281E 01	3.1914E 04
4.0000E 00	2.6855E 02	2.3935E 02	9.5871E 02	2.9193E 01	1.1547E 02	3.1885E 04
5.0000E 00	2.6855E 02	2.3913E 02	1.1978E 03	2.9412E 01	1.4488E 02	3.1855E 04
6.0000E 00	2.6855E 02	2.3891E 02	1.4368E 03	2.9632E 01	1.7452E 02	3.1825E 04
7.0000E 00	2.6855E 02	2.3869E 02	1.6754E 03	2.9854E 01	2.0437E 02	3.1796E 04
8.0000E 00	2.6855E 02	2.3847E 02	1.9139E 03	3.0078E 01	2.3445E 02	3.1766E 04
9.0000E 00	2.6855E 02	2.3824E 02	2.1522E 03	3.0304E 01	2.6475E 02	3.1735E 04
1.0000E 01	2.6855E 02	2.3801E 02	2.3902E 03	3.0531E 01	2.9528E 02	3.1705E 04
1.1000E 01	2.6855E 02	2.3779E 02	2.6280E 03	3.0760E 01	3.2604E 02	3.1674E 04
1.2000E 01	2.6855E 02	2.3755E 02	2.8655E 03	3.0991E 01	3.5704E 02	3.1643E 04

Fig. 3.44   Output from the mortgage program of Fig. 3.43.

MONTH = 0.0. It is important that MONTH be used in this statement. Using TIME in place of MONTH will cause the program to ignore the IF statement and give incorrect results. When a CSMP program variable is renamed, the newly assigned variable should be used in the program.

This example problem was programmed using macro functions simply to illustrate the use of such functions. Obviously, the problem can be written without using a macro and Fig. 3.45 shows one solution using straightforward statements. The output of this program is identical in all respects to the program which uses the macros.

In summary, CSMP can be used for problem solving without integration. The advantage of using CSMP versus FORTRAN will often be that DO loops and subscripted variables are not necessary. This leads to a less demanding effort on the part of the programmer, especially if the programmer is not an expert in using FORTRAN. One disadvantage of using CSMP in problems of this type is that generally slightly more computing time (CPU time) is required.

Example 3.12

The occasion may arise when the solution of a differential equation of the form

$$a_1 \frac{d^3x}{dt^3} + a_2 \frac{d^2x}{dt^2} + a_3 \frac{dx}{dt} + a_4 x = b_1 \frac{d^3r}{dt^3} + b_2 \frac{d^2r}{dt^2} + b_3 \frac{dr}{dt} + b_4 r \quad (3.30)$$

is desired. If we assume all zero initial conditions and all constant coefficients, applying Laplace transforms yields

$$\frac{X(s)}{R(s)} = \frac{b_1 s^3 + b_2 s^2 + b_3 s + b_4}{a_1 s^3 + a_2 s^2 + a_3 s + a_4} \quad (3.31)$$

This type of problem cannot be handled directly using CSMP system macros such as REALPL and CMPXPL.

The task in this example is to develop a macro which can be used for solving a ratio of polynomials up to the fifth-power with the form

$$\frac{X(s)}{R(s)} = \frac{b_1 s^5 + b_2 s^4 + b_3 s^3 + b_4 s^2 + b_5 s + b_6}{a_1 s^5 + a_2 s^4 + a_3 s^3 + a_4 s^2 + a_5 s + a_6} \quad (3.32)$$

Define,

$$\frac{Y(s)}{R(s)} = \frac{1}{a_1 s^5 + a_2 s^4 + a_3 s^3 + a_4 s^2 + a_5 s + a_6} \quad (3.33)$$

```
*****          EXAMPLE 3-11 WITHOUT USING MACROS        *****
*   BALANC   = BALANCE TO PAY ON THE LOAN
*   MONINT   = INTEREST FOR THE CURRENT MONTH
*   TOTINT   = TOTAL INTEREST PAID TO DATE
*   MPAY     = AMOUNT PAID ON LOAN FOR THE CURRENT MONTH
*   EQUITY   = CURRENT AMOUNT PAID-OFF ON THE LOAN
*   IRM      = THE MONTHLY INTEREST RATE (IN DECIMAL FORM)
*   PAY      = THE MONTHLY PAYMENT (A CONSTANT)


RENAME TIME = MONTH


INITIAL
PARAMETER P = 32000.0, I = 9.0, PERIOD = 300.0
INCON       PSTINT = 0.0, TOTINT = 0.0, MONINT = 0.0,...
            EQUITY = 0.0, PASPAY = 0.0
BALANC = P
IRM = I/1200.0
S = (1.0 + IRM)**PERIOD
PAY = (P*IRM*S)/(S - 1.0)


DYNAMIC
NOSORT
     IF(MONTH.EQ.0.0) GO TO 7
     MONINT = IRM*BALANC
     MPAY = PAY - MONINT
     EQUITY = MPAY + PASPAY
     BALANC  = P - EQUITY
     TOTINT = MONINT + PSTINT
     PASPAY = EQUITY
     PSTINT = TOTINT
   7 CONTINUE
TIMER FINTIM =300.0, PRDEL = 1.0, DELT = 1.0, OUTDEL = 10.0
PRINT PAY, MONINT, TOTINT,MPAY,EQUITY,BALANC
TITLE  OUTPUT FOR EXAMPLE 3-11 WITHOUT MACROS
END
STOP
ENDJOB
```

Fig. 3.45 Program for calculating mortgage payments without using macros.

and

$$\frac{X(s)}{Y(s)} = b_1 s^5 + b_2 s^4 + b_3 s^3 + b_4 s^2 + b_5 s + b_6 \tag{3.34}$$

In the time domain, Eq. (3.33) becomes

$$a_1 \frac{d^5 y}{dt^5} + a_2 \frac{d^4 y}{dt^4} + a_3 \frac{d^3 y}{dt^3} + a_4 \frac{d^2 y}{dt^2} + a_5 \frac{dy}{dt} + a_6 y = r \tag{3.35}$$

Now define

$$y_1 = y$$
$$\frac{dy_1}{dt} = \frac{dy}{dt} = y_2$$
$$\frac{dy_2}{dt} = \frac{d^2 y}{dt^2} = y_3$$

$$\frac{dy_3}{dt} = \frac{d^3y}{dt^3} = y_4$$

$$\frac{dy_4}{dt} = \frac{d^4y}{dt^4} = y_5$$

$$\frac{dy_5}{dt} = \frac{d^5y}{dt^5} = \frac{1}{a_1}[r - a_6 y_1 - a_5 y_2 - a_4 y_3 - a_3 y_4 - a_2 y_5]$$

From Eq. (3.34), after going to the time domain and making appropriate substitutions, we have

$$x = b_6 y_1 + b_5 y_2 + b_4 y_3 + b_3 y_4 + b_2 y_5 + b_1 \frac{dy_5}{dt} \tag{3.36}$$

where

$$\frac{dy_5}{dt} = \frac{d^5y}{dt^5} \tag{3.37}$$

An appropriate macro for these equations is

```
MACRO XOUT = TRANSF (BN, AD, IN)
   Y1 = INTGRL (0.0, Y2)
   Y2 = INTGRL (0.0, Y3)
   Y3 = INTGRL (0.0, Y4)
   Y4 = INTGRL (0.0, Y5)
   Y5 = INTGRL (0.0, Y5DOT)
 Y5DOT = (1.0/AD(1))*(IN−AD(6)*Y1−AD(5)*Y2−AD(4)*Y3−AD(3)* . . .
         Y4−AD(2)*Y5)
  XOUT = BN(6)*Y1+BN(5)*Y2+BN(4)*Y3+BN(3)*Y4+BN(2)* . . .
         Y5+BN(1)*Y5DOT
ENDMAC
```

The parameters for the polynomials can be placed in the macro using TABLE and STORAGE. For this case,

```
STORAGE BNUM1(6), ADEN1(6)
TABLE   BNUM1(1–6) = N1, N2, N3, N4, N5, N6, . . .
        ADEN1(1–6) = D1, D2, D3, D4, D5, D6, . . .
```

where BNUM1 and ADEN1 are dummy names for the numerator and denominator coefficients.

Care must be exercised in using the TABLE with this general fifth-order system. For example, if the particular transfer function is

$$\frac{2s^2 + 4s + 3}{9s^3 + 6s^2 + 5s + 8} \tag{3.38}$$

the polynomial should be viewed as

$$\frac{0s^3 + 2s^2 + 4s + 3}{9s^3 + 6s^2 + 5s + 8} \tag{3.39}$$

The parameters should be read in as

```
TABLE BNUM1(1–6) = 0.0, 2.0, 4.0, 3.0, 0.0, 0.0, . . .
      ADEN1(1–6) = 9.0, 6.0, 5.0, 8.0, 0.0, 0.0
```

$$\text{where } N1 = 0.0 \quad D1 = 9.0$$
$$N2 = 2.0 \quad D2 = 6.0$$
$$N3 = 4.0 \quad D3 = 5.0$$
$$N4 = 3.0 \quad D4 = 8.0$$
$$N5 = 0.0 \quad D5 = 0.0$$
$$N6 = 0.0 \quad D6 = 0.0$$

As an illustration of using this MACRO in a system, consider the diagram in Fig. 3.46. The program, which invokes the macro twice, is given in Fig. 3.47.

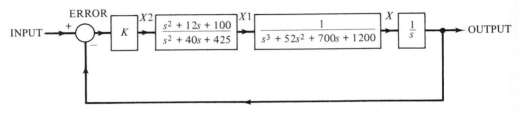

Fig. 3.46 System diagram for Example 3.12, illustrating the use of a macro function.

Note that BNUM1, ADEN1, and X2 replace N, D, and IN respectively for one reference to the macro. On the second reference, BNUM2, ADEN2, and X1 replace N, D, and IN. This explains the reason for naming the STORAGE and TABLE values BNUM1, BNUM2, ADEN1, and ADEN2.

The guidelines and examples which have been presented in this section on generating macro functions should serve as a good foundation for the beginning user. The scope of a problem will usually determine whether a special macro function should be written. Obviously, when properly used, a macro function can save considerable programming effort.

The power of the macro is further increased when used with procedure functions. The following section will show how this is accomplished.

PROCEDURE FUNCTION

As in the macro, a procedure function offers another way for users to define their own simulation functions. The major differences between these two translation control statements are (a) the macro is called within the program in a way that is analogous to calling a subroutine, the procedure function is a single entity within the program and is not called; (b) macro structure statements are sorted by the sorting algorithm while structure statements of the procedure function are not sorted. Therefore, the statements of a procedure function can be so written as to utilize FORTRAN logical and branching statements as well as subscripted variables.

```
************************************************************
*    A MACRO MUST BE PLACED AT THE FRONT OF THE DECK    *
************************************************************
MACRO   XOUT = TRANSF(BN, AD, IN)
    Y1  = INTGRL(0.0, Y2)
    Y2  = INTGRL(0.0, Y3)
    Y3  = INTGRL(0.0, Y4)
    Y4  = INTGRL(0.0, Y5)
    Y5  = INTGRL(0.0, Y5DOT)
Y5DOT = (1.0/AD(1))*(IN-AD(6)*Y1-AD(5)*Y2-AD(4)*Y3-AD(3)*Y4-AD(2)*Y5)
  XOUT = BN(6)*Y1+BN(5)*Y2+BN(4)*Y3+BN(3)*Y4+BN(2)*Y5+BN(1)*Y5DOT
ENDMAC


************************************************************
*   NOTE THAT STORAGE SIZE OF ALL ELEMENTS ARE SO INDICATED ON   *
*                   THE SINGLE FOLLOWING CARD                     *
************************************************************

STORAGE   BNUM1(6), ADEN1(6), BNUM2(6), ADEN2(6)

************************************************************
* THE PARTICULAR VALUES OF THE STORAGE LOCATIONS ARE GIVEN *
* USING "TABLE".  NOTE THAT TABLE IS TYPED ONLY ONCE AND    *
*          AND THE ... FEATURE IS USED FOR CONTINUATION     *
************************************************************

TABLE     BNUM1(1-6) = 1.0, 12.0, 100.0, 3*0.0,...
          ADEN1(1-6) = 1.0, 40.0, 425.0, 3*0.0,...
          BNUM2(1-6) = 3*0.0,1.0, 2*0.0,...
          ADEN2(1-6) = 1.0, 52.0, 700.0, 1200.0, 2*0.0

PARAMETER    K = 3000.0
INPUT = STEP(0.0)
ERROR = INPUT - OUTPUT
X2 = K*ERROR
X1 = TRANSF(BNUM1, ADEN1, X2)
X  = TRANSF(BNUM2, ADEN2, X1)
OUTPUT = INTGRL(0.0, X)
TIMER  FINTIM = 20.0,   OUTDEL = 0.4
PRTPLT OUTPUT
LABEL  OUTPUT OF SYSTEM WITH DOUBLE INVOCATION OF
LABEL  TRANSFER FUNCTION MACRO  - EXAMPLE 3-12
END
STOP
ENDJOB
```

Fig. 3.47 Program showing the use of a transfer function macro.

Except for a single important difference, one can achieve the same programming objectives with a nosort section as with a procedure function. This important difference is that a nosort section forms a division between sorted sections of the program. All statements occurring before a nosort section remain above the section. Likewise, statements made after a nosort section always remain after the section. However, statements within a procedure function are never sorted but the function itself is treated as a single entity and moved around within the overall program as required by the sorting algorithm.

The basic structure of a procedure function is given in the following example.

```
PROCEDURE     MASS = MVARY (SLOPE, SWITCH, MASSO)
   IF (TIME. GE. SWITCH) GO TO 14
   MASS = MASSO * (1.0 + SLOPE*TIME)
   GO TO 8
14 MASS = MASSO * (1.0 + SLOPE*SWITCH)
 8 CONTINUE
ENDPROCEDURE
```

Suppose for a given problem that SLOPE $= -0.2$ and SWITCH $= 3.0$. We assume these values have been entered previously on a parameter card. A condition for branching is set up by a logical IF statement involving TIME and SWITCH. The procedure function for this case will determine the time varying MASS as given in Fig. 3.48. This procedure function can be located essentially anywhere in the DYNAMIC segment of a simulation requiring the time-varying mass. The sort algorithm will place the procedure function in the correct position relative to other program statements. The time-varying characteristics of the mass can easily be changed by inserting new values for MASSO, SWITCH and SLOPE on parameter cards.

Although many of the structure format statements for using a procedure function are similar to a macro, the procedure accomplishes a different purpose and the following rules should be observed.

1 The first card of the procedure contains the word PROCEDURE followed by at least one blank space. A simple illustration is

   PROCEDURE JX, AMP = XFUN(XUP, PAR1, SIGNAL)

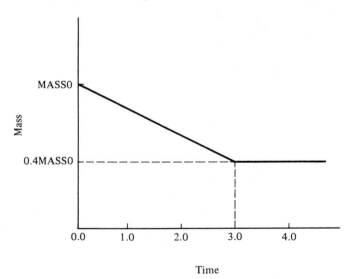

Fig. 3.48 Diagram of time-varying mass used in PROCEDURE illustration.

2　Unlike a macro, the arguments of a procedure cannot be functions, expressions, or numbers. They must be names defined elsewhere in the program.

3　As noted above, the first card of a procedure contains the word PROCEDURE. The very last card contains the single word, ENDPROCEDURE. All statements describing the procedure function are placed between the PROCEDURE card and the ENDPROCEDURE card. Since CSMP takes action on the first five characters of these words, the user may replace PROCEDURE by PROCED, and ENDPROCEDURE by ENDPRO.

4　As with the macro function, variables defined within a procedure are not available as output data by means of PRTPLT, PRINT, PREPARE or RANGE unless they are output names on the procedure definition card. In the following example,

```
PARAM CONST1 = 0.5, CONST2 = 0.1
PROCEDURE X1, X2 = READY(CONST1, CONST2)
    X1 = CONST1*EXP**(−3.0*TIME)
    X2 = CONST2*X1**2
    X3 = INTGRL(0.0, X1)
ENDPROCEDURE
TIMER FINTIM = 1.0, PRDEL = 0.02
PRINT X1, X2
END
STOP
ENDJOB
```

X3 is not available as output data; however, output is available for X1 and X2. Changing the definition card to

```
PROCEDURE X1, X2, X3 = READY (CONST1, CONST2)
```

will make X3 available as output data.

5　The procedure function may be located in both INITIAL and DYNAMIC segments. (Since the TERMINAL segment executes statements in the order they appear, a procedure function would not serve a useful purpose in this segment.) The procedure may not be placed in a nosort section.

6　Procedure functions may be placed within a macro. Even though the macro statements are sorted, the statements contained within each procedure will not be sorted except as a single block.

7　Procedure functions within a macro may not be used to invoke other macros. A macro, however, may invoke another macro which itself may contain one or more procedure functions.

8　A procedure function cannot be placed within another procedure function.

9　If the user desires all statements in a macro to be procedural then a single card containing the word **PROCEDURAL** should be placed immediately after the macro definition card. For example, the procedure function given previously in this section may be changed to a macro as follows:

```
MACRO    MASS = MVARY (SLOPE, SWITCH, MASSO)
         PROCEDURAL
         IF (TIME.GE.SWITCH) GO TO 14
         MASS = MASSO*(1.0 + SLOPE*TIME)
         GO TO 8
     14  MASS = MASSO*(1.0 + SLOPE + SWITCH)
      8  CONTINUE
         ENDMAC
```

Note that an ENDPROCEDURE card is not to be used in this case. Restrictions placed upon the arguments of MVARY, when used as a procedure function, are now removed and consequently the arguments now conform to those of a macro.

Most of the guidelines presented here on using the procedure function are applied in the following rather comprehensive problem.

Example 3.13

   The purpose of this problem is to illustrate how CSMP can be applied to simulating four-way traffic flow at a signal light. The simulation makes use of the procedure function as well as several other program features. These additional features include the mode integrator and the logic functions AND and NOT.
   The first step of the problem is to develop a reasonable model for traffic flow. In the diagram in Fig. 3.49 we assume that traffic flows from east to west, west to east, north to south and south to north. The assumption is made that cars do not turn at the light. The justification for making this assumption is simply to avoid the added complication in the problem. Turning traffic could be included with the addition of ten to twelve program statements.
   The first consideration is to establish the flow of traffic. One must recognize that the spacing between cars is a random variable even with a given density of traffic flow. The one exception is when the through-put (flow) is maximum. For example, in a 20 mph speed zone, we assume maximum through-put when the distance between each car is constant and each car is traveling at the maximum speed limit. This situation is shown in Fig. 3.50.
   To simplify the flow modeling we assume each car approaching the intersection is traveling at the same velocity but with a random arrival. Intuitively we know that even with random spacing between cars, the density or compactness of the traffic varies during the time of day. For this problem we assume the traffic density is as shown in Fig. 3.51. The intended interpretation here is that a relative density of one represents maximum traffic through-put as discussed earlier.
   The problem becomes one of producing a string of random traffic having a given density. This can be accomplished as follows. Define an impulse train by

$$\text{ETRIG} = \text{IMPULS(0.0, EPERID)}$$

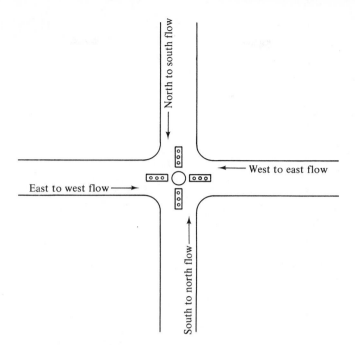

Fig. 3.49 Diagram showing intersection geometry at the traffic light.

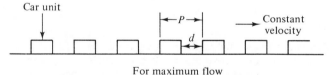

For maximum flow
- Each unit moves with the same velocity.
- The space, d, between each car is the same.
- The period of each unit is P sec.

Fig. 3.50 Spacing of cars for maximum flow rate.

where the pulse train starts at 0.0 and has a spacing of EPERID sec. The spacing EPERID is set equal to the period, P, corresponding to maximum traffic flow shown earlier in Fig. 3.50. Next a random-variable sequence of numbers between 0 and 1 is made available from the random-number generator, RNDGEN.

Thus   ERAND = RNDGEN(3)

where ERAND is the selected name of the sequence and 3 is a seed value (any odd integer). Now define a time-varying function, ESIG, such that

$$ESIG = EDENSE - ERAND$$

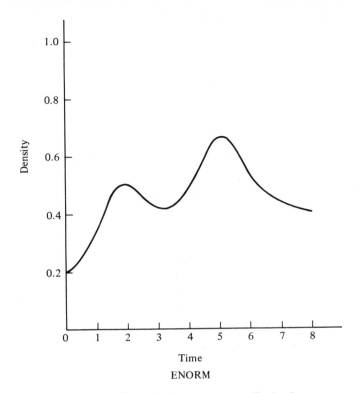

Fig. 3.51 Normalized east to west traffic density.

where EDENSE is the value of the traffic density (from Fig. 5.51) at a particular time. ESIG will be a random number between -1 and 1. If EDENSE is greater than ERAND, ESIG will be positive and this signifies that a car should be generated at this time. Conversely, if ESIG is negative, no car is generated. By using the logical AND function with ETRIG and ESIG as arguments, a trigger for a pulse generator is formed. The width of the pulse from the generator can be set to the length of a car. The statements required are

$$ESINK = AND(ESIG, ETRIG)$$
$$EFLOW = PULSE(EWIDTH, ESINK)$$

where EWIDTH is a parameter (the car length) and EFLOW is the flow of traffic. AND is a logical function defined by

$$Y = AND(X1, X2)$$
$$Y = 1 \text{ if } X1, X2 > 0$$
$$Y = 0 \text{ otherwise}$$

By combining the above statements, one can write a procedure function for generating EFLOW as follows.

PROCEDURE   EFLOW = EFORM(EWIDTH, EPERID)
        ETRIG = IMPULS(0.0, EPERID)
        ERAND = RNDGEN(3)
        EDENSE = AFGEN(ENORM, TIME)
        ESIG = EDENSE − ERAND
        ESINK = AND(ESIG, ETRIG)
        EFLOW = PULSE(EWIDTH, ESINK)
ENDPRO

The value of EDENSE is determined by using the linear interpolater, AFGEN, where the function is defined by the data of Fig. 3.51. Thus,

FUNCTION   ENORM = (0.0,0.2),(0.5,0.25),(1.0,0.34), . . .
        (1.5,0.46),(2.0,0.5),(2.5,0.45),(3.0,0.42),(3.5,0.43), . . .
        (4.0,0.48),(4.5,0.58)

Use of the AFGEN was previously explained in Chap. 1. At this point in the simulation development the user should write a short program and verify that EFORM is correctly generated. A short program for this purpose is given in Fig. 3.52 and the resulting output is presented in Fig. 3.53.

```
FUNCTION ENORM = (0.0,0.2),(.5,.25),(1.,.34),(1.5,.46),(2.,.5),...
       (2.5,.45),(3.,.42),(3.5,.43),(4.,.48),(4.5,.58),(5.,.66)

PARAM EWIDTH = 0.01, EPERID = 0.02

*****************************************************************************
*     NOTE ON THE FOLLOWING PROCEDURE TRANSLATION CONTROL CARD         *
*     THAT ALL VARIABLES REQUESTED FOR PRINTING MUST BE LISTED AS      *
*                OUTPUTS  OF THE PROCEDURE FUNCTION                     *
*****************************************************************************

PROCEDURE  ETRIG, EDENSE, ERAND, ESIG, EFLOW = EFORM(EWIDTH,EPERID)
       ETRIG = IMPULS(0.0, EPERID)
       ERAND = RNDGEN(3)
       EDENSE = AFGEN(ENORM, TIME)
       ESIG = EDENSE − ERAND
       ESINK = AND(ESIG, ETRIG)
       EFLOW = PULSE(EWIDTH,ESINK)
ENDPRO

TIMER FINTIM = 0.2, PRDEL = 0.002, DELT = 0.002
PRINT  ETRIG, EDENSE, ERAND, ESIG, EFLOW
TITLE I*******   TEST FOR RANDOM PULSE GENERATOR -   EXAMPLE 3-13 ******
END
STOP
ENDJOB
```

Fig. 3.52 Program for testing the random generation of a string of cars.

Having developed a reasonable representation of traffic flow, attention is next directed to the problem of determining (a) how many cars have approached the light; (b) how many cars have passed under the light; and (c) how many cars are waiting at any given instant of time. The total number of cars approaching the light can be determined by integrating the traffic flow and dividing by the width of a car pulse. This can be deter-

TIME	ETRIG	ECENSE	ERAND	ESIG	EFLOW
0.0	1.0000E 00	2.0000E-01	9.1557E-05	1.9991E-01	1.0000E 00
2.0000E-03	0.0	2.0020E-01	5.4933E-04	1.9965E-01	1.0000E 00
4.0000E-03	0.0	2.0040E-01	2.4720E-03	1.9793E-01	1.0000E 00
6.0000E-03	0.0	2.0060E-01	9.8878E-03	1.9071E-01	1.0000E 00
8.0000E-03	0.0	2.0080E-01	3.7079E-02	1.6372E-01	1.0000E 00
1.0000E-02	0.0	2.0100E-01	1.3348E-01	6.7515E-02	1.0000E 00
1.2000E-02	0.0	2.0120E-01	4.6720E-01	-2.6600E-01	0.0
1.4000E-02	0.0	2.0140E-01	6.0182E-01	-4.0042E-01	0.0
1.6000E-02	0.0	2.0160E-01	4.0612E-01	-2.0452E-01	0.0
1.8000E-02	0.0	2.0180E-01	2.0407E-02	1.8139E-01	0.0
2.0000E-02	1.0000E 00	2.0200E-01	4.6732E-01	-2.6532E-01	0.0
2.2000E-02	0.0	2.0220E-01	6.2025E-01	-4.1805E-01	0.0
2.4000E-02	0.0	2.0240E-01	5.1562E-01	-3.1322E-01	0.0
2.6000E-02	0.0	2.0260E-01	5.1150E-01	-3.0890E-01	0.0
2.8000E-02	0.0	2.0280E-01	4.2840E-01	-2.2560E-01	0.0
3.0000E-02	0.0	2.0300E-01	9.6687E-01	-7.6387E-01	0.0
3.2000E-02	0.0	2.0320E-01	9.4563E-01	-7.4244E-01	0.0
3.4000E-02	0.0	2.0340E-01	9.7195E-01	-7.6855E-01	0.0
3.6000E-02	0.0	2.0360E-01	3.2096E-01	-1.1736E-01	0.0
3.8000E-02	0.0	2.0380E-01	1.7823E-01	2.5567E-02	0.0
4.0000E-02	1.0000E 00	2.0400E-01	1.8079E-01	2.3215E-02	1.0000E 00
4.2000E-02	0.0	2.0420E-01	4.8062E-01	-2.7642E-01	1.0000E 00
4.4000E-02	0.0	2.0440E-01	2.5664E-01	-5.2244E-02	1.0000E 00
4.6000E-02	0.0	2.0460E-01	2.1430E-01	-9.6995E-03	1.0000E 00
4.8000E-02	0.0	2.0480E-01	9.7600E-01	-7.7120E-01	1.0000E 00
5.0000E-02	0.0	2.0500E-01	9.2729E-01	-7.2229E-01	0.0
5.2000E-02	0.0	2.0520E-01	7.7977E-01	-5.7457E-01	0.0
5.4000E-02	0.0	2.0540E-01	3.3299E-01	-1.2759E-01	0.0
5.6000E-02	0.0	2.0560E-01	9.8003E-01	-7.7443E-01	0.0
5.8000E-02	0.0	2.0580E-01	8.8322E-01	-6.7742E-01	0.0
6.0000E-02	1.0000E 00	2.0600E-01	4.7910E-01	-2.7310E-01	0.0
6.2000E-02	0.0	2.0620E-01	9.2562E-01	-7.1942E-01	0.0
6.4000E-02	0.0	2.0640E-01	2.4180E-01	-3.5397E-02	0.0
6.6000E-02	0.0	2.0660E-01	1.2020E-01	8.6404E-02	0.0
6.8000E-02	0.0	2.0680E-01	5.4501E-01	-3.3821E-01	0.0
7.0000E-02	0.0	2.0700E-01	1.8827E-01	1.8727E-02	0.0
7.2000E-02	0.0	2.0720E-01	2.2458E-01	-1.7379E-02	0.0
7.4000E-02	0.0	2.0740E-01	6.5301E-01	-4.4561E-01	0.0
7.6000E-02	0.0	2.0760E-01	8.9688E-01	-6.8928E-01	0.0
7.8000E-02	0.0	2.0780E-01	5.0414E-01	-2.9634E-01	0.0
8.0000E-02	1.0000E 00	2.0800E-01	9.5295E-01	-7.4495E-01	0.0
8.2000E-02	0.0	2.0820E-01	1.8040E-01	2.7796E-02	0.0
8.4000E-02	0.0	2.0840E-01	5.0590E-01	-2.9750E-01	0.0
8.6000E-02	0.0	2.0860E-01	4.1178E-01	-2.0318E-01	0.0
8.8000E-02	0.0	2.0880E-01	9.1757E-01	-7.0877E-01	0.0
9.0000E-02	0.0	2.0900E-01	7.9938E-01	-5.9038E-01	0.0
9.2000E-02	0.0	2.0920E-01	5.3812E-01	-3.2892E-01	0.0
9.4000E-02	0.0	2.0940E-01	3.4321E-02	1.7508E-01	0.0
9.6000E-02	0.0	2.0960E-01	3.6282E-01	-1.5322E-01	0.0
9.8000E-02	0.0	2.0980E-01	8.6803E-01	-6.5823E-01	0.0
1.0000E-01	1.0000E 00	2.1000E-01	9.4281E-01	-7.3281E-01	0.0
1.0200E-01	0.0	2.1020E-01	8.4457E-01	-6.3437E-01	0.0
1.0400E-01	0.0	2.1040E-01	5.8216E-01	-3.7176E-01	0.0
1.0600E-01	0.0	2.1060E-01	8.9176E-01	-6.8116E-01	0.0
1.0800E-01	0.0	2.1080E-01	1.1116E-01	9.9635E-02	0.0

Fig. 3.53 The printer-plotted output from the program of a random pulse generator.

162

mined using the INTGRL function as

$$ELEFT = INTGRL(0.0,\ EFLOW)$$
$$ETOTAL = ELEFT/EWIDTH$$

where ETOTAL is the total traffic arriving from the east.

For simplicity, the state of the light is assumed to be either green or red with amber ignored. The number of cars passing under the light can now be determined using the mode integrator function whose characteristics are given in Table 3.4.

Table 3.4

Mode Integrator Function

CSMP Statement	*Mathematical Equivalent*
Y = MODINT(IC, X1, X2, X3)	$y(t) = \int x_3(t)\,dt + IC$ $X_1 > 0$, any X_2
	$y(t) = IC$ $X_1 \leq 0,\ X_2 > 0$
	$y(t) =$ last output $X_1 \leq 0,\ X_2 \leq 0$

The input signal X_3 should be the actual traffic passing under the light with X_1 the state of the light. Thus, if X_1 is greater than zero (the light is green), $y(t)$ increases by integrating X_3. When the light changes to red, X_1 equals zero and $y(t)$ holds the last output value. Any non-positive number for X_2 can be used. Expressing this in program variables gives

$$ERIGHT = MODINT(0.0,\ EWLITE,\ -1.0,\ EWX2)$$
$$EPASED = ERIGHT/EWIDTH$$

where EWLITE = state of the east-west light

      EWX2 = the pulses passing under the light

      EPASED = the total number of cars passed through the light

The traffic backed-up at any time on the east side is given by

$$EBAKUP = ELEFT - ERIGHT$$
$$EWAIT = EBAKUP/EWIDTH$$

where EWAIT is the actual traffic backed up or waiting on the east side.

The excitation signal, EWX2, must take the following into account. When the light changes from red to green, traffic will not move instantaneously. We assume the flow linearly increases with time up to the limit of maximum through-put. Once the maximum flow is reached, the excitation will remain at this level until EBAKUP reaches zero. Once EBAKUP reaches zero, EWX2 should be changed to EFLOW. Suppose we assume that after 8 car units have passed under the light, EWX2 reaches the maximum value of EWIDTH/EPERID.

A program which combines the generation of random traffic with traffic volumn in the east to west direction is given in Fig. 3.54. The correct state of EWX2 is determined by the procedure function EWX3 except for testing the state of EBAKUP. Logically, this test should also be made in the EWX3 procedure function but in attempting to do so an algebraic loop will be formed between ERIGHT, EBAKUP, and EWX3. This algebraic loop is avoided by testing the state of EBAKUP in a NOSORT segment.

```
FUNCTION ENORM = (0.0,0.2),(.5,.25),(1.,.34),(1.5,.46),(2.,.5),...
      (2.5,.45),(3.,.42),(3.5,.43),(4.,.48),(4.5,.58),(5.,.66)

PARAM EWIDTH = 0.012, EPERID = 0.024, CYCLE = 1.0, EBAKUP = 0.0

LITRIG = IMPULS (0.0,CYCLE)
EWLITE = PULSE(CYCLE/2.0,LITRIG)
ETRIG = IMPULS(0.0, EPERID)

PROCEDURE  EFLCW = EFORM(ETRIG)
      ERAND = RNDGEN(3)
      EDENSE = AFGEN(ENCRM, TIME)
      ESIG = EDENSE - ERAND
      ESINK = AND(ESIG, ETRIG)
      EFLOW = PULSE(EWIDTH,ESINK)
ENDPRO

NOSORT
IF(EBAKUP.GT.0.0) EWX2 = EWX3
IF(EBAKUP.LE.0.0) EWX2 =EFLOW
SORT

ELEFT = INTGRL(0.0,EFLOW)
ERIGHT = MODINT(0.0,EWLITE,-1.0,EWX2)
EBAKUP = ELEFT-ERIGHT

PROCEDURE  EWX3 = GENRAL(EWLITE,EFLOW,CYCLE,EWIDTH,EPERID)
      IF(TIME.LT.(CYCLE/2.0)) GO TO 19
      IF(EWLITE.EQ.0.0) GO TO 21
      IF(EPASLT.EQ.0.0) OLDTIM = TIME
      EDELT = TIME - OLDTIM
      IF(EDELT.GE.(8.0*EPERID)) GO TO 23
      EWX3 = ((EWIDTH)/(8.0*(EPERID**2)))*EDELT
      EPASLT = EWLITE
      GO TO 27
 21   EWX3 = 0.0
      EPASLT = 0.0
      GO TO 27
 19   EPASLT = 0.0
      IF(EWLITE.GT.0.0) EWX3 = EFLOW
      IF(EWLITE.GE.0.0) EWX3 = 0.0
      GO TO 27
 23   EWX3 = EWIDTH/EPERID
 27   CONTINUE
ENDPRO

ETOTAL = ELEFT/EWIDTH
EPASED = ERIGHT/EWIDTH
EWAIT = EBAKUP/EWIDTH

TIMER FINTIM = 1.0, OUTDEL = 0.02, PRDEL = 0.02
PRTPLT ETOTAL, EPASED, EWAIT
LABEL   TEST FOR TRAFFIC MODEL - EXAMPLE 3-13
END
STOP
ENDJOB
```

Fig. 3.54 Traffic simulation program for east to west traffic.

This same basic model applies to west-east, north-south, and south-north traffic and in a natural way invites the use of macro functions.

A simpler program could have been written to illustrate the use of the procedure function. However, most meaningful real-world simulations require combinations of several CSMP features. The intention here has been to show how the procedure function can typically be used as an element in a more comprehensive simulation.

Fortran Subprograms

There are two forms of subprograms that can be used in CSMP programs, FORTRAN function and subroutine. The function subprogram can be used whenever the output is a single variable. A subroutine must always be used when there are multiple outputs.

The MACRO and PROCEDURE functions essentially serve the same purpose as subprograms and are generally preferred. However, subprograms do have certain advantages not found in the macro and procedure functions. The following is a listing of the most important advantages.

1 FORTRAN subprograms that are stored in the computer's system library can be called directly by CSMP programs.

2 The statements of a subprogram are not processed by the CSMP translator. Consequently, larger models can be developed by using subprograms.

3 Subprograms used in CSMP programs are compatible in FORTRAN programs. Therefore, subprograms can be interchanged between CSMP and FORTRAN programs.

General Guidelines

The rules for writing a subprogram for the use in a CSMP program are exactly the same as for a FORTRAN program. If needed, the reader should refer to specialized texts on FORTRAN[2-5] for a complete discussion on this topic. Implementation of subprograms in CSMP is somewhat different than in FORTRAN. Special considerations that must be followed are outlined by the following general rules.

1 All subprograms must follow the STOP card and furthermore, the last subprogram must be terminated by the ENDJOB card.

2 The following CSMP functional elements cannot be used in subprograms: INTGRL, DERIV, DELAY, ZHOLD, IMPL, MODINT, REALPL, LEDLAG, CMPXPL, HSTRSS, STEP, IMPULS, PULSE, RAMP, SINE, GAUSS, RNDGEN, and any user-defined macro function.

3 A FORTRAN function subprogram can be invoked in all segments and sections of a CSMP program. This includes invocations in macro and procedure functions. Using input variables P, Q, and R, a statement for calling the function XYZ is

$$X = XYZ(P, \ R, \ Q)$$

4 A FORTRAN subroutine can be called by two different methods. The CALL statement which is used in FORTRAN programs is only allowed in nosort sections and procedure functions. An example of a CALL statement for the subroutine named ABC is the following.

$$\text{CALL} \quad \text{ABC}(P, Q, R, X, Y)$$

If the output variables of the subroutine are on the right-hand side of the argument list of the subroutine, a second method for calling a subroutine can be used. This method can be used in all sections and segments, including macro and procedure functions. An example of this type of statement for the same ABC subroutine is shown below.

$$\text{X, Y} = \text{ABC}(P, Q, R)$$

It should be reemphasized that the above statement can only be used when the outputs X and Y are the last variables in the argument list of the subroutine.

Another limitation for using the above form for calling a subroutine is that it can only be used when there are two or more output variables.

5 When calling a FORTRAN subroutine that is stored in computer memory, the CALL form must be used in a nosort section or procedure function.

6 If the output of a subprogram includes subscripted variables, the calling statement in the CSMP program must be in a nosort section of a procedure function. All subscripted variables used as inputs or outputs must be declared on either a STORAGE or DIMENSION card.

The following example program illustrates the use of a FORTRAN subroutine to calculate the roots of a quadratic equation of the following form;

$$Ax^2 + Bx + C = 0 \qquad (3.40)$$

In this program the inputs to the subroutine are the coefficients A, B, and C. The outputs returned by the subroutine are the real part of the solution: X1REAL and X2REAL; and the imaginary part of the solution, X1IM and X2IM.

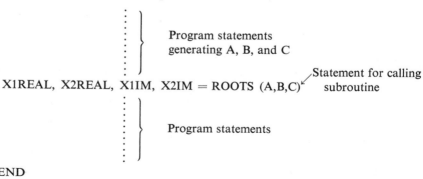

```
         ⋮ ⎫
         ⋮ ⎪  Program statements
         ⋮ ⎬  generating A, B, and C
         ⋮ ⎪
         ⋮ ⎭                           ╱Statement for calling
X1REAL, X2REAL, X1IM, X2IM = ROOTS (A,B,C)     subroutine
         ⋮ ⎫
         ⋮ ⎪  Program statements
         ⋮ ⎬
         ⋮ ⎭
END
STOP
```

```
    SUBROUTINE  ROOTS  (A,B,C,XIREAL,X2REAL,X1IM,X2IM)
    IF(B*B − 4.0*A*C) 1,2,3
  1 X1REAL = −B/(2.0*A)
    X2REAL = X1REAL
    X1IM    = SQRT(4.0*A*C−B*B)/(2.0*A)
    X2IM    = −X1IM
    GO TO 4
  2 X1REAL = −B/(2.0*A)
    X2REAL = X1REAL
    X1IM    = 0.0
    X2IM    = 0.0
    GO TO 4
  3 X1REAL = (−B+SQRT(B*B−4.0*A*C))/(2.0*A)
    X2REAL = (−B−SQRT(B*B−4.0*A*C))/(2.0*A)
    X1IM    = 0.0
    X2IM    = 0.0
  4 CONTINUE
    RETURN
    END
ENDJOB
```

The statement used in the above example program for calling the subroutine can be used in all segments and sections of the program. The following statement can also be used to call the subroutine, but it must be used in a nosort section or procedure function.

CALL ROOTS(A, B, C, X1REAL, X2REAL, X1IM, X2IM)

Example 3.14

One method for solving a set of simultaneous differential equations is to solve for the highest derivative of each variable and then successively integrate to obtain the total answer. This approach sometimes requires solving a set of simultaneous algebraic equations. As an example, consider the following problem,

$$\ddot{x}_1 + (t^2 + 4)\ddot{x}_2 \qquad = -0.5\dot{x}_1 - 1.7x_1 + 0.3x_2 + 0.21x_3^3 + \sin(t) \qquad (3.41)$$

$$5\ddot{x}_1 \qquad + 24\ddot{x}_3 = 0.017tx_1 - 5\dot{x}_2 - x_2 + 0.1x_3 + 1.6\dot{x}_3 \qquad (3.42)$$

$$56\ddot{x}_1 + 2t\ddot{x}_2 + 100\ddot{x}_3 = \dot{x}_1\dot{x}_2 - x_3 - 2.4\dot{x}_3 + e^{-x_1/23} \qquad (3.43)$$

where all initial conditions are zero.

The method of solution is to first solve for $\ddot{x}_1$, $\ddot{x}_2$, and $\ddot{x}_3$ as functions of the other variables that appear in the equations. This requires the solution of a set of simultaneous linear algebraic equations as shown by Eq. (3.44).

$$XDD = A^{-1}Y \qquad (3.44)$$

where A, XDD, and Y are defined by the following matrices.

$$A = \begin{bmatrix} 1 & t^2 + 4 & 0 \\ 5 & 0 & 24 \\ 56 & 2t & 100 \end{bmatrix} \tag{3.45}$$

$$XDD = \begin{bmatrix} \ddot{x}_1 \\ \ddot{x}_2 \\ \ddot{x}_3 \end{bmatrix} \tag{3.46}$$

$$Y = \begin{bmatrix} -0.5\dot{x}_1 - 1.7x_1 + 0.3x_2 + 0.21x_3^3 + \sin(t) \\ 0.017tx_1 - 5\dot{x}_2 - x_2 + 0.1x_1 + 1.6\dot{x}_3 \\ \dot{x}_1\dot{x}_2 - x_3 - 2.4\dot{x}_3 + e^{-x_1/23} \end{bmatrix} \tag{3.47}$$

Since the elements of both the A and Y matrices are time dependent, the XDD vector must be calculated at each integration step. Most computer installations have programs available for solving matrix relationships of the form given by Eq. (3.44). At many centers these subroutines are on-line and can be called from storage. In this example, a subroutine which uses the Gauss-Jordan elimination method with complete pivotal searching is employed. The subroutine SOLVE is called from the CSMP program. Within the subroutine SOLVE is another subroutine named INVERT. Figure 3.55 shows a listing of the CSMP program and the FORTRAN subroutines for solving the set of differential equations.

Because A is a double-subscripted variable, the FORTRAN DIMENSION statement must be used. The FORTRAN LOGICAL instruction is a FORTRAN specification statement and is used to declare that SING is a logical variable. If the matrix A is singular, the logical variable SING is true. In this example, the matrix is never singular and hence the solution of the equation set is given in Fig. 3.56.

Since the variable-step Runge-Kutta integration method is used in this example, subroutine SOLVE is called four times for each integration step. If this procedure is used to solve an equation set that has a large number of variables, a large amount of computer time will be required for the multiple solution of the XDD vector.

It should be noted that the only way to call the subroutine SOLVE is by using the CALL SOLVE instruction in a nosort or procedure section. The following statement cannot be used since the output variable (XDD) is not the last argument of the subroutine.

XDD = SOLVE(A, 3, 3, 3, Y, SING)

Note that there is only one output variable. This is another reason why the above form for calling the subroutine SOLVE cannot be used.

Special consideration must be made when using subprograms where the output depends on past values of output and past and present values of input.

Memory

This statement is required for subprograms that contain memory functions. A memory function is one where the output depends only on past values of input and output. As an example, consider the following statement.

MEMORY ABC(13)

The above instruction specifies that 13 storage locations are allocated for memory functions which are contained in the subprogram ABC. The MEMORY

```
TITLE   PROGRAM TO DEMONSTRATE THE USE OF A SUBROUTINE AND TO SOLVE
TITLE    A SYSTEM OF SIMULTANEOUS DIFFERENTIAL EQUATIONS
/        DIMENSION  A(3,3), XDD(3), Y(3)
/        LOGICAL SING
  FIXED  I, J
NOSORT
  DO 1 I = 1,3
  DO 1 J = 1,3
1 A(I,J) = 0.0
  A(1,1) = 1.0
  A(1,2) = TIME*TIME + 4.0
  A(2,1) = 5.0
  A(2,3) = 24.0
  A(3,1) = 56.0
  A(3,2) = 2.0*TIME
  A(3,3) = 100.0
  Y(1) = -0.5*XD1 - 1.7*X1 + 0.3*X2 + 0.21*(X3**3) + SIN(TIME)
  Y(2) = 0.017*X1*TIME - 5.0*XD2 - X2 + 0.1*X3 + 1.6*XD3
  Y(3) = XD1*XD2 - X3 - 2.4*XD3 + EXP(-X1/23.0)
  CALL SOLVE(A,3,3,3,Y,XDD,SING)
SORT
      XD1 = INTGRL(0.0,XDD(1))
      XD2 = INTGRL(0.0,XDD(2))
      XD3 = INTGRL(0.0,XDD(3))
      X1  = INTGRL(0.0,XD1)
      X2  = INTGRL(0.0,XD2)
      X3  = INTGRL(0.0,XD3)
  PRINT X1, X2, X3
  TIMER FINTIM = 0.2, PRDEL = 0.01
END
STOP
C   THE FOLLOWING TWO SUBROUTINES WILL SOLVE A SET OF SIMULTANEOUS
C      LINEAR EQUATIONS.
       SUBROUTINE SOLVE(A,MA,NA,N,B,X,SING)
       REAL A(MA,NA),B(N),X(N)
       LOGICAL SING
       CALL INVERT(A,N,MA,SING)
       IF(SING)RETURN
       DO 10 I=1,N
       X(I)=0.0
       DO 10 J=1,N
  10   X(I)=X(I)+A(I,J)*B(J)
       RETURN
       END
       SUBROUTINE INVERT(A,ORDER,DIM A, SING)
       INTEGER ORDER,DIM A
       REAL A(DIM A,1),B(100),C(100)
       N=ORDER
       LOGICAL SING
       INTEGER IP(100),IQ(100)
       DO 1 K=1,N
       PIVOT=0.
       DO 100 I=K,N
       DO 100 J=K,N
       IF(ABS(A(I,J))-ABS(PIVOT))100,100,101
  101  PIVOT=A(I,J)
       IP(K)=I
       IQ(K)=J
  100  CONTINUE
       IF(PIVOT)102,900,102
  102  IPK=IP(K)
       IQK=IQ(K)
       IF(IPK-K)200,299,200
  200  DO 201 J=1,N
       Z=A(IPK,J)
       A(IPK,J)=A(K,J)
  201  A(K,J)=Z
  299  CONTINUE
       IF(IQK-K)300,399,300
```

Fig. 3.55 CSMP program and subroutines used to solve a set of differential equations.

```
300     DO 301 I=1,N
        Z=A(I,IQK)
        A(I,IQK)=A(I,K)
301     A(I,K)=Z
399     CONTINUE
        DO 400 J=1,N
        IF(J-K)403,402,403
402     B(J)=1./PIVOT
        C(J)=1.
        GOTO 404
403     B(J)=-A(K,J)/PIVOT
        C(J)=A(J,K)
404     A(K,J)=0.
400     A(J,K)=0.
        DO 405 I=1,N
        DO 405 J=1,N
405     A(I,J)=A(I,J)+C(I)*B(J)
1       CONTINUE
        K=N
        DO 500 KDUM=1,N
        IQK=IQ(K)
        IPK=IP(K)
        IF(IPK-K)501,502,501
501     DO 503 I=1,N
        Z=A(I,IPK)
        A(I,IPK)=A(I,K)
503     A(I,K)=Z
502     IF(IQK-K)504,500,504
504     DO 506 J=1,N
        Z=A(IQK,J)
        A(IQK,J)=A(K,J)
506     A(K,J)=Z
500     K=K-1
        SING = .FALSE.
        RETURN
900     SING = .TRUE.
        RETURN
        END
ENDJOB
```

Fig. 3.55 (Continued)

PROGRAM TO DEMONSTRATE THE USE OF A SUBROUTINE AND TO SOLVE   RKS   INTEGRATION
A SYSTEM OF SIMULTANEOUS DIFFERENTIAL EQUATIONS

TIME	X1	X2	X3
0.0	0.0	C.C	0.0
1.0000E-02	1.4215E-06	-3.1430E-07	-2.9598E-07
2.0000E-02	5.6849E-06	-1.0927E-06	-1.1833E-06
3.0000E-02	1.2790E-05	-2.0889E-06	-2.6614E-06
4.0000E-02	2.2737E-05	-3.0570E-06	-4.7307E-06
5.0000E-02	3.5528E-05	-3.7516E-06	-7.3923E-06
6.0000E-02	5.1166E-05	-3.9282E-06	-1.0648E-05
7.0000E-02	6.9655E-05	-3.3428E-06	-1.4501E-05
8.0000E-02	9.1001E-05	-1.7521E-06	-1.8953E-05
9.0000E-02	1.1521E-04	1.0862E-06	-2.4010E-05
1.0000E-01	1.4229E-04	5.4138E-06	-2.9676E-05
1.1000E-01	1.7225E-04	1.1471E-05	-3.5957E-05
1.2000E-01	2.0510E-04	1.9498E-05	-4.2859E-05
1.3000E-01	2.4085E-04	2.9733E-05	-5.0389E-05
1.4000E-01	2.7952E-04	4.2413E-05	-5.8555E-05
1.5000E-01	3.2111E-04	5.7774E-05	-6.7367E-05
1.6000E-01	3.6565E-04	7.6053E-05	-7.6833E-05
1.7000E-01	4.1314E-04	9.7482E-05	-8.6964E-05
1.8000E-01	4.6361E-04	1.2229E-04	-9.7770E-05
1.9000E-01	5.1708E-04	1.5072E-04	-1.0926E-04
2.0000E-01	5.7356E-04	1.8299E-04	-1.2146E-04

Fig. 3.56 PRINT output for the solution of the set of differential equations.

card should appear in the deck before the first reference to the subprogram. Examples of some CSMP memory elements are the following

INTGRL
DELAY
REALPL
CMPXPL

History

This statement is used with subprograms where functions are used in which the output depends on past and present values of output and present values of input. The following instruction will reserve 8 history storage locations for the subprogram ABC, and 17 for the subprogram DEF.

HISTORY ABC(8) DEF(17)

The HISTORY statement must appear before any reference to the subprogram. Some CSMP history functions are:

DERIV   IMPULS   RST
GAUSS   PULSE   STEP
HSTRSS   RNDGEN   ZHOLD

A more comprehensive discussion of MEMORY and HISTORY can be found in References (1 and 6).

Implicit Functions

For certain types of simulations, the output of CSMP will state that an algebraic loop is formed between a specified list of variables and that the program cannot be executed. This problem arises whenever a loop does not contain at least one memory function.

As noted in the previous section, there are two types of memory functions defined in CSMP. If we consider a function to be characterized as having an output and input, the following applies.

(a) A MEMORY function is one whose output depends only on past values of the output and input.

(b) A HISTORY function is one whose output depends on past values of the output and input and in addition, the present value of the input.

There are four MEMORY functions in CSMP: INTGRL, REALPL, CMPXPL, and DELAY. Each loop of a simulation must contain either at least one of these functions or a user-defined memory function to avoid an algebraic loop. However, loops are not always obvious in a simulation.

Consider the simple circuit shown in Fig. 3.57(a). If the output is the voltage

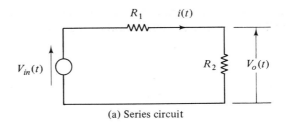

(a) Series circuit

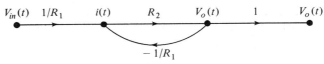

(b) Flow graph for the series circuit

Fig. 3.57   Series circuit and proposed flowgraph.

across R_2 and the input is the independent source $V_{in}(t)$, a set of equations describing the network is

$$V_o(t) = i(t) R_2 \tag{3.48}$$

$$i(t) = \frac{V_{in}(t)}{R_1} - \frac{V_o(t)}{R_1} \tag{3.49}$$

A flow graph for the two equations is given in Fig. 3.57(b). Since the flow graph contains a feedback loop without a memory function, programming these equations will result in an algebraic loop. An algebraic loop can often be circumvented by either rewriting the simulation equations or using the CSMP implicit function.

Formulation of Implicit Functions

An equation of the form $F(x) = 0$ is an implicit function. The equation for $F(x)$ can be written as

$$x = f(x) \tag{3.50}$$

where $F(x) \neq f(x)$. As an example, suppose

$$F(x) = x^2 + 3x - 4 = 0 \tag{3.51}$$

One way of expressing the variable x is

$$x = f(x) = \tfrac{1}{3}(4 - x^2) \tag{3.52}$$

The CSMP function

$$X = IMPL(XO, \ ERROR, \ FOFX) \tag{3.53}$$

will often find an x which satisfies the general equation, $x = f(x)$, and in particular may satisfy Eq. (3.52). In the IMPL function arguments,

        XO = an initial guess for x as specified by the user

   ERROR = the error specified by the user in finding x

    FOFX = the right hand side of Eq. (3.50)

A general rule for using the IMPL function is that the expressions for FOFX must appear after X = IMPL (XO, ERROR, FOFX) and the last such expression must specify FOFX. Furthermore, the variable X must appear at least once on the right side of the equations defining FOFX. As applied to Eq. (3.52) one may write either

           X = IMPL (XO, ERROR, FOFX)
           A = 4.0/3.0
           B = −(X∗∗2)/3.0
           FOFX = A + B

or

           X = IMPL (XO, ERROR, FOFX)
           FOFX = (4.0 − X∗∗2)/3.0

which illustrates that the user is not required to split-up FOFX.

    CSMP uses a Wegstein's accelerated convergence algorithm to calculate x_{n+1} using x_{n-1}, x_n, $f(x_{n-1})$ and $f(x_n)$ with a user-specified initial guess for x.[6] The test for satisfying the user-specified ERROR is based on

$$\left| \frac{x_{n+1} - x_n}{x_{n+1}} \right| \leq \text{ERROR} \qquad \text{for } |x_{n+1}| > 1 \qquad (3.54)$$

or

$$|x_{n+1} - x_n| \leq \text{ERROR} \qquad \text{for } |x_{n+1}| \leq 1 \qquad (3.55)$$

    Generally there will be more than one way of expressing $x = f(x)$. For the case, $F(x) = x^2 + 3x - 4 = 0$, two ways of writing x are

$$x = \frac{4 - x^2}{3} = f(x) \qquad (3.56)$$

and

$$x = \pm\sqrt{4 - 3x} = f(x) \qquad (3.57)$$

The IMPL function results will often depend on the form selected for $f(x)$ as well as the initial guess. The polynomial, $x^2 + 3x - 4 = 0$, has roots at $x = 1$ and $x = -4$. Suppose Eq. (3.56) is used to represent $f(x)$. The convergence of the implicit algorithm may be observed by printing out the variables x and $f(x)$ each time the program passes through the implicit loop. A short program for this purpose is given in Fig. 3.58. With ERROR = 0.05, the following values of X were obtained for the indicated initial values of XO.

```
PARAM XO = -3.0, ERROR = 0.05
NOSORT
X = IMPL(XO,ERROR,FCFX)
A = 4.0/3.0
B = -(X**2)/3.0
FOFX1 = A + B
        WRITE(6,1) X,FOFX1
    1   FORMAT(2E20.10)
FOFX = A + B
TIMER FINTIM = 1.0, DELT = 0.25, PRDEL = 0.25
PRINT X
TITLE   ILLUSTRATION OF IMPLICIT FUNCTION
END
STOP
ENDJOB
```

Fig. 3.58 Program for printing the values of x and $f(x)$ during the implicit loop convergence process.

Initial Guess XO	Solution X
0.5	1
2.5	1
−1.0	1
−3.0	−4
−6.0	−4

Rather than present a lengthy discussion on how convergence takes place, the reader is encouraged to submit the previous program of Fig. 3.58 and observe the trial values of X and FOFX1 as the program converges to the final value of X.

If X is expressed by

$$x = -\sqrt{4 - 3x} \qquad (3.58)$$

solutions for X, using ERROR $= 0.05$, are as follows

Initial Guess XO	Solution X
3.0	−4
−1.0	−4
−2.0	−4
−6.0	−4

We observe that for XO $= -1$, X converges to $+1$ in one case and -4 in the other.

Corresponding to the program in Fig. 3.58 a portion of subroutine Update will give

```
30000  X = IMPL( 1, XO,ERROR,FOFX)
       IF(NALARM) 30002, 30002, 30001
30001  CONTINUE
       A = 4.0/3.0
       B = −(X**2)/3.0
       FOFX1 = A + B
       WRITE(6,1) X, FOFX1
1      FORMAT(2E20.10)
       FOFX = A + B
       GO TO 30000
30002  CONTINUE
```

On first entering the IMPL routine, NALARM is set equal to 1 and X equal to XO. The value of FOFX is determined and the program returns to 30000. When the error criteria is satisfied, NALARM is set to zero and X set equal to the last value of FOFX. The program goes to 30002 and continues in the simulation. If the error criteria is not satisfied after 100 iterations, NALARM is set to −1 and the run is terminated with a message stating that 100 iterations have been exceeded.

The entire implicit function may be considered as a functional block with X as the output variable. Occasions may arise in which intermediate variables defining the implicit function are required for calculations elsewhere in a program. In such cases, extreme care must be exercised to avoid erroneous results. If at all possible, the desired intermediate variables should be recalculated outside the implicit loop. However, it may be possible to define the intermediate variables as outputs from either a procedure function or a user-defined macro which contains the entire implicit loop. Example 3.16 illustrates the use of a macro for this purpose.

Rules and guidelines for using the IMPL function can be summarized as follows.

1 As many statements as desired may be used to define $f(x)$; however, the output name in the last statement of the definition must be identical to the third argument, FOFX, of the IMPL statement. The output name for FOFX cannot be the output of an expression containing an INTGRL block nor can it be the output of a user defined macro, procedure or user-supplied routine. The last statement in the implicit loop cannot be continued.

2 The implicit variable must appear on the right side of an equal sign at least once in the statements defining FOFX.

3 An implicit loop can be defined within a macro or procedure provided the entire set of required statements is contained within the macro or procedure.

4 An implicit loop cannot be defined within another implicit loop.

Example 3.15

Consider the circuit diagram of Fig. 3.57(a) which was earlier described by the equations

$$V_o(t) = i(t)R_2 \tag{3.59}$$

$$i(t) = \frac{V_{in}(t)}{R_1} - \frac{V_o(t)}{R_1} \tag{3.60}$$

A simple CSMP program for the solution of $i(t)$ and the resulting output is

```
PARAM R1 = 8.0, R2 = 2.0
VIN = 10.0*STEP(0.0)
VO  = IA*R2
IA  = VIN/R1 - VO/R1
TIMER FINTIM = 1.0, DELT = 0.25, PRDEL = 0.25
PRINT IA
END
STOP
ENDJOB
```

SIMULATION INVOLVES AN ALGEBRAIC LOOP CONTAINING THE
FOLLOWING   ELEMENTS   VIN   VO   IA
********PROBLEM CAN NOT BE EXECUTED********

The algebraic loop can be broken by using the implicit function shown below.

```
PARAM R1 = 8.0, R2 = 2.0, IAO = 2.0, ERROR = 0.05
VIN = 10.0*STEP(0.0)
IA  = IMPL(IAO, ERROR, FOFIA)
A   = VIN/R1
VO  = IA*R2
B   = -VO/R1
FOFIA = A + B
TIMER FINTIM = 1.0, DELT = 0.25, PRDEL = 0.25
PRINT IA
END
STOP
ENDJOB
```

The output from the PRINT request is

TIME	IA
0.0	1.0000E 00
2.5000E-01	1.0000E 00
5.0000E-01	1.0000E 00
7.5000E-01	1.0000E 00
1.0000E 00	1.0000E 00

The initial value of IA is specified on the PARAM card as IAO = 2.0 and the acceptable error as ERROR = 0.05. Note that FOFIA is the last statement used in defining IA and also corresponds to the third argument in the IMPL function.

Example 3.16

The diagram of a feedback system which does not contain a memory function is shown in Fig. 3.59. An unaware user may write the following program only to find that the simulation involves an algebraic loop.

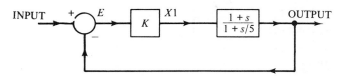

Fig. 3.59 Diagram for Example 3.16. (This diagram does not contain a memory function.)

```
PARAM K = 0.6
INPUT = STEP(0.0)
E        = INPUT − OUTPUT
X1       = K*E
OUTPUT = LEDLAG(1.0, 1.0/5.0, X1)
METHOD RKSFX
TIMER FINTIM = 2.0, DELT = 0.05, PRDEL = 0.1
PRINT OUTPUT†
END
STOP
ENDJOB
```

The algebraic loop can be broken by making OUTPUT an implicit function in the following manner.

```
PARAM OUTO = 0.2, ERROR = 0.05, K = 0.6
INPUT   = STEP(0.0)
OUTPUT = IMPL(OUTO, ERROR, FOFOUT)
E          = INPUT − OUTPUT
X1         = K*E
FOFOUT = LEDLAG(1.0, 1.0/5.0, X1)
METHOD RKSFX
TIMER FINTIM = 2.0, DELT = 0.05, PRDEL = 0.1
PRINT OUTPUT†
END
STOP
ENDJOB
```

The response of the system has a discontinuity at time equals zero and hence a fixed-step integration method is used to avoid computational difficulties with the variable-step integration. The printed output from the program is shown in Fig. 3.60. One can easily verify by analytical methods that the values for OUTPUT are correct.

†OUTPUT is acceptable for S/360 CSMP but should be changed to another variable for CSMP III.

PRINTED OUTPUT FOR LEDLAG ALGEBRAIC LOOP

TIME	OUTPUT
0.0	7.5000E-01
1.0000E-01	6.8202E-01
2.0000E-01	6.2637E-01
3.0000E-01	5.8080E-01
4.0000E-01	5.4350E-01
5.0000E-01	5.1296E-01
6.0000E-01	4.8795E-01
7.0000E-01	4.6747E-01
8.0000E-01	4.5071E-01
9.0000E-01	4.3699E-01
1.0000E 00	4.2575E-01
1.1000E 00	4.1655E-01
1.2000E 00	4.0902E-01
1.3000E 00	4.0285E-01
1.4000E 00	3.9780E-01
1.5000E 00	3.9367E-01
1.6000E 00	3.9029E-01
1.7000E 00	3.8752E-01
1.8000E 00	3.8525E-01
1.9000E 00	3.8339E-01
2.0000E 00	3.8187E-01

Fig. 3.60 Printer output from the algebraic loop containing the LEDLAG function.

Now suppose that one desires to evaluate the integral of E^2 which is often a meaningful index of performance. For this problem, this can be accomplished simply by adding the statement, ISA = INTGRL(0.0,E**2), outside the implicit loop. As an alternate approach, E may be expressed as the output of a macro and incorporated into a program as follows.

```
MACRO E, OUTPUT = CHECK(OUTO, ERROR, K, INPUT)
    OUTPUT = IMPL(OUTO, ERROR, FOFOUT)
    E       = INPUT - OUTPUT
    X1      = K*E
    FOFOUT = LEDLAG(1.0, 1.0/5.0, X1)
ENDMAC
PARAM OUTO = 0.2, ERROR = 0.05, K = 0.6
INPUT = STEP(0.0)
E, OUTPUT = CHECK(OUTO, ERROR, K, INPUT)
ISA = INTGRL(0.0, E**2)
METHOD RKSFX
TIMER FINTIM = 2.0, DELT = 0.05, PRDEL = 0.1
PRINT E, ISA, OUTPUT
END
STOP
ENDJOB
```

In summary, the implicit function can often be used for breaking an algebraic loop. The user however should be aware that the specified initial value and the user-selected form of $f(x)$ can influence the solution. Normally, the user will know the general results expected from the simulation. Therefore, if an unexpected (or unwanted) solution is obtained when using the IMPL function, one should try several different initial conditions and perhaps change the form of $f(x)$.

REFERENCES

1. *System/360 Continuous System Modeling Program* System Manual Y20-0111, IBM Corporation, Data Processing Division, White Plains, N.Y.

2. MCCRACKEN, DANIEL D., *A Guide to FORTRAN IV Programming*. New York: John Wiley and Sons, Inc., 1972.

3. COOPER, LAURA, and MARILYN SMITH, *Standard FORTRAN: A Problem-Solving Approach*. Boston: Houghton-Mifflin Company, 1973.

4. HOLDEN, HERBERT L., *Introduction to FORTRAN IV*, New York: The MacMillan Company, 1970.

5. ANDERSON, DECIMA M., *Computer Programming FORTRAN IV*, New York: Appleton-Century-Crofts, 1966.

6. *Continuous System Modeling Program III (CSMP III) Program Reference Manual* SH-19-7001-2, Program Number 5734-X59, IBM Corporation, Data Processing Division, White Plains, N.Y.

7. SAGE, ANDREW P., *Optimum Systems Control*, Englewood Cliffs, N.J.: Prentice-Hall, Inc., 1968.

PROBLEMS

1 Using the END and/or CONTINUE card, solve for the time-history of the value of a bank account over a four-year period. One hundred dollars is deposited at the beginning of the first year at the interest rate of 5% per year. At the beginning of the second year the interest rate is increased to 6%, and at the beginning of the third year the interest rate is 6.75%. The fourth year, the interest rate is 7.5%. Assume that money is continuously compounded according to the following formula.

$$\frac{d\$}{dt} = i\$$$

where i = yearly interest rate (0.05, 0.06, 0.0675, 0.075)

Answer:

At end of fourth year, $\$ = 128.72$

2 A linear transfer function that can be used to approximate the response of a human operator to a visual command in turning a steering wheel is shown below. The input is a unit-step input.

$$\theta_{IN}(s) \longrightarrow \boxed{\frac{(1.6s+1)(0.11s+1)e^{-0.2s}}{(1.2s+1)(0.15s+1)(0.0035s^2+0.084s+1)}} \longrightarrow \theta_o(s)$$

Determine the response of the man to a unit-step input. Note that the $e^{-0.2s}$ term represents a pure time delay of 0.2 sec.

Fig. P3.2

Answer:

$$\text{At TIME} = 1.0, \theta_0 = 1.1874$$

3 A real estate purchase of $42,000 is financed for 25 years at an interest rate of 9.5%. Write a CSMP program to calculate the monthly payments, the amount paid in interest each month, and the amount paid on the principal each month.

Answer:

Monthly payment = $366.95, at the tenth month interest = $329.97, principal = $36.98

4 Consider the system represented in the following diagram.

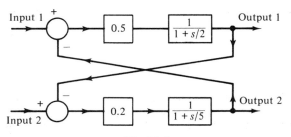

Fig. P3.4

If Input 1 and Input 2 are unit-steps applied at $t = 0$, find the responses of Output 1 and Output 2.

Answer:

$$\text{At TIME} = 1.2, \text{Output 1} = 0.39523, \text{Output 2} = 0.12559$$

5 The location of a hole is given by the X and Y dimensions. Assume that the X dimension varies between 11.900 in. and 12.100 in. with a uniform distribution and Y dimension has a Gaussian distribution with a mean of 5.000 inches and a standard deviation of 0.100 inches.

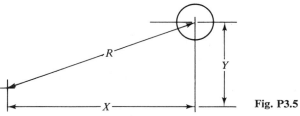

Fig. P3.5

How many parts in a production run of 1000 will have a radius R outside the tolerance of $R = 13.000 \pm 0.100$?

Answer:

135 parts

6 Use subscripted variables and the array form of the integral statement to solve the following set of equations that describe a mechanical system. The system is initially at rest.

$$m_1\ddot{x}_1 + (k_1 + k_2)x_1 + c\dot{x}_1 - k_2x_2 - c\dot{x}_2 = 0$$

$$m_2x_2 + (k_2 + k_3)x_2 + c\dot{x}_2 - k_2x_1 - c\dot{x}_1 - k_3x_3 = 0$$

$$m_3\ddot{x}_3 + k_3x_3 - k_3x_2 = 12\sin(4.0t)$$

where $m_1 = 2.3,\ m_2 = 3.4,\ m_3 = 0.9$

$k_1 = 19.0,\ k_2 = 45.0,\ k_3 = 12.0,\ c = 0.8$

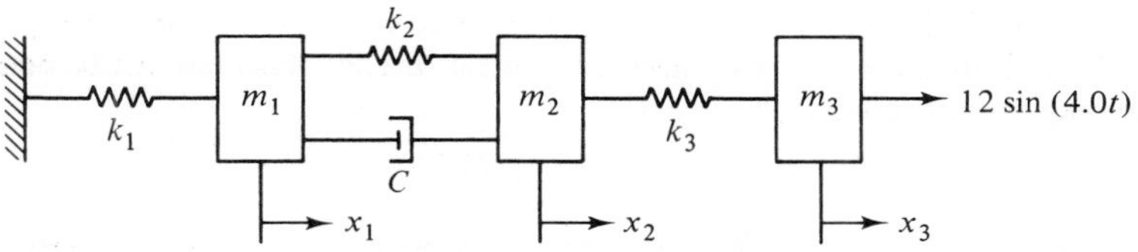

Fig. P3.6

Answer:

At TIME = 2.0, $x_1 = -0.2075$, $x_2 = -0.2707$, $x_3 = 0.4517$

7 The gas pressure acting across a small caliber bullet varies with the distance traveled by the bullet down the barrel of the gun.

The following table gives this pressure-distance relationship for a 26 in. barrel.

X, in.	Pressure lb/in.^{2}
0	5,500
1.0	13,000
2.0	22,500
3.0	27,700
4.0	30,200
5.0	29,000
6.0	27,500
8.0	22,000
10.0	16,000
12.0	11,100
15	6,200
18	4,100
21	3,750
26	3,540

Pressure

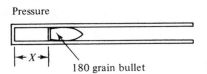

180 grain bullet   **Fig. P3.7**

Assume the diameter of the bullet is 0.300 in. and weighs 180 gr (0.0234 lb). The frictional drag on the barrel is to be considered small compared to the gas force.

Use Newton's second law to find the time required for the bullet to leave the barrel and the muzzle velocity. Compare answers using the NLFGEN and AFGEN arbitrary function generators.

$$m\ddot{x} = PA$$

Answers:

$$\text{Using AFGEN} \quad v = 27546 \text{ in./sec}$$
$$\text{Using NLFGEN} \quad v = 27512 \text{ in./sec}$$

8 An automobile costing \$4000 is financed for 36 months with monthly payments of \$130.50. The annual effective compound interest rate for this loan can be determined by solving the expression

$$I_{\text{annual}} = (1 + i)^{12} - 1$$

where I_{annual} = annual effective compound interest rate expressed in decimal form (not per cent)

i = interest rate per month expressed in decimal form

The solution for i (rate per month) can be found from

$$a_{n/i} = \frac{(1 + i)^n - 1}{i(1 + i)^n}$$

where in this case

$$a_{n/i} = \frac{4000}{130.50} = 30.651$$

and $n = 36$. An implicit expression for i can therefore be written as

$$i = \frac{(1 + i)^n - 1}{30.651(1 + i)^n}$$

Develop a CSMP program using the IMPL function to solve for i. Include a statement in your program for finding I_{annual}. Use PRINT to give the values of I_{annual} and i. What effect does the initial guess for i have on the solution for i? Use $i = 0.01$ and $i = 5.0$ as initial values and error = 0.00001.

Answers:

For i (initial) = 0.01, i = .0089613, I_{annual} = .11298
For i (initial) = 5.0, same answers

9 Use the multiple run capability of a PARAMETER card to solve Mathieu's equation for the following values of a

$$a = 0.5, 1.5, 3.0, 5.0$$
$$b = 2.0, \omega = 3.0$$

Mathieu's equation is

$$\frac{d^2y}{dt^2} + (a - 2b \cos \omega t)y = 0$$

$$y(0) = \dot{y}(0) = 2.0$$

Answers:

At TIME = 1.0		
$y = 4.979$	$a = 0.5$
$y = 3.669$	$a = 1.5$
$y = 2.046$	$a = 3.0$
$y = 0.4326$	$a = 5.0$

10 The figure below shows a tapered beam with an axial load of 10,000 lb and a uniform load of 72 lb/in. The differential equation describing the deflection of the beam is

$$EI\frac{d^2y}{dx^2} = -M \qquad M = \text{moment due to uniform lead} + p_y$$

Using the **CALL RERUN** statement, find the deflection of the beam.

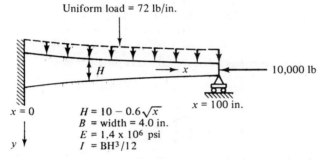

Uniform load = 72 lb/in.

H

x

10,000 lb

$x = 0$

$x = 100$ in.

$H = 10 - 0.6\sqrt{x}$
$B = \text{width} = 4.0$ in.
$E = 1.4 \times 10^6$ psi
$I = BH^3/12$

y

Fig. P3.10

Answer:

$$Y_{max} = 0.388 \text{ in.} \qquad \text{at } X = 66 \text{ in.}$$

11 The acceleration indicated by an accelerometer is given by the following formula.

$$A = \omega_n^2 z$$

The equation of motion of the seismic mass is

$$\ddot{z} + 2\zeta\omega_n\dot{z} + \omega_n^2 z = \ddot{y}$$

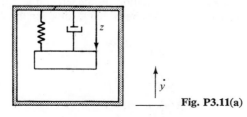

z

$\dot{y}$

Fig. P3.11(a)

Using the multiple run capability of the PARAMETER and END cards, find the indicated acceleration reading for the four combinations of natural frequencies and damping ratios. The input acceleration is shown below.

1. $\omega_n = 10$ rad/sec $\zeta = 0.1$
2. $\omega_n = 10$ rad/sec $\zeta = 0.7$
3. $\omega_n = 100$ rad/sec $\zeta = 0.1$
4. $\omega_n = 100$ rad/sec $\zeta = 0.7$

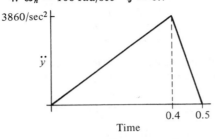

Fig. P3.11(b)

Answers:

Case 1 At TIME $= 0.4$, $A = 4054.0$
Case 2 At TIME $= 0.4$, $A = 2430.0$
Case 3 At TIME $= 0.4$, $A = 3839.0$
Case 4 At TIME $= 0.4$, $A = 3725.0$

12 A steel cantilever beam has a distributed load as shown below.

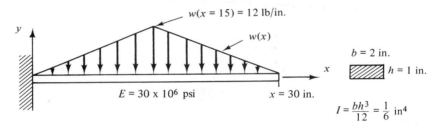

Fig. P3.12

The equation which describes the shape is the following fourth order differential equation.

$$\frac{d^4y}{dx^4} = -\frac{w(x)}{EI}$$

$w(x) =$ density of the distributed load

Using the following boundary condition on the left end and the AFGEN function to represent the density of the distributed load, solve for the deflected shape of the beam.

$$y(0) = \frac{dy(0)}{dx} = 0$$

$$\frac{d^2y(0)}{dx^2} = -\frac{2700}{EI}$$

$$\frac{d^3y(0)}{dx^3} = \frac{180}{EI}$$

Answer:

$$\text{At } x = 30, y = -0.111 \text{ in.}$$

13 The approximate dimensions of the cross-section of a railroad rail are given by the following table.

y	t
0	6
0.5	6
1	1.7
1.3	1.0
3.0	0.65
4.7	0.8
4.9	0.95
5.0	1.2
5.3	2.9
6.4	2.9
6.55	1.0

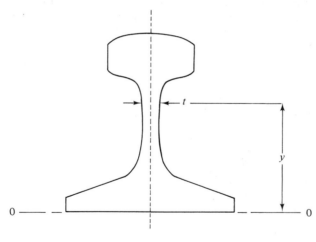

Fig. P3.13

Use the AFGEN function to approximate the shape of the rail and find the center of mass and the moment of inertia about line O–O. The moment of inertia is given by the following formula.

$$I = \int y^2 t \, dy$$

Answer:

$$y_{cg} = 2.87 \text{ in.} \qquad I = 171.9 \text{ in.}^4$$

14 The transfer function of a system is given by

$$\frac{Y(s)}{U(s)} = \frac{12}{(s+2)(s+6)}$$

A unit-step input is applied to the system. It is desired to sample the output signal, $y(t)$, every 0.04 sec as shown in the diagram below.

$$\frac{U}{\text{Step input}} \longrightarrow \boxed{\frac{12}{(s+2)(s+6)}} \longrightarrow \frac{Y}{T = 0.04}$$

Write a CSMP program which outputs the sampled signal. Use the **IMPULS** function for modeling the sampler.

Answer:

$y = 0.76307$ at TIME $= 0.72$ using variable-step integration.

15 The following equations describe the motion of a top spinning about a frictionless pivot point. The only external torque is due to gravity.

$$\ddot{\theta} = \frac{I'\dot{\psi}^2 \sin\theta \cos\theta - I(\dot{\phi} + \dot{\psi}\cos\theta)\dot{\psi}\sin\theta + WD\sin\theta}{I'}$$

$$\ddot{\psi} = \frac{-2I'\dot{\theta}\dot{\psi}\cos\theta + I(\dot{\phi} + \dot{\psi}\cos\theta)\dot{\theta}}{I'\sin\theta}$$

$$\ddot{\phi} = \dot{\psi}\dot{\theta}\sin\theta - \ddot{\psi}\cos\theta$$

I = moment of inertia of top about spin axis = 0.001 lb-in-sec^2
I' = moment of inertia of top about a line perpendicular to the spin axis and through the pivot point = 0.04 lb-in-sec^2
W = weight = 0.8 lb
D = 2.0 in.

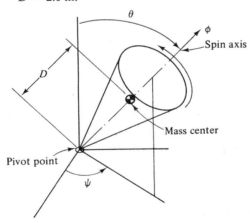

Fig. P3.15

Find the motion of the top for the following initial conditions. Note the singularity at $\theta = 0$.

$\psi = \dot{\psi} = 0$, $\dot{\phi} = 500$ radians/sec, $\theta = 1.0$ radians, $\dot{\theta} = 0.5$ radians/sec

Answer:

At TIME = 1.0 sec, $\theta = 63.83°$, $\psi = 200.55°$, $\dot{\phi} = 499.31$ rad/sec

16 The difference equation for a low pass filter can be expressed as

$$y(kT) = (1 - \alpha)x(kT) + \alpha y[(k-1)T]$$

for α less than 1. For this case select $\alpha = 0.8$ and $T = 0.1$. The input signal, $x(t)$, is given by

$$x(t) = t + 0.5 \sin(6\pi t)$$

Write a CSMP program which (a) gives the input signal $x(t)$ every 0.1 sec; and (b) gives the filtered signal $y(t)$ every 0.1 sec.†

Answer:

$$\text{At TIME} = 4.1, \; x(t) = 4.5755, \; y(t) = 3.7666$$

17 A classical problem in the calculus of variation is to determine the path of quickest descent–the brachistochoe problem. This problem involves finding the curve between two points that a sliding particle would descend in the minimum time. The exact curve is a cycloid.

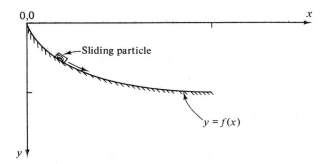

Fig. P3.17

The time required for a mass to slide down a curve under uniform gravity is given by the following expression.

$$t = \frac{1}{\sqrt{2g}} \int_0^x \sqrt{\frac{1 + y'^2}{y}} \, dx$$

Using Euler's equation, the above integral can be minimized by the solution to the following differential equations

$$y' = \sqrt{\frac{c}{y} - 1}$$

The value of $c = 30.32$ will yield a solution that will pass through the following point: $x = 25$, $y = 25.9$. Using this value of c, solve the above equation to determine the shape of the curve. Note that in integrating the above expression for y', it is necessary to add a small number (10^{-6}) to y to avoid dividing by zero at the start of the integration. Also, it is necessary to use a small step size at the beginning of the integration. Use the CONTINUE card to increase the step size of a fixed-step size integration method (RKSFX) after $x = 0.5$.

Answer:

$$\text{At } x = 25, \; y = 26.01$$

†See Problem 18, Chap. 4 for additional comments.

18 Consider the first-order differential equation given by

$$\frac{dy(t)}{dt} + 3y(t) = u(t)$$

$$y(0) = 0$$

where $u(t)$ is a unit-step defined to be 1 for t greater than or equal to zero and zero for t less than zero. One approach for numerical solution to this equation is to use the Tustin approximation for differentiation. In such case

$$\frac{dy(t)}{dt} \longrightarrow \frac{2}{T}\left[\frac{z-1}{z+1}\right]Y(z)$$

where T is the time between numerical iterations and z is the z-transform variable. Using this approximetion one writes

$$\frac{2}{T}\left[\frac{z-1}{z+1}\right]Y(z) + 3Y(z) = U(z)$$

which yields

$$\frac{Y(z)}{U(z)} = K_1 \frac{(1+z^{-1})}{(1+K_2 z^{-1})}$$

$$K_1 = \frac{T}{2+3T}, \qquad K_2 = \frac{3T-2}{3T+2}$$

Using the left shifting property of z-transforms gives the difference equation

$$y(kT) = K_1[u(kT) + u((k-1)T] - K_2 y((k-1)T)$$

(a) Solve the original differential equation using the standard INTGRL function of CSMP. Use trapezoidal integration (i.e., METHOD TRAPZ) with DELT = 0.1.
(b) Write a CSMP program which solves the difference equation. Use $T = 0.1$. Compare the results with that of part (a).

Answers:
   At TIME = 1.0, $y(t)$ = 0.31578 [part (a)], $y(t)$ = 0.31711 [part (b)]

19 The steady-state current for a 60 hertz-120 volt input to the circuit below is given by the following expression.

$$i = 120\left[\frac{R_2^2 + L^2\omega^2}{L^2\omega^2(R_1+R_2)^2 + (R_1 R_2)^2}\right]^{1/2} \qquad \omega = 120\pi \text{ rad/sec}$$

The phase angle of the current can be expressed as

$$\phi = \tan^{-1}\left(\frac{L\omega}{R_2}\right) - \tan^{-1}\left[\frac{L\omega(R_1+R_2)}{R_1 R_2}\right]$$

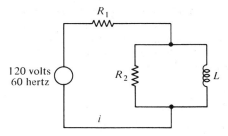

Fig. P3.19

The values of the resistances and inductance with the appropriate tolerances and standard deviations are:

$$R_1 = 10 \text{ ohms } (\pm 10\%), \qquad \sigma_{R_1} = 0.33 \text{ ohms}$$
$$R_2 = 60 \text{ ohms } (\pm 20\%), \qquad \sigma_{R_2} = 4.0 \text{ ohms}$$
$$L = 0.18 \text{ henries } (\pm 20\%), \quad \sigma_L = 0.012 \text{ henries}$$

Use the GAUSS function to calculate the current and phase angle for gaussian distributed random values of R_1, R_2, and L. Find the average and standard deviation of both the current and phase angle for 300 different combinations of parameters.

An expression for the standard deviation is:

$$\sigma_x = \left[\frac{n \sum_{i=1}^{n} x_i^2 - \left(\sum_{i=1}^{n} x_i \right)^2}{n(n-1)} \right]^{1/2}$$

Answer:

$$i_{\text{ave}} = 2.294 \text{ amps}, \quad \phi_{\text{ave}} = -34.29°$$
$$\sigma_i = 0.104 \text{ amps} \qquad \sigma_\phi = 2.264°$$

20 The purpose of this problem is to illustrate how a macro is invoked within another macro. Consider the general system diagram shown in Fig. P3.20-1. Both subsystem *A* and subsystem *B* can be described by a state equation of the form

$$\begin{bmatrix} \dot{x}_1(t) \\ \dot{x}_2(t) \end{bmatrix} = \begin{bmatrix} a_{11} & a_{12} \\ a_{21} & a_{22} \end{bmatrix} \begin{bmatrix} x_1(t) \\ x_2(t) \end{bmatrix} + \begin{bmatrix} b_1 \\ b_2 \end{bmatrix} r(t)$$

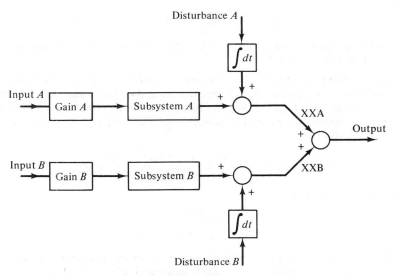

Fig. P3.20-1

where $x_1(0) = \text{IC1}$ and $x_2(0) = \text{IC2}$. The program in Fig. P3.20-2 is a proposed method for writing a single MACRO to describe both subsystems *A* and *B* and a single MACRO to describe both **XXA** and **XXB**. Verify that the program is correct.

```
MACRO Y = STATE(A11,A12,A21,A22,B1,B2,R,IC1,IC2)
  X1DOT = A11*X1 + A12*X2 + B1*R
  X2DOT = A21*X1 + A22*X2 + B2*R
  X1 = INTGRL(IC1,X1DOT)
  X2 = INTGRL(IC2,X2DOT)
  Y = X1
ENDMAC

* SECOND MACRO FOLLOWS
MACRO OUTX = SIMULA(INPUT,GAIN,DIST,S11,S12,...
       S21,S22,SB1,SB2,SIC1,SIC2)
  XX = STATE(S11,S12,S21,S22,SB1,SB2,GAIN*INPUT,SIC1,SIC2)
  QQ = INTGRL(0.0,DIST)
  OUTX = XX + QQ
ENDMAC

* PARAMETERS FOR THE SUBSYSTEMS FOLLOW
PARAMETER SAA11 = 0.0, SAA12 = 1.0, SAA21 = -2.0, SAA22 = -3.0,...
          SAB1 = 0.0, SAB2 = 1.4, SAGAIN = 0.5, SAIC1 = 0.0,SAIC2 = 0.0
PARAMETER SBA11 = 0.0, SBA12 = 1.0, SBA21 = -4.0, SBA22 = -5.0,...
          SBB1 = 0.0, SBB2 = 1.0, SBGAIN = 2.1, SBIC1 = 0.0, SBIC2 = 0.0


INA = STEP(0.0)
DISTA = 0.2*SIN(400.0*TIME)
INB = STEP(0.1)
DISTB = EXP(-0.1*TIME)
XXA = SIMULA(INA,SAGAIN,DISTA,SAA11,SAA12,SAA21,...
      SAA22,SAB1,SAB2,SAIC1,SAIC2)
XXB = SIMULA(INB, SBGAIN, DISTB, SBA11, SBA12, SBA21,...
      SBA22, SBB1, SBB2, SBIC1, SBIC2)
OUTPUT = XXA + XXB
TIMER FINTIM = 2.0, OUTDEL = 0.04
PRTPLT OUTPUT(XXA,XXB)
END
STOP
```

Fig. P3.20-2

Find the response of the system for the inputs, disturbances and system parameters as given in Figure P3.20-2.

21   Many problems in control theory formerly solved by classical methods are now formulated as an optimal control problem using state space analysis. This problem illustrates the use of CSMP for simulation of a time optimal control problem in which the control is expressed in closed form. The state equations for a particular system are given by

$$\begin{bmatrix} \dot{x}_1(t) \\ \dot{x}_2(t) \end{bmatrix} = \begin{bmatrix} 0 & 1 \\ 1 & -\alpha \end{bmatrix} \begin{bmatrix} x_1(t) \\ x_2(t) \end{bmatrix} + \begin{bmatrix} 0 \\ 1 \end{bmatrix} u(t)$$

with initial conditions

$$x_1(0) = x_{10}$$
$$x_2(0) = x_{20}$$

The problem is to find the closed-loop control $u(t)$ such that for any initial set, $x(0)$, the system response time for state $x(t)$ is minimized. Figure P3.21-1 gives a diagram which expresses the essence of this problem.

The solution for the control $u(t)$ can be expressed as[7]

$$u(t) = -\text{sign}\left[x_1 + \frac{1}{\alpha} x_2 - \left(\frac{\text{sign}(x_2)}{\alpha^2}\right)(ln(1 + \alpha |x_2|)) \right]$$

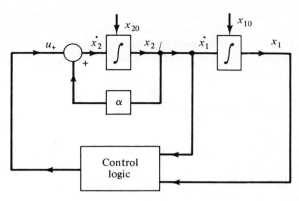

Fig. P3.21-1

which is maximum effort in that $u(t)$ will be either plus or minus one. The solution for the control logic is obtained from the argument of the sign function and is thus given by

$$x_1 + \frac{1}{\alpha} x_2 - \left(\frac{\text{sign}(x_2)}{\alpha^2}\right)(1n(1 + \alpha |x_2|)) = 0$$

This equation defines the switching curve as shown in Fig. P3.21-2.

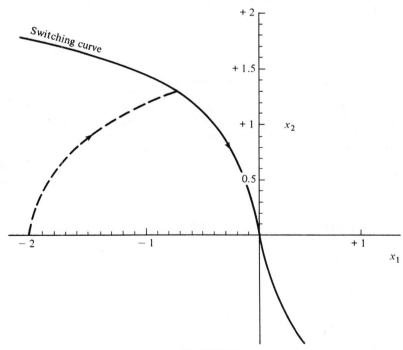

Fig. P3.21-2

The program listing for simulating this system is given in Fig. P3.21-3. The complexity of writing a single statement for $u(t)$ can be avoided by grouping the terms of the switching curve. A conventional FORTRAN statement for the sign of a function is used in the simulation along with standard CSMP simulation statements. The FORTRAN statement

$$y = \text{sign}\ (P_1, P_2)$$

takes the value of P_1 and multiplies by the sign of P_2.

```
PARAM ALPHA = 0.4
INCON ICX1 = -2.0, ICX2 = 0.0
X1DOT = X2
X1 = INTGRL(ICX1, X1DOT)
X2DOT = -ALPHA*X2 + U
X2 = INTGRL(ICX2, X2DOT)
GAMMA = ALOG(1.0 + ALPHA*ABS(X2))
BETA = SIGN(1.0, X2)/(ALPHA*ALPHA)
RHO = X1 + (X2)/ALPHA
U = -SIGN(1.0, RHO - BETA*GAMMA)
TIMER FINTIM = 3.0, OUTDEL = 0.06
PRTPLT X1(U,X2)
LABEL STATE RESPONSE OF X1 AND X2 FOR TIME OPTIMAL
CONTROL
END
STOP
ENDJOB
```

Fig. P3.21-3   Program listing for time optimal control.

The time optimal response, starting from any point in the x_1, x_2 plane should respond under control of maximum effort $u(t)$ until the system trajectory reaches the switching curve. At this point the sign of the control is reversed and the system should follow the ideal switching curve to the origin. Use the program above with ICX1 = -2 and ICX2 = 0 and find the system response for $x_1(t)$ and $x_2(t)$. Use a FINTIM of 3.0 and an OUTDEL of 0.06. Verify that the response intersects the switching curve and then proceeds to the origin.

Answer:

At TIME = 2.1, X1 = -0.36077, X2 = 0.05055, U = -1.0

4

FURTHER APPLICATION
OF CSMP

The types of problems which can be simulated with CSMP are practically unlimited. Consequently, examples in Chaps. 2 and 3 were rather general and specifically selected to illustrate basic features. For the most part, these earlier chapters present those attributes of the program which are required in the majority of simulations. The material in Chap. 4 is somewhat more limited in scope but nevertheless relates to important engineering applications.

One normally views CSMP as a tool for obtaining the transient or dynamic response of a system of equations. In the first part of this chapter, however, two methods are given for obtaining the steady-state frequency response of a system. One of these methods relies on direct application of the system transfer function: the Laplace variable s is replaced by $j\omega$ and the response is obtained as a function of ω. The second method uses transient response information to find the frequency response.

The second topic of the chapter deals with particular block diagram forms of transfer functions and illustrates how these diagrams can be used to formulate the system state-variable equations. An example is given in which the state equations are solved using CSMP.

Real-time digital control has become increasingly important with recent technological development in the minicomputer and microprocessor field. Therefore, a section is included which gives the basic concepts for simulating direct digital control. The material covers sample and hold simulation, recursive digital control algorithms, and the digital three-mode (PID) controller. Anyone involved with real-time digital control will find this material to be helpful in evaluating the performance of proposed computer control systems.

Although CSMP is normally viewed as a software package for simulating continuous systems, the program also has the capability of simulating digital logic. The final section in this chapter presents an introduction to logic simulation and thus provides a method for investigating the performance of purposed configurations without actually building hardware realizations.

Frequency Response

The concept of frequency response is well known in most scientific fields, particularly physics and engineering.[1-4] Basically this concept arises when considering an nth order, linear, differential equation with constant coefficients. In particular, if such an equation is written as

$$\frac{d^n x(t)}{dt^n} + a_{n-1} \frac{d^{n-1} x(t)}{dt^{n-1}} + \cdots + a_1 \frac{dx(t)}{dt} + a_0 x(t)$$

$$= \frac{b_m d^m f(t)}{dt^m} + b_{m-1} \frac{d^{m-1} f(t)}{dt^{m-1}} + \cdots + b_1 \frac{df(t)}{dt} + b_0 f(t) \qquad (4.1)$$

and $f(t) = F \sin \omega t$, the steady-state solution for $x(t)$ can be expressed as

$$x(t) = |X(j\omega)| \sin(\omega t + \phi) \qquad (4.2)$$

where

$$|X(j\omega)| = |A(\omega) + jB(\omega)| \qquad (4.3)$$

and

$$\phi = \tan^{-1}[B(\omega)/A(\omega)] \qquad (4.4)$$

The question arises as to whether CSMP can be effectively used to calculate the frequency response of a system. The answer is a qualified yes. We have seen that if the transfer function or differential equation of a system is known, the response of the system to any forcing function can be determined using INTGRL, REALPL, LEDLAG, CMPXPL, and user-defined transfer function macros. If the forcing function is selected as a sine wave, the response of the system in steady state will yield the $|X(j\omega)|$ and ϕ. This, however, is a rather awkward approach in that CSMP must go through the transient portion of the solution for each frequency before reaching steady state. The transient calculations are wasted computer time with respect to determining the steady-state solution. Thus, even though CSMP can be used to determine the frequency response, excessive computer time is required if the procedure described above is followed. Only in rare instances, such as calculating the response for one or two frequencies, could one justify using this procedure.

While the above method should generally be avoided, there are other approaches to finding the frequency response using CSMP. Two methods are given below. For the first method a macro is developed and the response determined without using integrators. The second method uses an array of integrators and is applicable to finding the frequency response directly from measured time-domain data.

Frequency Response Without Integration

Consider a transfer function of the form

$$G(s) = \frac{N_6 s^5 + N_5 s^4 + N_4 s^3 + N_3 s^2 + N_2 s + N_1}{D_6 s^5 + D_5 s^4 + D_4 s^3 + D_3 s^2 + D_2 s + D_1} \qquad (4.5)$$

Replacing s by $j\omega$, multiplying out, and collecting real and imaginary terms gives

$$G(j\omega) = \frac{(N_5 \omega^4 - N_3 \omega^2 + N_1) + j(N_6 \omega^5 - N_4 \omega^3 + N_2 \omega)}{(D_5 \omega^4 - D_3 \omega^2 + D_1) + j(D_6 \omega^5 - D_4 \omega^3 + D_2 \omega)} \qquad (4.6)$$

Define

$$
\begin{aligned}
X_1 &= (N_5 \omega^4 - N_3 \omega^2 + N_1) \\
X_2 &= (N_6 \omega^5 - N_4 \omega^3 + N_2 \omega) \\
X_3 &= (D_5 \omega^4 - D_3 \omega^2 + D_1) \\
X_4 &= (D_6 \omega^5 - D_4 \omega^3 + D_2 \omega)
\end{aligned}
\qquad (4.7)
$$

giving

$$G(j\omega) = \frac{X_1 + jX_2}{X_3 + jX_4} \qquad (4.8)$$

Rationalizing the expression for $G(j\omega)$ yields

$$G(j\omega) = \frac{(X_1 X_3 + X_2 X_4) + j(X_2 X_3 - X_1 X_4)}{(X_3^2 + X_4^2)} \qquad (4.9)$$

We therefore have from Eq. (4.8)

$$|G(j\omega)| = \left[\frac{X_1^2 + X_2^2}{X_3^2 + X_4^2}\right]^{1/2} \qquad (4.10)$$

and from (4.9)

$$\phi = \underline{/G(j\omega)} = \tan^{-1}\left[\frac{X_2 X_3 - X_1 X_4}{X_1 X_3 + X_2 X_4}\right] \qquad (4.11)$$

To compute the frequency response, various values of ω must be used to calculate X_1, X_2, X_3, and X_4.

The values assigned to ω will normally depend upon the intended application. Typically, linear increments of ω are used for Nyquist plots whereas logarithmic increments are used for Bode diagrams. The frequency increments will be equally spaced on a log scale provided,

$$\omega_i = \omega_{min}(10)^{i\Delta} \qquad i = 0, 1, 2, \ldots, m \qquad (4.12)$$

where m = the total number of steps over the frequency range

     ω_{min} = starting value of ω

     Δ = 1/(the number of desired points per decade)

     ω_i = ω for the ith point

The natural stepping variable of CSMP is the program variable TIME. The equations for the frequency response developed above are, however, not functions of time. Nevertheless, the index $i\Delta$ can be replaced by TIME if TIME is incremented by 0, Δ, 2Δ, 3Δ, ..., $m\Delta$. When integration is not specified in CSMP, the increments of TIME can be controlled by the value of DELT given on the TIMER card. The following example illustrates how these observations may be used for obtaining the frequency response.

Example 4.1

The transfer function of a system is given by

$$G(s) = \frac{25}{s^2 + 3s + 25} \tag{4.13}$$

The frequency response is desired over the range $\omega = 0.1$ to $\omega = 10$ radians/sec with 20 equally spaced log steps per decade.

A program for this purpose is given in Fig. 4.1. In this program a procedural macro is used to calculate $|G(j\omega)|$ and $\underline{/G(j\omega)}$. The coefficients of the transfer function are placed into the program using the STORAGE and TABLE features. Frequency is incremented by the equation

$$\text{OMEGA} = \text{WMIN}*(10.0**(\text{TIME})) \tag{4.14}$$

```
MACRO FASE, MAG = FREQ(N,D,OMEGA)
PROCEDURAL
    X1 = N(5)*(OMEGA**4)-N(3)*(OMEGA**2)+N(1)
    X2 =   N(6)*(OMEGA**5)-N(4)*(OMEGA**3)+N(2)*OMEGA
    X3 = D(5)*(OMEGA**4)-D(3)*(OMEGA**2)+D(1)
    X4 =   D(6)*(OMEGA**5)-D(4)*(OMEGA**3)+D(2)*OMEGA
    MAG = (SQRT(X1*X1+X2*X2))/(SQRT(X3*X3+X4*X4))
    CONV = 57.29578
    XREAL = X1*X3 + X2*X4
    XIMAG = X2*X3 - X1*X4
    FASE = (ATAN2(XIMAG,XREAL))*CONV
    IF(FASE)2, 2, 3
3    FASE = FASE - 360.0
2    CONTINUE
ENDMAC
INITIAL
     PARAM WMIN = 0.1, WMAX = 10.0
DYNAMIC
STORAGE  NUM(6), DEN(6)
TABLE  NUM(1-6) = 25.0, 5*0.0,...
       DEN(1-6) = 25.0, 3.0, 1.0, 3*0.0
NOSORT
OMEGA = WMIN*(10.0**(TIME))
PHASE, VALUE = FREQ(NUM,DEN,OMEGA)
FINISH OMEGA = WMAX
*   SET OUTDEL EQUAL TO ONE OVER THE NUMBER OF STEPS PER DECADE
TIMER FINTIM =.4.0, OUTDEL = 0.05, DELT = 0.05
PRTPLT VALUE(OMEGA,PHASE), PHASE(OMEGA,VALUE)
LABEL  OUTPUT FOR EXAMPLE 4-1
END
STOP
ENDJOB
```

Fig. 4.1 Frequency response program listing using a procedural macro.

The value of TIME is incremented as the smallest value of DELT and OUTDEL. Termination of OMEGA is governed by FINTIM or the finish statement, FINISH = WMAX. One must be careful in specifying FINTIM in this case. When TIME = FINTIM, OMEGA will be $(\omega_{min})10^{FINTIM}$. For FINTIM = 4 the maximum OMEGA is 10,000 times ω_{min} which gives 1000 for $\omega_{min} = 0.1$. Obviously, in this case the FINISH statement will terminate the run.

The form of output can be selected by the user. The customary Bode diagram will be obtained if printer-plots of magnitude and phase are requested. As shown on the output in Fig. 4.2, TIME is still listed in the first column and should be ignored except to observe that its increments vary according to the value of OUTDEL. The actual value of OMEGA can be listed on the plots by specifying it as an argument of the requested PRTPLT variable. For example

PRTPLT VALUE (OMEGA)

The program was written for any transfer function with up to a maximum of a fifth-order numerator and denominator. The macro could be extended to a more general case, for example a tenth-order numerator and denominator, but at the sacrifice of using more CPU time.

Frequency Response Using Array Integrators

A transfer function is defined as the ratio of the Laplace transform of the output to the Laplace transform of the input when the system has zero initial conditions. As noted in the previous section, a transfer function can be defined only for linear, time-invariant, systems. The frequency-response function can be determined from the transfer function by replacing s with $j\omega$. We can then write

$$G(j\omega) = \frac{X(j\omega)}{F(j\omega)} = \frac{\int_{-\infty}^{+\infty} x(t)e^{-j\omega t}\, dt}{\int_{-\infty}^{+\infty} f(t)e^{-j\omega t}\, dt} \tag{4.15}$$

where $X(j\omega)$ and $F(j\omega)$ are the Fourier transforms of $x(t)$ and $f(t)$ respectively. These integrals are defined provided they are absolutely convergent. This is shown by

$$\int_{-\infty}^{+\infty} |x(t)e^{-j\omega t}|\, dt = \int_{-\infty}^{+\infty} |x(t)|\, dt < \infty \tag{4.16}$$

and similarly for the integral involving $f(t)$. If $x(t)$ and $f(t)$ are zero for $t < 0$, Eq. (4.15) becomes

$$G(j\omega) = \frac{\int_{0}^{\infty} x(t)e^{-j\omega t}\, dt}{\int_{0}^{\infty} f(t)e^{-j\omega t}\, dt} \tag{4.17}$$

Evaluating these integrals gives another method for finding the frequency response. From the standpoint of a numerical integration one would never attempt to evaluate the integral over the range $(0, \infty)$. Thus, from practical considerations, the

OUTPUT FOR EXAMPLE 4-1

PAGE   1

VALUE VERSUS TIME

MINIMUM
2.3506E-01

MAXIMUM
1.7459E 00

TIME	VALUE	OMEGA	PHASE
0.0	1.0003E 00	1.0000E-01	-6.8779E-01
5.0000E-02	1.0004E 00	1.1220E-01	-7.7178E-01
1.0000E-01	1.0005E 00	1.2589E-01	-8.6606E-01
1.5000E-01	1.0007E 00	1.4125E-01	-9.7187E-01
2.0000E-01	1.0008E 00	1.5849E-01	-1.0907E 00
2.5000E-01	1.0010E 00	1.7783E-01	-1.2240E 00
3.0000E-01	1.0013E 00	1.9953E-01	-1.3738E 00
3.5000E-01	1.0016E 00	2.2387E-01	-1.5419E 00
4.0000E-01	1.0021E 00	2.5119E-01	-1.7309E 00
4.5000E-01	1.0026E 00	2.8184E-01	-1.9432E 00
5.0000E-01	1.0033E 00	3.1623E-01	-2.1819E 00
5.5000E-01	1.0041E 00	3.5481E-01	-2.4504E 00
6.0000E-01	1.0052E 00	3.9811E-01	-2.7525E 00
6.5000E-01	1.0066E 00	4.4668E-01	-3.0929E 00
7.0000E-01	1.0083E 00	5.0119E-01	-3.4766E 00
7.5000E-01	1.0105E 00	5.6234E-01	-3.9098E 00
8.0000E-01	1.0132E 00	6.3096E-01	-4.3997E 00
8.5000E-01	1.0166E 00	7.0794E-01	-4.9546E 00
9.0000E-01	1.0210E 00	7.9433E-01	-5.5850E 00
9.5000E-01	1.0266E 00	8.9125E-01	-6.3033E 00
1.0000E 00	1.0336E 00	1.0000E 00	-7.1250E 00
1.0500E 00	1.0426E 00	1.1220E 00	-8.0697E 00
1.1000E 00	1.0541E 00	1.2589E 00	-9.1626E 00
1.1500E 00	1.0688E 00	1.4125E 00	-1.0437E 01
1.2000E 00	1.0877E 00	1.5849E 00	-1.1938E 01
1.2500E 00	1.1121E 00	1.7783E 00	-1.3728E 01
1.3000E 00	1.1439E 00	1.9953E 00	-1.5896E 01
1.3500E 00	1.1856E 00	2.2387E 00	-1.8573E 01
1.4000E 00	1.2405E 00	2.5119E 00	-2.1958E 01
1.4500E 00	1.3122E 00	2.8184E 00	-2.6368E 01
1.5000E 00	1.4086E 00	3.1623E 00	-3.2311E 01
1.5500E 00	1.5290E 00	3.5481E 00	-4.0618E 01
1.6000E 00	1.6616E 00	3.9811E 00	-5.2539E 01
1.6500E 00	1.7459E 00	4.4668E 00	-6.9360E 01
1.7000E 00	1.6627E 00	5.0118E 00	-9.0452E 01
1.7500E 00	1.3794E 00	5.6234E 00	-1.1143E 02
1.8000E 00	1.0402E 00	6.3095E 00	-1.2804E 02
1.8500E 00	7.6003E-01	7.0794E 00	-1.3978E 02
1.9000E 00	5.5637E-01	7.9432E 00	-1.4797E 02
1.9500E 00	4.1224E-01	8.9125E 00	-1.5384E 02
2.0000E 00	3.0950E-01	9.9999E 00	-1.5820E 02
2.0500E 00	2.3506E-01	1.1220E 01	-1.6155E 02

Fig. 4.2 Printer-plot of the frequency response amplitude using log ω steps.

198

range must be limited to a finite practical upper bound, t_f. The expression for $G(j\omega)$ can be viewed as

$$G(j\omega) = \frac{\int_0^{t_f} x(t)e^{-j\omega t}\,dt + \int_{t_f}^{\infty} x(t)e^{-j\omega t}\,dt}{\int_0^{t_f} f(t)e^{-j\omega}\,dt + \int_{t_f}^{\infty} f(t)e^{-j\omega t}\,dt} \qquad (4.18)$$

If the integration process is terminated after t_f, error in $G(j\omega)$ will result according to the remaining integrals evaluated from t_f to ∞.[5]

Let the input to a system be a pulse of duration $t_f/2$ where the t_f selected is large enough to cover the range of the transient response shown in Fig. 4.3. The

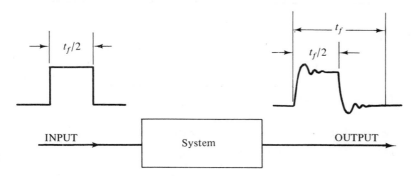

Fig. 4.3 Typical time domain signals for determining frequency response.

input signal is not required to be an ideal pulse but should contain sufficiently high harmonics to cover the frequency range of interest for $G(j\omega)$. $G(jw)$ can be written as

$$G(j\omega) = \frac{\int_0^{t_f} x(t)\cos(\omega t)\,dt - j\int_0^{t_f} x(t)\sin(\omega t)\,dt}{\int_0^{t_f} f(t)\cos(\omega t)\,dt - j\int_0^{t_f} f(t)\sin(\omega t)\,dt} \qquad (4.19)$$

Suppose the frequency response is desired at the frequency points

$$\omega_i = \omega_1, \omega_2, \omega_3, \ldots, \omega_{19}, \omega_{20}, \omega_{21}$$

Determining $G(j\omega_i)$ requires that 84 integrals be evaluated. In particular, if

$$G(j\omega) = \frac{A(j\omega_i) - jB(j\omega_i)}{C(j\omega_i) - jD(j\omega_i)} \qquad (4.20)$$

then

$$A(j\omega_i) = \int_0^{t_f} x(t)\cos(\omega_i t)\,dt \qquad i = 1, 2, \ldots, 21 \qquad (4.21)$$

and similar expressions prevail for $B(j\omega_i)$, $C(j\omega_i)$, and $D(j\omega_i)$. Observe that

$$A(j\omega_i) = \text{INTGRL}[IC, x(t)\cos(\omega_i t)] \qquad (4.22)$$

and thus 21 integrals of this form, one for each ω_i, could be programmed.

A series of integrals of this form can best be handled using the integrator array feature of CSMP. This type of integration was presented earlier in Example 3.3. The following example will show how these concepts can be programmed in CSMP for finding the frequency response.†

Example 4.2

Find the pulse response and frequency response for the system shown in Fig. 4.4.

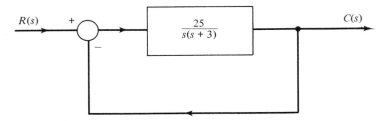

Fig. 4.4   System diagram for Example 4.2.

We observe that the closed-loop transfer function is given by

$$\frac{C(s)}{R(s)} = \frac{25}{s^2 + 3s + 25} \tag{4.23}$$

which is identical to the transfer function in Example 4.1.

The first problem to resolve is the width of the pulse test function. Comparing the above transfer function to the standard second-order form, $s^2 + 2\zeta\omega_n s + \omega_n^2$, gives $\omega_n = 5$ and $\zeta = 0.3$. The $\pm 1.8\%$ settling time is $t_s = 4/\zeta\omega_n = 2.67$ sec. To insure small error for $G(j\omega)$ we select a pulse width of 4.2 sec.

If we now apply a pulse R given by

$$\begin{aligned} R &= Q1 - Q2 \\ Q1 &= STEP(0.0) \\ Q2 &= STEP(4.2) \end{aligned} \tag{4.24}$$

then $c(t) = x(t)$ will be the response of the system with $R = r(t) = f(t)$ as the input.

An ordinary CSMP program can be written for the system in Fig. 4.4 to find $c(t)$. This $c(t)$ plays the role of $x(t)$ in the earlier mathematical discussion.

A program for finding the pulse response and the frequency response from $\omega = 0.1$ to $\omega = 10$ radians/sec at ten equal log steps per decade is shown in Fig. 4.5. The following key points are noted.

Frequency W(I) is calculated in the initial segment using a conventional DO loop. The program determines the pulse response, XX, of the system. The time-varying XX is passed on to a procedure function where the time-varying XA(I), XB(I), XC(I), and XD(I) (I = 1, 2, . . . , 21) are determined. The procedure function returns XA(I), XB(I), XC(I), and XD(I) to the arguments of the array integrators A(I), B(I), C(I), and D(I).

†Slight modifications are required for using the array integrator in CSMP III. See Chap. 5.

```
*     ILLUSTRATION OF FREQUENCY RESPONSE USING ARRAY INTEGRATION
/        DIMENSION A(21),B(21),C(21),D(21),AIC(21),BIC(21),CIC(21),DIC(21),
/     $ XA(21),XB(21),XC(21),XD(21)
/        EQUIVALENCE (A1,A(1)),(B1,B(1)),(C1,C(1)),(D1,D(1)),
/     $(AIC1,AIC(1)),(BIC1,BIC(1)),(CIC1,CIC(1)),(DIC1,DIC(1)),
/     $ (XA1,XA(1)),(XB1,XB(1)),(XC1,XC(1)),(XD1,XD(1))
FIXED I,ITOTAL
STORAGE W(21), GAIN(21), PHASE(21)
**


INITIAL
PARAMETER XDECAD = 10.0, NUMDEC = 2.0, WMIN = 0.1
NOSORT
ITOTAL = NUMDEC*XDECAD + 1.0
X = 0.0
      DO 20 I = 1, ITOTAL
      X = 1.0/XDECAD + X
   20 W(I) = WMIN*(10.0**(X))
**


DYNAMIC
Q1 = STEP(0.0)
Q2 = STEP(4.2)
R = Q1 - Q2
E = R - XX
X1 = (25./3.0)*E
X2 = REALPL(0.0,1.0/3.0,X1)
XX = INTGRL(0.0,X2)
A1 = INTGRL(AIC1,XA1,21)
B1 =  INTGRL(BIC1,XB1,21)
C1 = INTGRL(CIC1,XC1,21)
D1 =  INTGRL(DIC1,XD1,21)
PROCEDURE XA1,XB1,XC1,XD1 = FORCE(W,XX,R,ITOTAL)
      DO 1 I = 1,ITOTAL
      XA(I) = XX*COS(W(I)*TIME)
      XB(I) = XX*SIN(W(I)*TIME)
      XC(I) = R*COS(W(I)*TIME)
   1  XD(I) = R*SIN(W(I)*TIME)
ENDPRO
*


TERMINAL
      CONV = 57.29578
      DO 3 I = 1,ITOTAL
      GAIN(I) = (SQRT(A(I)**2 + B(I)**2))/(SQRT(C(I)**2 + D(I)**2))
       PHASE(I) = (ATAN2(-B(I),A(I))-ATAN2(-D(I),C(I)))*CONV
      IF(PHASE(I))9,9,8
   8 PHASE(I) = PHASE(I) - 360.0
   9 CONTINUE
   3 WRITE(6,10) W(I),GAIN(I), PHASE(I)
  10 FORMAT(1H0,8H OMEGA =,E12.5,3X,11HAMPLITUDE =,2X,E12.5
     $,3X,7HPHASE =,2X,F10.5)
**

TIMER FINTIM = 10.0, OUTDEL = 0.2, DELT = 0.001, PRDEL = 0.1
PRTPLT XX
LABEL  OUTPUT FOR EXAMPLE 4-2 ARRAY
END
STOP
ENDJOB
```

Fig. 4.5 CSMP program listing for finding frequency response with integrator array.

In the terminal segment conventional FORTRAN statements are used to calculate

$$\text{GAIN(I)} = \sqrt{\frac{A(I)^2 + B(I)^2}{C(I)^2 + D(I)^2}} \tag{4.25}$$

$$\text{PHASE(I)} = \tan^{-1}\left(\frac{-B(I)}{A(I)}\right) - \tan^{-1}\left(\frac{-D(I)}{C(I)}\right) \tag{4.26}$$

```
OUTPUT FOR EXAMPLE 4-2 ARRAY                                    PAGE   1

                     MINIMUM              XX      VERSUS TIME      MAXIMUM
                    -3.6443E-01                                   1.3647E 00
   TIME        XX       I                                              I
0.0          0.0        ----------+
2.0000E-01   3.8141E-01 -------------------+
4.0000E-01   1.0186E 00 ------------------------------------------+
6.0000E-01   1.3555E 00 --------------------------------------------------+
8.0000E-01   1.2945E 00 -----------------------------------------------+
1.0000E 00   1.0573E 00 -------------------------------------------+
1.2000E 00   8.8749E-01 -------------------------------------+
1.4000E 00   8.7212E-01 ------------------------------------+
1.6000E 00   9.5215E-01 ---------------------------------------+
1.8000E 00   1.0292E 00 ------------------------------------------+
2.0000E 00   1.0513E 00 -------------------------------------------+
2.2000E 00   1.0279E 00 ------------------------------------------+
2.4000E 00   9.9580E-01 ----------------------------------------+
2.6000E 00   9.8107E-01 ---------------------------------------+
2.8000E 00   9.8608E-01 ---------------------------------------+
3.0000E 00   9.9846E-01 ----------------------------------------+
3.2000E 00   1.0063E 00 -----------------------------------------+
3.4000E 00   1.0063E 00 -----------------------------------------+
3.6000E 00   1.0019E 00 ----------------------------------------+
3.8000E 00   9.9819E-01 ----------------------------------------+
4.0000E 00   9.9741E-01 ----------------------------------------+
4.2000E 00   9.9877E-01 ----------------------------------------+
4.4000E 00   6.1897E-01 --------------------------------+
4.6000E 00  -1.7633E-02 ---------+
4.8000E 00  -3.5483E-01 +
5.0000E 00  -2.9445E-01 --+
5.2000E 00  -5.7717E-02 --------+
5.4000E 00   1.1226E-01 --------------+
5.6000E 00   1.2801E-01 ---------------+
5.8000E 00   4.8140E-02 -----------+
6.0000E 00  -2.9031E-02 ---------+
6.2000E 00  -5.1352E-02 --------+
6.4000E 00  -2.8103E-02 ---------+
6.6000E 00   4.1044E-03 ----------+
6.8000E 00   1.8961E-02 -----------+
7.0000E 00   1.4011E-02 ----------+
7.2000E 00   1.6021E-03 ----------+
7.4000E 00  -6.3230E-03 ----------+
7.6000E 00  -6.3044E-03 ----------+
7.8000E 00  -1.9344E-03 ----------+
8.0000E 00   1.8041E-03 ----------+
8.2000E 00   2.6104E-03 ----------+
8.4000E 00   1.2482E-03 ----------+
8.6000E 00  -3.6340E-04 ----------+
8.8000E 00  -9.9750E-04 ----------+
9.0000E 00  -6.5606E-04 ----------+
9.2000E 00  -1.4780E-05 ----------+
9.4000E 00   3.4776E-04 ----------+
9.6000E 00   3.0644E-04 ----------+
9.8000E 00   7.1827E-05 ----------+
1.0000E 01  -1.0674E-04 ----------+
```

Fig. 4.6 Pulse response of the system for Example 4.2.

The program output for GAIN(I) and PHASE(I) are neither printer-plotted nor printed in the usual CSMP form but are listed using FORTRAN Write and Format statements.

The pulse response of the system is given in Fig. 4.6 where we note that the response overshoot is approximatley 36%, as expected for $\zeta = 0.3$. We also note that for all practical purposes the system has reached steady state in 4 sec.

Frequency response information is given in column format in Fig. 4.7. Comparing this output with the results of Example 4.1 (Fig. 4.2) shows excellent agreement. The decibel amplitude could easily be added to the output listing by inserting the statement,

$$\text{DBGAIN(I)} = 20.0*\text{ALOG(GAIN(I))} \tag{4.27}$$

along with slight modifications in the Write and Format statements.

```
        PROBLEM DURATION 0.0        TO   1.0000E 01


VARIABLE        MINIMUM         TIME        MAXIMUM           TIME
XX            -3.6443E-01    4.9000E 00    1.3647E 00      7.0000E-01

   OMEGA = 0.12589E 00    AMPLITUDE =  0.10005E 01    PHASE =    -0.86600

   OMEGA = 0.15849E 00    AMPLITUDE =  0.10008E 01    PHASE =    -1.09073

   OMEGA = 0.19953E 00    AMPLITUDE =  0.10013E 01    PHASE =    -1.37399

   OMEGA = 0.25119E 00    AMPLITUDE =  0.10021E 01    PHASE =    -1.73112

   OMEGA = 0.31623E 00    AMPLITUDE =  0.10033E 01    PHASE =    -2.18208

   OMEGA = 0.39811E 00    AMPLITUDE =  0.10052E 01    PHASE =    -2.75266

   OMEGA = 0.50119E 00    AMPLITUDE =  0.10083E 01    PHASE =    -3.47651

   OMEGA = 0.63096E 00    AMPLITUDE =  0.10132E 01    PHASE =    -4.39891

   OMEGA = 0.79433E 00    AMPLITUDE =  0.10210E 01    PHASE =    -5.58496

   OMEGA = 0.10000E 01    AMPLITUDE =  0.10336E 01    PHASE =    -7.12580

   OMEGA = 0.12589E 01    AMPLITUDE =  0.10540E 01    PHASE =    -9.16152

   OMEGA = 0.15849E 01    AMPLITUDE =  0.10874E 01    PHASE =   -11.93899

   OMEGA = 0.19953E 01    AMPLITUDE =  0.11440E 01    PHASE =   -15.89472

   OMEGA = 0.25119E 01    AMPLITUDE =  0.12405E 01    PHASE =   -21.95506

   OMEGA = 0.31623E 01    AMPLITUDE =  0.14086E 01    PHASE =   -32.29756

   OMEGA = 0.39810E 01    AMPLITUDE =  0.16619E 01    PHASE =   -52.54179

   OMEGA = 0.50118E 01    AMPLITUDE =  0.16627E 01    PHASE =   -90.46185

   OMEGA = 0.63095E 01    AMPLITUDE =  0.10394E 01    PHASE =  -128.03969

   OMEGA = 0.79432E 01    AMPLITUDE =  0.55605E 00    PHASE =  -147.91357

   OMEGA = 0.99999E 01    AMPLITUDE =  0.30965E 00    PHASE =  -158.23978

   OMEGA = 0.12589E 02    AMPLITUDE =  0.17989E 00    PHASE =  -164.20155
```

Fig. 4.7 Frequency response information for Example 4.2 using the integrator array.

```
*      ILLUSTRATION CF FREQUENCY RESPONSE USING ARRAY INTEGRATION
/      DIMENSION A(21),B(21),C(21),D(21),AIC(21),BIC(21),CIC(21),DIC(21),
/    $ XA(21),XB(21),XC(21),XD(21)
/      EQUIVALENCE (A1,A(1)),(B1,B(1)),(C1,C(1)),(D1,D(1)),
/    $(AIC1,AIC(1)),(BIC1,BIC(1)),(CIC1,CIC(1)),(DIC1,DIC(1)),
/    $ (XA1,XA(1)),(XB1,XB(1)),(XC1,XC(1)),(XD1,XD(1))
FIXED I,ITOTAL
STORAGE W(21), GAIN(21), PHASE(21)
FUNCTION  SIGNAL = (0.0,0.0),(.2,.3814),(.4,1.019),(.6,1.355),...
  (.8,1.295),(1.0,1.057),(1.2,.8875),(1.4,.872),(1.6,.952),(1.8,1.03),...
  (2.0,1.051),(2.2,1.028),(2.4,.996),(2.6,.981),(2.8,.986),...
  (3.0,.998),(3.2,1.00),(3.4,1.00),(3.6,1.00),(3.8,1.00),...
  (4.00,1.00),(4.2,.617),(4.4,-.0182),(4.6,-.354),(4.8,-.294),...
  (5.0,-.057),(5.2,.112),(5.4,.128),(5.6,.0478),(5.8,-.029),...
  (6.0,-.051),(6.2,-.0279),(6.4,.004),(6.6,.0189),(6.8,.0139),...
  (7.0,.0015),(7.2,-.0063),(7.4,-.006),(7.6,-.0019),(7.8,.0018),...
  (8.0,.0026),(8.2,.00123),(8.4,-.0004),(8.6,-.0009),(8.8,-.0006),...
  (9.0,0.0),(9.2,0.0),(9.4,0.0),(9.6,0.0),(9.8,0.0),(10.0,0.0)
**


INITIAL
PARAMETER XDECAD = 10.0, NUMDEC = 2.0, WMIN = 0.1
NOSORT
ITOTAL = NUMDEC*XDECAD + 1.0
X = 0.0
     DO 20 I = 1, ITOTAL
     X = 1.0/XDECAD + X
 20 W(I) = WMIN*(10.0**(X))
**


DYNAMIC
Q1 = STEP(0.0)
Q2 = STEP(4.2)
R = Q1 - Q2
XX =  AFGEN(SIGNAL,TIME)
A1 = INTGRL(AIC1,XA1,21)
B1 =  INTGRL(BIC1,XB1,21)
C1 = INTGRL(CIC1,XC1,21)
D1 =  INTGRL(DIC1,XD1,21)
PROCEDURE XA1,XB1,XC1,XD1 = FORCE(W,XX,R,ITOTAL)
     DO 1 I = 1,ITOTAL
     XA(I) = XX*COS(W(I)*TIME)
     XB(I) = XX*SIN(W(I)*TIME)
     XC(I) = R*COS(W(I)*TIME)
  1  XD(I) = R*SIN(W(I)*TIME)
ENDPRO
*
METHCD RKSFX


TERMINAL
TERMINAL
     CONV = 57.29578
     DO 3 I = 1,ITOTAL
     GAIN(I) = (SQRT(A(I)**2 + B(I)**2))/(SQRT(C(I)**2 + D(I)**2))
      PHASE(I) = (ATAN2(-B(I),A(I))-ATAN2(-D(I),C(I)))*CONV
     IF(PHASE(I))9,9,8
   8 PHASE(I) = PHASE(I) - 360.0
   9 CONTINUE
   3 WRITE(6,10) W(I),GAIN(I), PHASE(I)
  10 FORMAT(1H0,8H OMEGA =,F12.5,3X,11HAMPLITUDE =,2X,E12.5
    $,3X,7HPHASE =,2X,F10.5)
**


TIMER FINTIM = 10.0, OUTDEL = 0.2, DELT = 0.005, PRDEL = 0.1
PRTPLT XX
LABEL  ARRAY INTEGRATION USING NUMERICAL FUNCTION EXAMPLE 4-2
END
STOP
```

Fig. 4.8 CSMP program listing for finding frequency response
from pulse response using the integrator array.

The CPU time for running this program on the 360/65 was approximately 40 sec. The multiple integrations account for the rather large CPU time as compared to Example 4.1 which used only 10 sec for twice as many frequency points. Obviously, if the system transfer function is known, the method of Example 4.1 is preferred over the array integration method.

The real advantage of the array-integrator method exhibits itself in the following light. There are many practical systems where direct frequency response measurements are extremely difficult and uneconomical to obtain. Making such measurements on chemical process plants and metal-rolling mills is practically out of the question. However, the pulse response of many systems can be directly and economically measured. If this time domain data is coupled to the array-integrator program via either AFGEN or NLFGEN then the system frequency response can be obtained.

To illustrate this point, suppose the time response data of Fig. 4.7 is used to define a function.

FUNCTION SIGNAL = (0.0, 0.0), (0.2, 0.3814), . . .

$$\cdot$$
$$\cdot$$
$$\cdot$$

from which the signal **XX** is obtained by:

XX = AFGEN (SIGNAL,TIME)

The program in Fig. 4.8 has incorporated these changes by simply removing the original system-transfer function. The resulting frequency response information is shown in Fig. 4.9. The response compares favorably to the "exact" answers and deviations can be attributed to interpolation inaccuracy.

In summary, this section shows two methods for obtaining a system frequency response using CSMP. The first method applies directly to those systems with known transfer functions. The second method can be applied to systems with known transfer functions and, moreover, has its greatest value in determining the frequency response from pulse-response information as measured directly from the physical system.

State Variable Formulation from Transfer Functions

The occasion often arises when one wishes to express the dynamics of a system in state-variable form rather than the customary transfer function. The state equations for a linear, time-invariant, system can be written as

$$\dot{\mathbf{x}}(t) = A\mathbf{x}(t) + B\mathbf{u}(t) \tag{4.28}$$

$$\mathbf{y}(t) = C\mathbf{x}(t) + D\mathbf{u}(t) \tag{4.29}$$

where $\mathbf{x}(t)$ = n-dimensional column vector called the system state vector,

   $\mathbf{y}(t)$ = q-dimensional column vector called the system output vector,

   $\mathbf{u}(t)$ = m-dimensional column vector called the system input vector,

and A, B, C, and D have appropriate matrix dimensions.

OMEGA = 0.10000E 00	AMPLITUDE =	0.99969E 00	PHASE =	-0.68448
OMEGA = 0.12589E 00	AMPLITUDE =	0.99986E 00	PHASE =	-0.86155
OMEGA = 0.15849E 00	AMPLITUDE =	0.10001E 01	PHASE =	-1.08507
OMEGA = 0.19953E 00	AMPLITUDE =	0.10006E 01	PHASE =	-1.36697
OMEGA = 0.25119E 00	AMPLITUDE =	0.10013E 01	PHASE =	-1.72244
OMEGA = 0.31623E 00	AMPLITUDE =	0.10024E 01	PHASE =	-2.17153
OMEGA = 0.39811E 00	AMPLITUDE =	0.10041E 01	PHASE =	-2.73948
OMEGA = 0.50119E 00	AMPLITUDE =	0.10069E 01	PHASE =	-3.45968
OMEGA = 0.63096E 00	AMPLITUDE =	0.10113E 01	PHASE =	-4.37542
OMEGA = 0.79433E 00	AMPLITUDE =	0.10183E 01	PHASE =	-5.54671
OMEGA = 0.10000E 01	AMPLITUDE =	0.10297E 01	PHASE =	-7.05657
OMEGA = 0.12589E 01	AMPLITUDE =	0.10486E 01	PHASE =	-9.01148
OMEGA = 0.15849E 01	AMPLITUDE =	0.10623E 01	PHASE =	-15.71143
OMEGA = 0.19953E 01	AMPLITUDE =	0.11272E 01	PHASE =	-16.04182
OMEGA = 0.25119E 01	AMPLITUDE =	0.12128E 01	PHASE =	-21.98824
OMEGA = 0.31623E 01	AMPLITUDE =	0.14504E 01	PHASE =	-33.11356
OMEGA = 0.39810E 01	AMPLITUDE =	0.15816E 01	PHASE =	-52.59436
OMEGA = 0.50118E 01	AMPLITUDE =	0.15293E 01	PHASE =	-89.78917
OMEGA = 0.63095E 01	AMPLITUDE =	0.74014E 00	PHASE =	-128.34552
OMEGA = 0.79432E 01	AMPLITUDE =	0.42237E 00	PHASE =	-149.91252
OMEGA = 0.99999E 01	AMPLITUDE =	0.20528E 00	PHASE =	-155.85031

Fig. 4.9 Frequency response information obtained by Fourier transform.

Consider a system represented by the transfer function

$$\frac{C(s)}{R(s)} = \frac{a_2 s^2 + a_1 s + a_0}{s^4 + b_3 s^3 + b_2 s^2 + b_1 s + b_0} \tag{4.30}$$

A block diagram which represents this transfer function is given in Fig. 4.10. The following equations can be written from the diagram.

$$\dot{x}_1(t) = x_2(t)$$
$$\dot{x}_2(t) = x_3(t)$$
$$\dot{x}_3(t) = x_4(t)$$
$$\dot{x}_4(t) = r(t) - b_0 x_1(t) - b_1 x_2(t) - b_2 x_3(t) - b_3 x_4(t)$$

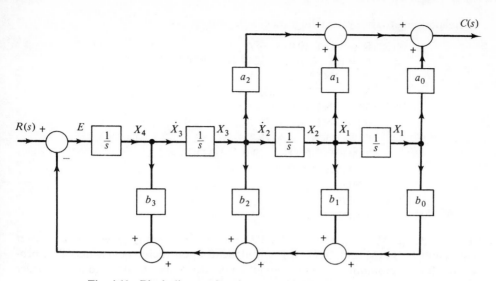

Fig. 4.10 Block diagram for phase variable state equations.

and

$$c(t) = a_0 x_1 + a_1 x_2 + a_2 x_3$$

Expressing these equations in matrix form gives

$$\begin{bmatrix} \dot{x}_1(t) \\ \dot{x}_2(t) \\ \dot{x}_3(t) \\ \dot{x}_4(t) \end{bmatrix} = \begin{bmatrix} 0 & 1 & 0 & 0 \\ 0 & 0 & 1 & 0 \\ 0 & 0 & 0 & 1 \\ -b_0 & -b_1 & -b_2 & -b_3 \end{bmatrix} \begin{bmatrix} x_1(t) \\ x_2(t) \\ x_3(t) \\ x_4(t) \end{bmatrix} + \begin{bmatrix} 0 \\ 0 \\ 0 \\ 1 \end{bmatrix} r(t) \qquad (4.31)$$

and

$$c(t) = [a_0 \quad a_1 \quad a_2 \quad 0] \begin{bmatrix} x_1(t) \\ x_2(t) \\ x_3(t) \\ x_4(t) \end{bmatrix} \qquad (4.32)$$

Comparing with Eqs. (4.28) and (4.29) we have

$$A = \begin{bmatrix} 0 & 1 & 0 & 0 \\ 0 & 0 & 1 & 0 \\ 0 & 0 & 0 & 1 \\ -b_0 & -b_1 & -b_2 & -b_3 \end{bmatrix} \qquad B = \begin{bmatrix} 0 \\ 0 \\ 0 \\ 1 \end{bmatrix} \qquad C = [a_0 \quad a_1 \quad a_2 \quad 0]$$

and $\mathbf{x}(t) = [x_1(t), x_2(t), x_3(t), x_4(t)]^T$, $\mathbf{y}(t) = c(t)$, $\mathbf{u}(t) = r(t)$.

Assuming the input to the system in Fig. 4.10 is a unit step, a CSMP program corresponding to the system diagram can be written as

$$
\begin{aligned}
&R = \text{STEP } (0.0) \\
&E = R - BO*X1 - B1*X2 - B2*X3 - B3*X4 \\
&X4 = \text{INTGRL } (0.0, \ E) \\
&X3 = \text{INTGRL } (0.0, \ X4) \\
&X2 = \text{INTGRL } (0.0, \ X3) \\
&X1 = \text{INTGRL } (0.0, \ X2) \\
&C = AO*X1 + A1*X2 + A2*X3 \\
&\text{TIMER FINTIM} = 3.0, \ \text{OUTDEL} = 0.1 \\
&\text{PRTPLT } X1, \ X2, \ X3, \ X4, \ C \\
&\text{END} \\
&\text{STOP} \\
&\text{ENDJOB}
\end{aligned}
$$

The constants A0, A1, A2, B0, B1, B2, and B3 are usually given on parameter cards and FINTIM and OUTDEL will depend upon these coefficients. This procedure for programming a transfer function can easily be extended to higher-order polynomials. The state variable formulation in this particular form is commonly known as the phase variable representation.[6,7,8]

This illustration shows that a set of state equations can always be written for a system transfer function. The state-representation for a transfer function is not unique. In fact, it can be proven that for a given transfer function there exist infinitely many state representations.

Another method for expressing the transfer function of Eq. (4.30) is given in Fig. 4.11. Using the output of each integration (the $1/s$ terms) as states, one can

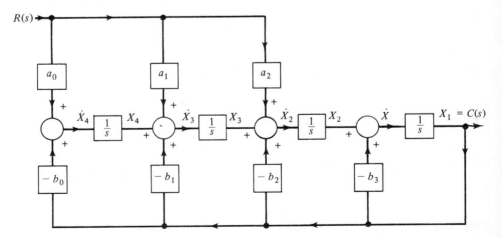

Fig. 4.11   Alternate block diagram form for state variable representation.

easily write a set of equations corresponding to Eqs. (4.28) and (4.29). This is done in an exercise at the end of this chapter.

Digital Control Systems

Digital computers have become an important method for controlling feedback systems. The purpose of this section is to illustrate how CSMP can be used to simulate the digital control process.

A simplified diagram showing the role of the digital computer in a feedback system is given in Fig. 4.12. The analog output of the system is converted to a binary representation by the analog to digital (A/D) converter. The system input must be converted to a compatible binary representation. The computer subtracts the binary output from the binary input to form a binary error signal. The error signal becomes the "input" to a properly selected control algorithm.

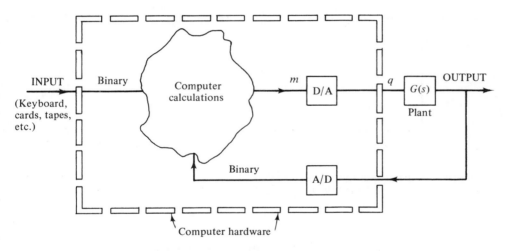

Fig. 4.12   Basic digital feedback control system.

Assume that algorithm calculations are made every T sec to form the binary signal, m, as shown in the diagram. This binary signal is converted to an analog signal by the digital to analog converter (D/A) to produce the signal q which is applied to the system. The design problem is to determine the algorithm which satisfies system specifications. The functions carried out by the computer hardware in Fig. 4.12 can be equivalently represented by the more straightforward model as given in Fig. 4.13. The problem of selecting the control algorithm reduces to one of finding the function $D(z)$ where z is the z-transform variable. The major attention in this section will be given to the simulation problem, since methods for finding $D(z)$ are given in the literature.[9,10,11]

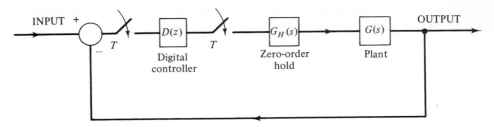

Fig. 4.13   Simplified diagram for digital control.

Example 4.3

The system shown in Fig. 4.14 does not include a controller $D(z)$. It is desired, how-ever, to investigate the effect of the sampling process on the system stability. Ideal sam-pling, followed by a zero-order hold, can be simulated as follows. If the input to the ideal

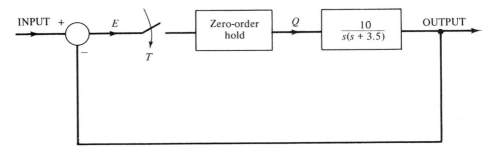

Fig. 4.14   System diagram with sample and hold. (Example 4.3.)

sampler is E, then the output of the zero-order hold can be represented as Q. This process can be simulated using the following CSMP statements.

$$A1 = \text{IMPULS}(0.0, \ T)$$
$$Q = \text{ZHOLD}(A1, \ E)$$

The reader will recall that $A1 = \text{IMPULS}(P1, P2)$ produces a unit-impulse train starting at P1 with pulses occurring every P2 sec. Properties of the zero-order hold are defined in Table 4.1. (Also, see Appendix I)

Table 4.1

Properties of the Zero-Order Hold

CSMP Form	Function	
$Q = \text{ZHOLD}(A1, \ E)$	$Q(t) = E$	$A1 > 0$
	$Q(t) = \text{last output}$	$A1 \leq 0$
	$Q(t) = 0$	$t = \text{initial } t$

Now suppose we desire to find the step response of the system for sampling times of $T = 0.3$ and $T = 0.8$. A simple program for this purpose is given below.

```
PARAM  T = (0.3, 0.8)
INPUT = STEP(0.0)
E = INPUT − OUTPUT
A1 = IMPULS(0.0, T)
Q = ZHOLD(A1, E)
X1 = (10.0/3.5)*Q
X2 = REALPL(0.0, 1.0/3.5, X1)
OUTPUT = INTGRL(0.0, X2)
METHOD RKSFX
TIMER  FINTIM = 6.0, OUTDEL = 0.24, DELT = 0.01
PRTPLT OUTPUT
LABEL STEP RESPONSE OF SYSTEM WITH SAMPLE AND HOLD
END
STOP
ENDJOB
```

A fixed-step integration (RKSFX) is selected for this example to insure that an integration is performed when the sampling switch closes. If a variable-step integration is used, the integration will not generally occur at the sampling time but rather when KEEP = 1. The resulting response will not be representative of uniform sampling every T sec. When using the sample and hold, variable integration can be selected but either OUTDEL or PRDEL should be chosen to be a sub-multiple of the sampling interval. Fixed-step integration is recommended for simulating discrete data systems. This is particularly important when a digital controller is placed in the system.

The responses in Fig. 4.15 show the effect of the sampling process on the system stability. The continuous system response has approximately 12% overshoot while the overshoot becomes larger as T in the sampling process is increased.

Example 4.4

The purpose of this example is to illustrate a procedure for finding the time response of a closed-loop feedback system which contains a digital controller, $D(z)$, and a zero-order hold. The basic block diagram of the system is shown in Fig. 4.16. This system is identical to the one given in Example 4.3 except the gain is raised from 10 to 70 and a digital controller is added to the system. A procedure for selecting the controller is given in reference 11.

First consider the simulation of $D(z)$ followed by the zero-order hold. For this example the particular $D(z)$ is given by

$$D(z) = 0.1 \frac{z - 0.981}{z - 0.998}$$

If this function is represented in a more general form as

$$D(z) = \frac{a_0 + a_1 z^{-1}}{1 + b_1 z^{-1}}$$

we can easily employ the phase variable or rectangular programming methods given in the previous section by replacing s by z. The block diagram for the rectangular form is given in Fig. 4.17. The controller and the zero-order hold can be programmed as a MACRO using the following CSMP statements:

$$\text{MACRO E2} = \text{COMP(SAMTIM, TX, A0, A1, B1, E1)}$$
$$\text{TX} = \text{IMPULS(0.0, SAMTIM)}$$
$$\text{EZ} = \text{TX}*\text{E1}$$
$$\text{Y0} = \text{A0}*\text{EZ} + \text{P1}$$
$$\text{Y1} = \text{A1}*\text{EZ} - \text{B1}*\text{Y0}$$
$$\text{P1} = \text{DELAY(1000, SAMTIM, Y1)}$$
$$\text{E2} = \text{ZHOLD(TX, Y0)}$$
$$\text{ENDMAC}$$

This macro is completely general for any $D(z)$ of the form so prescribed and can easily be extended to higher-order $D(z)$.

A block diagram of the total system is shown in Fig. 4.18 and is used to identify the inputs and outputs of the various blocks within the CSMP program. Figure 4.19 gives the program listing corresponding to the block diagram. Figure 4.20 gives the step response, $c(t)$, of the compensated system. Compare this response to that obtained earlier in Example 4.3. The gain was 10 but has now been raised to 70. Note also that fixed-step integration is used with $\text{DELT} = 0.02$ to insure that integration is performed at the sampling interval of $\text{T} = 0.1$.

While the method for simulating $D(z)$ given in this example is completely general for any digital controller, it is not directly the method followed in programming a control algorithm for real-time digital control. We recall that the operator z^{-1} was simulated by

```
        STEP RESPONSE OF CONTINUOUS SYSTEM TO COMPARE WITH SAMPLE AND HOLD  PAGE   1

                                  MINIMUM           OUTPUT VERSUS TIME            MAXIMUM
                                    0.0                                         1.1215E 00
            TIME          OUTPUT     I                                             I
          0.0           0.0          +
          1.8000E-01    1.2934E-01   -----+
          3.6000E-01    4.0196E-01   --------------------+
          5.4000E-01    6.8711E-01   -----------------------------------+
          7.2000E-01    9.1221E-01   --------------------------------------------------+
          9.0000E-01    1.0526E 00   -----------------------------------------------------------+
          1.0800E 00    1.1151E 00   ---------------------------------------------------------------+
          1.2600E 00    1.1215E 00   ----------------------------------------------------------------+
          1.4400E 00    1.0964E 00   -------------------------------------------------------------+
          1.6200E 00    1.0605E 00   ---------------------------------------------------------+
          1.8000E 00    1.0272E 00   ------------------------------------------------------+
          1.9800E 00    1.0031E 00   ---------------------------------------------------+
          2.1600E 00    9.8956E-01   --------------------------------------------------+
          2.3400E 00    9.8478E-01   -------------------------------------------------+
          2.5200E 00    9.8579E-01   -------------------------------------------------+
          2.7000E 00    9.8966E-01   --------------------------------------------------+
          2.8800E 00    9.9414E-01   ---------------------------------------------------+
          3.0600E 00    9.9789E-01   ---------------------------------------------------+
          3.2400E 00    1.0004E 00   ----------------------------------------------------+
          3.4200E 00    1.0016E 00   ----------------------------------------------------+
          3.6000E 00    1.0019E 00   ----------------------------------------------------+
          3.7800E 00    1.0016E 00   ----------------------------------------------------+
          3.9600E 00    1.0011E 00   ----------------------------------------------------+
          4.1400E 00    1.0005E 00   ----------------------------------------------------+
          4.3200E 00    1.0001E 00   ----------------------------------------------------+
          4.5000E 00    9.9988E-01   ---------------------------------------------------+
          4.6800E 00    9.9977E-01   ---------------------------------------------------+
          4.8600E 00    9.9977E-01   ---------------------------------------------------+
          5.0400E 00    9.9982E-01   ---------------------------------------------------+
          5.2200E 00    9.9989E-01   ---------------------------------------------------+
          5.4000E 00    9.9995E-01   ---------------------------------------------------+
          5.5800E 00    1.0000E 00   ---------------------------------------------------+
          5.7600E 00    1.0000E 00   ---------------------------------------------------+
```

(a)
Continuous
system

Fig. 4.15 System step-response for Example 4.3.

```
                                                              PAGE   1
                          MINIMUM           OUTPUT VERSUS TIME        MAXIMUM
                           0.0              T     = 3.0000E-01        1.8433E 00
          TIME     OUTPUT     I                                       I
          0.0      0.0        +
          2.4000E-01  2.2180E-01  ------+
          4.8000E-01  6.6318E-01  ------------------+
          7.2000E-01  1.0824E 00  ------------------------------+
          9.6000E-01  1.3189E 00  -------------------------------------+
          1.2000E 00  1.3460E 00  --------------------------------------+
          1.4400E 00  1.2343E 00  ----------------------------------+
          1.6800E 00  1.0719E 00  -----------------------------+
          1.9200E 00  9.3795E-01  -------------------------+
          2.1600E 00  8.7641E-01  ------------------------+
          2.4000E 00  8.8489E-01  ------------------------+
          2.6400E 00  9.3330E-01  -------------------------+
          2.8800E 00  9.9045E-01  ---------------------------+
          3.1200E 00  1.0314E 00  ----------------------------+
          3.3600E 00  1.0453E 00  ----------------------------+
          3.6000E 00  1.0368E 00  ----------------------------+
          3.8400E 00  1.0175E 00  ---------------------------+
          4.0800E 00  9.9823E-01  ---------------------------+
          4.3200E 00  9.8635E-01  --------------------------+
          4.5600E 00  9.8408E-01  --------------------------+
          4.8000E 00  9.8875E-01  --------------------------+
          5.0400E 00  9.9594E-01  ---------------------------+
          5.2800E 00  1.0022E 00  ---------------------------+
          5.5200E 00  1.0054E 00  ---------------------------+
          5.7600E 00  1.0053E 00  ---------------------------+
          6.0000E 00  1.0032E 00  ---------------------------+
```

(b)
Sample
and hold,
$T = 0.3$ sec

```
                                                              PAGE   1
                          MINIMUM           OUTPUT VERSUS TIME        MAXIMUM
                           0.0              T     = 8.0000E-01        1.8433E 00
          TIME     OUTPUT     I                                       I
          0.0      0.0        +
          2.4000E-01  2.2180E-01  ------+
          4.8000E-01  7.0724E-01  ---------------------+
          7.2000E-01  1.3065E 00  --------------------------------------+
          9.6000E-01  1.7891E 00  -------------------------------------------------------+
          1.2000E 00  1.8172E 00  ---------------------------------------------------------+
          1.4400E 00  1.6271E 00  ----------------------------------------------------+
          1.6800E 00  1.3451E 00  ---------------------------------------+
          1.9200E 00  1.0454E 00  ------------------------------+
          2.1600E 00  7.4309E-01  ----------------------+
          2.4000E 00  4.3961E-01  ------------+
          2.6400E 00  3.6110E-01  ----------+
          2.8800E 00  5.4558E-01  ----------------+
          3.1200E 00  8.4359E-01  ------------------------+
          3.3600E 00  1.1343E 00  ---------------------------------+
          3.6000E 00  1.2845E 00  -------------------------------------+
          3.8400E 00  1.3667E 00  ---------------------------------------+
          4.0800E 00  1.4058E 00  ----------------------------------------+
          4.3200E 00  1.2947E 00  -------------------------------------+
          4.5600E 00  1.0894E 00  -------------------------------+
          4.8000E 00  8.4347E-01  ------------------------+
          5.0400E 00  7.0573E-01  --------------------+
          5.2800E 00  7.0726E-01  --------------------+
          5.5200E 00  7.6892E-01  ----------------------+
          5.7600E 00  8.6170E-01  -------------------------+
          6.0000E 00  9.8047E-01  ---------------------------+
```

(c)
Sample
and hold,
$T = 0.8$ sec

Fig. 4.15. (*Continued*)

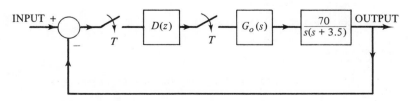

Fig. 4.16　System diagram with a digital controller. (Example 4.4.)

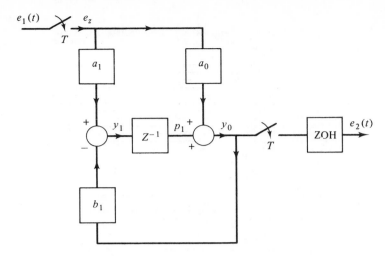

Fig. 4.17 Block diagram representation for the realization of a first-order digital controller.

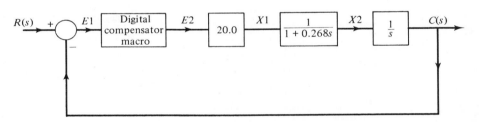

Fig. 4.18 Total simulation diagram for Example 4.4.

```
MACRO    E2 = COMP(SAMTIM, TX, A0, A1, B1, E1)
         TX = IMPULS(0.0, SAMTIM)
         EZ = TX*E1
         YO = A0*EZ + P1
         Y1 = A1*EZ - B1*YO
         P1 = DELAY(1000, SAMTIM,Y1)
         E2 = ZHOLD(TX, YO)
ENDMAC
*        PLANT STRUCTURE FOLLOWS
PARAMETER A0 = 0.1, A1 = -0.0981, B1 = -0.998, SAMTIM = .1
 C  = INTGRL(0.0,X2)
 X2 = REALPL(0.0,0.286,X1)
 X1 = 20.0*E2
 E2 = COMP(SAMTIM, TX, A0, A1, B1, E1)
 E1 = R - C
 R  = STEP(0.0)
METHOD RKSFX
TIMER FINTIM = 5.0, OUTDEL = 0.18, DELT = 0.02
PRTPLT C
LABEL  STEP RESPONSE FOR DIGITAL CONTROL SYSTEM EXAMP. 4-4
END
STOP
ENDJOB
```

Fig. 4.19 Program listing for a digital control system. (Example 4.4.)

```
STEP RESPONSE FOR DIGITAL CONTROL SYSTEM EXAMP. 4-4              PAGE   1
                        MINIMUM           C      VERSUS TIME       MAXIMUM
                          0.0                                     1.1883E 00
    TIME          C         I                                         I
 0.0,           0.0         +
 1.8000E-01     9.2493E-02  ---+
 3.6000E-01     3.0146E-01  ------------+
 5.4000E-01     5.4612E-01  ----------------------+
 7.2000E-01     7.7337E-01  -------------------------------+
 9.0000E-01     9.5472E-01  -----------------------------------------+
 1.0800E 00     1.0806E 00  -----------------------------------------------+
 1.2600E 00     1.1542E 00  ---------------------------------------------------+
 1.4400E 00     1.1851E 00  ------------------------------------------------------+
 1.6200E 00     1.1857E 00  ------------------------------------------------------+
 1.8000E 00     1.1681E 00  ----------------------------------------------------+
 1.9800E 00     1.1421E 00  --------------------------------------------------+
 2.1600E 00     1.1148E 00  ------------------------------------------------+
 2.3400E 00     1.0906E 00  ----------------------------------------------+
 2.5200E 00     1.0716E 00  ---------------------------------------------+
 2.7000E 00     1.0582E 00  --------------------------------------------+
 2.8800E 00     1.0499E 00  -------------------------------------------+
 3.0600E 00     1.0456E 00  -------------------------------------------+
 3.2400E 00     1.0439E 00  -------------------------------------------+
 3.4200E 00     1.0438E 00  -------------------------------------------+
 3.6000E 00     1.0443E 00  -------------------------------------------+
 3.7800E 00     1.0449E 00  -------------------------------------------+
 3.9600E 00     1.0451E 00  -------------------------------------------+
 4.1400E 00     1.0449E 00  -------------------------------------------+
 4.3200E 00     1.0441E 00  -------------------------------------------+
 4.5000E 00     1.0430E 00  -------------------------------------------+
 4.6800E 00     1.0416E 00  -------------------------------------------+
 4.8600E 00     1.0400E 00  -------------------------------------------+
```

Fig. 4.20 Step response of a system containing a digital controller.
(Example 4.4.)

using the CSMP DELAY function. In the following example the controller will be programmed using a recursive difference equation.

Example 4.5

The system diagram of a process to be controlled by a digital algorithm is shown in Fig. 4.21.[12] We note from the diagram that $D(z)$ is given by

$$D(z) = \frac{Q(z)}{E(z)} = \frac{0.19185(z - 0.56496)(z - 0.96585)}{(z - 0.99501)(z - 0.42857)} \tag{4.33}$$

This controller can be programmed as a difference equation using either a *direct, cascade,* or *parallel* realization.[11] Each of these methods will be briefly discussed.

Direct Realization. The function $D(z)$ can be expressed as

$$D(z) = \frac{Q(z)}{E(z)} = \frac{0.19185z^2 - 0.29368z + 0.10469}{z^2 - 1.42358z + 0.42643} \tag{4.34}$$

The standard procedure here is to divide the numerator and denominator by z^2 and cross-multiply. In terms of $Q(z)$ and $E(z)$ this yields

$$Q(z) = 0.19185E(z) - 0.29368z^{-1}E(z) + 0.10469z^{-2}E(z)$$
$$+ 1.42358z^{-1}Q(z) - 0.42643z^{-2}Q(z) \tag{4.35}$$

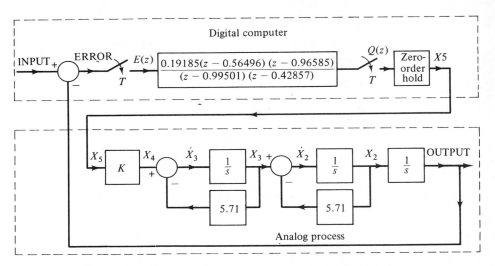

Fig. 4.21 System diagram with second-order digital controller. (Example 4.5.)

Using the real-translation property of z-transforms[13] gives

$$Q(nT) = 0.19185E(nT) - 0.29368E[(n-1)T] + 0.10469E[(n-2)T]$$
$$+ 1.42358Q[(n-1)T] - 0.42643Q[(n-2)T] \qquad (4.36)$$

where T is the sampling interval, $n = 0, 1, 2, \ldots$, and

$Q(nT)$ = the present value of the output Q at $t = nT$

$Q[(n-1)T]$ = the first past value of Q

$Q[(n-2)T]$ = the second past value of Q

$E(nT)$ = the present value of the input E at $t = nT$

$E[(n-1)T]$ = the first past value of E

$E[(n-2)T]$ = the second past value of E

For convenience in programming, the expression for $Q(nT)$ is written as

$$Q = 0.19185E - 0.29368E1 + 0.10469E2 + 1.42358Q1 - 0.42643Q2$$

The present value of E is always known in a process. After solving for Q, the coefficients of the algorithm are immediately updated by including the following statements;

$$Q2 = Q1$$
$$Q1 = Q$$
$$E2 = E1 \qquad (4.37)$$
$$E1 = E$$

```
INITIAL
        E1 = 0.0
        Q1 = 0.0
        E2 = 0.0
        Q2 = 0.0
        PARAM K = 326.
DYNAMIC
        INPUT = STEP(C.0)
        E = INPUT - OUT
        OUT = INTGRL(0.0, X2)
        X2 = INTGRL(C.0, X2D)
        X2D = X3 - 5.71*X2
        X3 = INTGRL(0.0, X3D)
        X3D = X4 - 5.71*X3
        X4 = K*X5
        X5 = ZHCLC(A1, C)
        A1 = IMPULS(C.0, C.1)
PROCEDURE Q = DUMMY(E,E1,E2,Q1,Q2,A1)
        IF(A1.NE.1.0) GO TO 10
        IF(KEEP.NE.1.C) GO TO 10
        Q = .19185*E-.29368*E1+.10469*E2+1.42358*Q1-.42643*Q2
        E2 = E1
        E1 = E
        Q2 = Q1
        Q1 = Q
10      CONTINUE
ENDPRO
METHOD RKSFX
TIMER FINTIM = 8.0, CUTDEL = 0.16, DELT = 0.01
PRTPLT OUT
LABEL  OUTPUT FOR EXAMPLE 4-5
END
STOP
ENDJOB
```

Fig. 4.22 Program listing for Example 4.5 using direct realization.

A CSMP program which incorporates this realization for $D(z)$ is given in Fig. 4.22. We note that the expressions for $D(z)$ are given as a procedure function since they must be executed sequentially. The statement

$$\text{IF(KEEP.NE.1.0) GO TO 10}$$

insures that the statements of the procedure function cannot be executed except when an integration is performed. Furthermore, when fixed-step integration is selected, the statement

$$\text{IF(A1.NE.1.0) go to 10}$$

allows the procedure function to be executed only at sampling instants nT where $n = 0, 1, 2, \ldots$ and T (sampling time) is an integral multiple of DELT. One will note that when a variable step-size integration method is used, integration will not normally occur at nT and the program will generally give misleading results.

Initial values for E1, E2, Q1 and Q2 are given in the INITIAL segment. It is important to observe that E2 is updated prior to E1 and Q2 is updated prior to Q1 in the procedure function.

The response of the system for a unit-step input applied at $t = 0$ is given in Fig. 4.23.

```
                              MINIMUM                 CUT   VERSUS TIME            MAXIMUM
                              C.0                                                  1.2328E 00
  TIME            OUT          I                                                    I
0.0              0.0           +
1.6000E-01       2.7389E-02   -+
3.2000E-01       1.4074E-01   -----+
4.8000E-01       3.1569E-01   ------------+
6.4000E-01       5.1267E-01   -------------------+
8.0000E-01       7.0248E-01   ----------------------------+
9.6000E-01       8.6790E-01   ------------------------------------+
1.12CCE 00       1.0009E 00   ---------------------------------------+
1.2800E 00       1.1001E 00   -----------------------------------------------+
1.4400E 00       1.1678E 00   -------------------------------------------------+
1.6000E 00       1.2088E 00   ---------------------------------------------------+
1.7600E 00       1.2286E 00   ----------------------------------------------------+
1.9200E 00       1.2326E 00   ----------------------------------------------------+
2.0800E 00       1.2257E 00   ----------------------------------------------------+
2.2400E 00       1.2120E 00   ---------------------------------------------------+
2.4000E 00       1.1947E 00   ---------------------------------------------------+
2.5600E 00       1.1761E 00   --------------------------------------------------+
2.7200E 00       1.1577E 00   -------------------------------------------------+
2.8800E 00       1.1406E 00   ------------------------------------------------+
3.0400E 00       1.1252E 00   -----------------------------------------------+
3.2000E 00       1.1118E 00   -----------------------------------------------+
3.3600E 00       1.1003E 00   ----------------------------------------------+
3.5200E 00       1.0905E 00   ----------------------------------------------+
3.6800E 00       1.0823E 00   ---------------------------------------------+
3.8400E 00       1.0752E 00   ---------------------------------------------+
4.0000E 00       1.0692E 00   --------------------------------------------+
4.1600E 00       1.0640E 00   --------------------------------------------+
4.3200E 00       1.0595E 00   --------------------------------------------+
4.4800E 00       1.0554E 00   -------------------------------------------+
4.6400E 00       1.0516E 00   -------------------------------------------+
4.8000E 00       1.0482E 00   -------------------------------------------+
4.9600E 00       1.0450E 00   -------------------------------------------+
5.1200E 00       1.0420E 00   -------------------------------------------+
5.2800E 00       1.0392E 00   -------------------------------------------+
5.4400E 00       1.0365E 00   -------------------------------------------+
5.6000E 00       1.0340E 00   ------------------------------------------+
5.7600E 00       1.0317E 00   ------------------------------------------+
5.9200E 00       1.0295E 00   ------------------------------------------+
6.0800E 00       1.0274E 00   ------------------------------------------+
6.2400E 00       1.0255E 00   ------------------------------------------+
6.4000E 00       1.0237E 00   ------------------------------------------+
6.5600E 00       1.0220E 00   ------------------------------------------+
6.7200E 00       1.0204E 00   ------------------------------------------+
6.8800E 00       1.0189E 00   ------------------------------------------+
7.0400E 00       1.0176E 00   ------------------------------------------+
7.2000E 00       1.0163E 00   ------------------------------------------+
7.3600E 00       1.0152E 00   ------------------------------------------+
7.5200E 00       1.0141E 00   ------------------------------------------+
7.6800E 00       1.0131E 00   ------------------------------------------+
7.8400E 00       1.0122E 00   ------------------------------------------+
8.0000E 00       1.0113E 00   ------------------------------------------+
```

Fig. 4.23 Unit-step response of the second-order digital control system. (Example 4.5.)

Cascade Realization. For cascade realization, $D(z)$ is expressed as

$$D(z) = \frac{Q(z)}{E(z)} = 0.19185\left[\frac{1 - 0.56496z^{-1}}{1 - 0.99501z^{-1}}\right]\left[\frac{1 - 0.96585z^{-1}}{1 - 0.42857z^{-1}}\right] \qquad (4.38)$$

where the grouping of the numerator and denominator terms within the brackets is arbitrary. An input/output relationship is assigned to each term of $Q(z)/E(z)$ by expressing

$$\frac{Q(z)}{E(z)} = \frac{Q(z)}{V(z)} \cdot \frac{V(z)}{S(z)} \cdot \frac{S(z)}{E(z)} \tag{4.39}$$

where

$$\frac{Q(z)}{V(z)} = 0.19185 \tag{4.40}$$

$$\frac{V(z)}{S(z)} = \left[\frac{1 - 0.56496z^{-1}}{1 - 0.99501z^{-1}}\right] \tag{4.41}$$

$$\frac{S(z)}{E(z)} = \left[\frac{1 - 0.96585z^{-1}}{1 - 0.42857z^{-1}}\right] \tag{4.42}$$

Each of these expressions can be written in difference equation form to give

$$S = E - 0.96585E1 + 0.42857S1$$
$$V = S - 0.56496S1 + 0.99501V1 \tag{4.43}$$
$$Q = 0.19185V$$

Again, after processing the above equations, the coefficients must be updated by setting

$$E1 = E$$
$$S1 = S \tag{4.44}$$
$$V1 = V$$

The simulation for the total system is the same as that given in Fig. 4.22 except the procedure function becomes

```
       PROCEDURE  Q = DUMMY(E, A1, E1, S1, V1)
              IF(A1.NE.1.0)  GO  TO  10
              IF(KEEP.NE.1.0)  GO  TO  10
              S = E - 0.96585*E1 + 0.42857*S1
              V = S - 0.56496*S1 + 0.99501*V1
              Q = 0.19185*V
              E1 = E
              S1 = S
              V1 = V
       10       CONTINUE
       ENDPRO
```

and in the INITIAL segment E1 = 0.0, S1 = 0.0, and V1 = 0.0.

Parallel Realization In the parallel realization method $D(z)$ is expanded in partial-fraction form. When the numerator and denominator of $D(z)$ are the same order, the denominator is first divided into the numerator, which for this example gives

$$D(z) = 0.19185 + \frac{(-0.02057z + 0.022875)}{(z - 0.99501)(z - 0.42857)} \tag{4.45}$$

In partial-fraction form, $D(z)$ becomes

$$D(z) = 0.19185 + \frac{A}{z - 0.99501} + \frac{B}{z - 0.42857} \tag{4.46}$$

where $A = 0.00425$ and $B = -0.02482$. We now write the function as

$$D(z) = \frac{Q(z)}{E(z)} = \frac{M(z)}{E(z)} + \frac{N(z)}{E(z)} + \frac{P(z)}{E(z)} \tag{4.47}$$

where:

$$\frac{M(z)}{E(z)} = 0.19185 \tag{4.48}$$

$$\frac{N(z)}{E(z)} = \frac{0.00425z^{-1}}{1 - 0.99501z^{-1}} \tag{4.49}$$

$$\frac{P(z)}{E(z)} = \frac{-0.02482z^{-1}}{1 - 0.42857z^{-1}} \tag{4.50}$$

These expressions lead directly to the difference equations

$$M = 0.19185E$$
$$N = 0.00425E1 + 0.99501N1$$
$$P = -0.02482E1 + 0.42857P1 \tag{4.51}$$
$$Q = M + N + P$$

with coefficient updating

$$E1 = E$$
$$M1 = M$$
$$N1 = N \tag{4.52}$$
$$P1 = P$$

The total simulation program is again identical to that of Fig. 4.22 except that the INITIAL segment contains the values E1 = 0.0, M1 = 0.0, N1 = 0.0, P1 = 0.0 and the procedure function is given by

```
PROCEDURE  Q = DUMMY(E, E1, M1, N1, P1, A1)
           IF(A1.NE.1.0) GO TO 10
           IF(KEEP.NE.1.0) GO TO 10
           M = 0.19185*E
           N = 0.00425*E1 + 0.99501*N1
           P = -0.02482*E1 + 0.42857*P1
           Q = M + N + P
           E1 = E
           M1 = M
           N1 = N
           P1 = P
     10  CONTINUE
     ENDPRO
```

The unit-step output response for each of the three methods is practically identical to the response given in Fig. 4.23. The reader may ask why three different methods are presented. When these algorithms are programmed for real-time digital control, two factors become important. First, the computer time required to execute the algorithm may be an important factor. If this is the sole consideration one would select the algorithm that requires the fewest number of additions and multiplications. Second, numerical roundoff and truncation may have a significant influence on the system response. In this case one might try each realization and select the one that gives the best results. Several investigators have found that parallel realization most often gives the best performance.

Example 4.6

Classical analog compensation techniques that rely heavily upon frequency response methods can be extended to digital control by implementing appropriate algorithms in the digital computer. The bases for this extension were given in the previous examples.

Obtaining frequency response information or developing models can be extremely time consuming and therefore is not always a luxury the control engineer can afford. This is particularly true in the process control industry. Normally one finds in this case that a three-mode controller is the most attractive compromise for improving system performance. The term three-mode stems from the nature of the controller in that one mode is directly proportional to the applied signal, one mode is proportional to the integral of the signal, and one mode is proportional to the derivative. Therefore the term, *proportional-integral-derivative* (PID) *controller* is often used rather than three-mode.

The diagram in Fig. 4.24 shows the representation of this controller as applied to a unity-negative feedback system. The continuous PID controller can be approximated on a digital computer using several forms of algorithms.[10,14] As a simplified version we can write

$$X_I(k) = X_I(k-1) + T*E(k) \tag{4.53}$$

$$X_D(k) = (1/T)(E(k) - E(k-1)) \tag{4.54}$$

where Eq. (4.53) corresponds to integration and (4.54) to differentiation. The output of the digital PID is then given by

$$Y(k) = K_p*E(k) + K_I(X_I(k-1) + T*E(k)) + K_D((1/T)(E(k) - E(k-1))) \tag{4.55}$$

We note that

$$T = \text{sampling time}$$

$$E(k) = \text{present value of the applied signal}$$

$$E(k-1) = \text{first past value of } E(k)$$

$$X_I(k) = \text{present estimate of the integral}$$

$$X_I(k-1) = \text{first past value of } X_I(k)$$

Consider the system shown in Fig. 4.25 where the sample switch and the zero-order-hold represent the digital computer without a control algorithm. With a sampling time of $T = 0.01$, the step response shown in Fig. 4.26 is practically identical to the continuous system (i.e. without the sampler and hold).

A program using the same basic system but including a digital representation of the PID controller is given in Fig. 4.27. The algorithm for the controller is contained in a

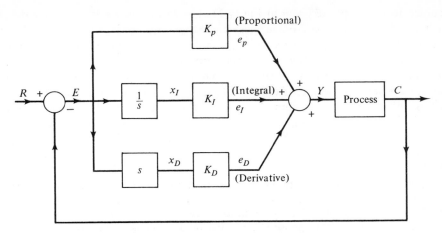

Fig. 4.24 Ideal representation of an ideal continuous PID controller.

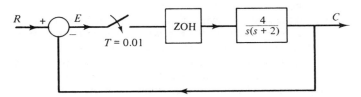

Fig. 4.25 System of Example 4.6 without a digital PID controller.

procedure function in the same way as given previously in Example 4.5. Earlier comments with respect to using fixed integration steps rather than variable steps also apply to this simulation.

Integration and differentiation for the controller are approximated by Eqs. (4.53) and (4.54), respectively. Using $K_p = 12$, $K_I = 20$, and $K_D = 4$ gives the step response shown in Fig. 4.28. We note the improvement in the system rise-time as well as a decrease in the overshoot when compared to the response without a controller.

In conclusion, this section has presented several methods for simulating a digital control system. There are many other ways to view the digital control system, for example, the digital implementation of a predictive controller.[15] Nevertheless, the methods presented here are fundamental and generally apply to more advanced techniques.

There is strong justification for learning the programming concepts presented in this section. Advances in computer technology, particularly in the microprocessor area, have reduced the cost for digital control by a factor of ten during the last decade. Continued improvement in computer technology, coupled with the flexibility offered by a digital controller, will in all likelihood result in a growing number of applications in this area.

```
                          MINIMUM                COUT   VERSUS TIME              MAXIMUM
                          0.0                                                    1.1628E 00
     TIME         COUT        I                                                      I
  0.0           0.0           +
  8.0000E-02    1.2118E-02    +
  1.6000E-01    4.5765E-02    -+
  2.4000E-01    9.6963E-02    ----+
  3.2000E-01    1.6191E-01    ------+
  4.0000E-01    2.3704E-01    ----------+
  4.8000E-01    3.1907E-01    ------------+
  5.6000E-01    4.0504E-01    -----------------+
  6.4000E-01    4.9235E-01    ---------------------+
  7.2000E-01    5.7873E-01    -------------------------+
  8.0000E-01    6.6229E-01    -----------------------------+
  8.8000E-01    7.4146E-01    --------------------------------+
  9.6000E-01    8.1503E-01    ------------------------------------+
  1.0400E 00    8.8209E-01    ---------------------------------------+
  1.1200E 00    9.4202E-01    -----------------------------------------+
  1.2000E 00    9.9446E-01    -------------------------------------------+
  1.2800E 00    1.0393E 00    ---------------------------------------------+
  1.3600E 00    1.0765E 00    ----------------------------------------------+
  1.4400E 00    1.1065E 00    -----------------------------------------------+
  1.5200E 00    1.1295E 00    ------------------------------------------------+
  1.6000E 00    1.1460E 00    -------------------------------------------------+
  1.6800E 00    1.1567E 00    --------------------------------------------------+
  1.7600E 00    1.1621E 00    --------------------------------------------------+
  1.8400E 00    1.1628E 00    --------------------------------------------------+
  1.9200E 00    1.1596E 00    --------------------------------------------------+
  2.0000E 00    1.1531E 00    -------------------------------------------------+
  2.0800E 00    1.1440E 00    -------------------------------------------------+
  2.1600E 00    1.1328E 00    ------------------------------------------------+
  2.2400E 00    1.1201E 00    -----------------------------------------------+
  2.3200E 00    1.1065E 00    ----------------------------------------------+
  2.4000E 00    1.0923E 00    ----------------------------------------------+
  2.4800E 00    1.0781E 00    ---------------------------------------------+
  2.5600E 00    1.0641E 00    --------------------------------------------+
  2.6400E 00    1.0507E 00    --------------------------------------------+
  2.7200E 00    1.0381E 00    -------------------------------------------+
  2.8000E 00    1.0264E 00    ------------------------------------------+
  2.8800E 00    1.0159E 00    ------------------------------------------+
  2.9600E 00    1.0065E 00    -----------------------------------------+
  3.0400E 00    9.9837E-01    ----------------------------------------+
  3.1200E 00    9.9147E-01    ----------------------------------------+
  3.2000E 00    9.8579E-01    ---------------------------------------+
  3.2800E 00    9.8129E-01    ---------------------------------------+
  3.3600E 00    9.7789E-01    ---------------------------------------+
  3.4400E 00    9.7552E-01    --------------------------------------+
  3.5200E 00    9.7408E-01    --------------------------------------+
  3.6000E 00    9.7346E-01    --------------------------------------+
  3.6800E 00    9.7356E-01    --------------------------------------+
  3.7600E 00    9.7427E-01    --------------------------------------+
  3.8400E 00    9.7548E-01    --------------------------------------+
  3.9200E 00    9.7709E-01    --------------------------------------+
  4.0000E 00    9.7901E-01    --------------------------------------+
```

Fig. 4.26 Step response for Example 4.6 without PID controller.

```
INITIAL
        T  =  0.01
        E1  =  0.0
        INTG1  =  0. 0
        KINTG  =  20.0
        KPROP  =  12.0
        KDERIV  =  4.0
DYNAMIC
        E  =  INPUT  -  RESP
        INPUT  =  STEP(0.0)
        RESP  =  INTGRL(0.0,X1)
        X1  =  REALPL(0.0,1.0/2.0,X2)
        X2  =  2.0*X3
        X3  =  ZHOLD(A1,Y)
        A1  =  IMPULS(0.0,T)
PROCEDURE Y  =  PIC(A1,E,E1,T,KPROP,KINTG,KDERIV,INTG1)
        IF(A1.NE.1.0)   GO  TO  10
        IF(KEEP.NE.1.0)    GO  TO  10
        EPROP  =  KPROP*E
        EDERIV  =  KDERIV*(1.0/T)*(E  -  E1)
        INTG  =  INTG1  +  T*E
        EINTG  =  KINTG*INTG
        Y  =  EPROP  +  EDERIV  +  EINTG
        E1  =  E
        INTG1  =  INTG
10      CONTINUE
ENDPRO
METHOD RKSFX
TIMER FINTIM = 4.0,  OUTDEL = 0.08,  DELT = 0.001
PRTPLT RESP
LABEL OUTPUT FOR EXAMPLE 4-6 WITH PID CONTROLLER
END
STOP
ENDJOB
```

Fig. 4.27   CSMP program listing for PID controller. (Example 4.6.)

Simulation of Digital Logic

The capability for simulating various digital logic functions is available in CSMP. We recall that functional blocks such as INTGRL are the key elements for simulating continuous systems. The same holds true for the digital logic case in that again standard blocks are available which make it possible to simulate practically any digital configuration.

Most of the functions related to logic simulation are given in Table 4.2. These functions are also given in Appendix I along with other special switching logic. As an introduction, example problems are given below that illustrate how these elements can be used in combination to perform useful logic tasks.

Example 4.7

This problem illustrates how a random-generated binary number (either 0, 1, 2, or 3) can be processed through digital logic so as to form the equivalent decimal representation of the number and thereby activate the proper edges of a Nixie light.

In Fig. 4.29(a), XL1 and XL2 form the binary representation of 0, 1, 2, and 3. E1 through E7 in 4.29(b) is a truth table for the edges of the light as defined in 4.29(c). The

		MINIMUM 0.0	RESP VERSUS TIME	MAXIMUM 1.0777E 00
TIME	RESP	I		I
0.0	0.0	+		
8.0000E-02	7.9619E-01	---------------------------------------+		
1.6000E-01	1.0108E 00	--+		
2.4000E-01	1.0625E 00	---+		
3.2000E-01	1.0755E 00	--+		
4.0000E-01	1.0777E 00	--+		
4.8000E-01	1.0758E 00	---+		
5.6000E-01	1.0718E 00	--+		
6.4000E-01	1.0664E 00	--+		
7.2000E-01	1.0601E 00	--+		
8.0000E-01	1.0533E 00	---+		
8.8000E-01	1.0463E 00	---+		
9.6000E-01	1.0393E 00	--+		
1.0400E 00	1.0326E 00	---+		
1.1200E 00	1.0263E 00	--+		
1.2000E 00	1.0205E 00	---+		
1.2800E 00	1.0152E 00	---+		
1.3600E 00	1.0106E 00	--+		
1.4400E 00	1.0066E 00	--+		
1.5200E 00	1.0032E 00	---+		
1.6000E 00	1.0004E 00	---+		
1.6800E 00	9.9818E-01	--+		
1.7600E 00	9.9646E-01	--+		
1.8400E 00	9.9519E-01	--+		
1.9200E 00	9.9431E-01	---------------------------------------+		
2.0000E 00	9.9379E-01	---------------------------------------+		
2.0800E 00	9.9355E-01	---------------------------------------+		
2.1600E 00	9.9356E-01	---------------------------------------+		
2.2400E 00	9.9375E-01	---------------------------------------+		
2.3200E 00	9.9309E-01	---------------------------------------+		
2.4000E 00	9.9454E-01	---------------------------------------+		
2.4800E 00	9.9506E-01	---------------------------------------+		
2.5600E 00	9.9562E-01	---------------------------------------+		
2.6400E 00	9.9620E-01	---------------------------------------+		
2.7200E 00	9.9677E-01	--+		
2.8000E 00	9.9732E-01	--+		
2.8800E 00	9.9784E-01	--+		
2.9600E 00	9.9832E-01	--+		
3.0400E 00	9.9875E-01	--+		
3.1200E 00	9.9913E-01	---+		
3.2000E 00	9.9946E-01	---+		
3.2800E 00	9.9974E-01	---+		
3.3600E 00	9.9997E-01	---+		
3.4400E 00	1.0001E 00	---+		
3.5200E 00	1.0003E 00	---+		
3.6000E 00	1.0004E 00	---+		
3.6800E 00	1.0004E 00	---+		
3.7600E 00	1.0005E 00	---+		
3.8400E 00	1.0005E 00	--- ----+		
3.9200E 00	1.0005E 00	---+		
4.0000E 00	1.0005E 00	---+		

Fig. 4.28 Step response of system with PID digital controller.
(Example 4.6.)

Table 4.2

Logic Functions for CSMP (Also, see Appendix I)

Statement Form		Function
$Y = RST \ (X_1, X_2, X_3)$	$Y = 0$	$X_1 > 0$
	$Y = 1$	$X_2 > 0, X_1 \leq 0$
Y_{N-1} = Previous state of flip-flop	$\left. \begin{array}{l} Y = 0 \\ Y = 1 \\ Y = 0 \\ Y = 1 \end{array} \right]$ $\begin{array}{l} X_1 \leq 0, \\ X_2 \leq 0, \end{array}$	$\left[\begin{array}{l} X_3 > 0, Y_{N-1} = 1 \\ X_3 > 0, Y_{N-1} = 0 \\ X_3 \leq 0, Y_{N-1} = 1 \\ X_3 \leq 0, Y_{N-1} = 1 \end{array} \right.$
RESETTABLE FLIP-FLOP		
$Y = COMPAR \ (X_1, X_2)$	$Y = 0$	$X_1 < X_2$
COMPARATOR	$Y = 1$	$X_1 \geq X_2$
$Y = AND \ (X_1, X_2)$	$Y = 1$	$X_1 > 0, X_2 > 0$
	$Y = 0$	OTHERWISE
$Y = NAND \ (X_1, X_2)$	$Y = 0$	$X_1 > 0, X_2 > 0$
NOT AND	$Y = 1$	OTHERWISE
$Y = IOR \ (X_1, X_2)$	$Y = 0$	$X_1 \leq 0, X_2 \leq 0$
INCLUSIVE OR	$Y = 1$	OTHERWISE
$Y = NOR \ (X_1, X_2)$	$Y = 1$	$X_1 \leq 0, X_2 \leq 0$
NOT OR	$Y = 0$	OTHERWISE
$Y = EOR \ (X_1, X_2)$	$Y = 1$	$X_1 \leq 0, X_2 > 0$
	$Y = 1$	$X_1 > 0, X_2 \leq 0$
EXCLUSIVE OR	$Y = 0$	OTHERWISE
$Y = NOT \ (X)$	$Y = 1$	$X \leq 0$
NOT	$Y = 0$	$X > 0$
$Y = EQUIV \ (X_1, X_2)$	$Y = 1$	$X_1 \leq 0, X_2 \leq 0$
	$Y = 1$	$X_1 > 0, X_2 > 0$
EQUIVALENT	$Y = 0$	OTHERWISE

logic statements defining E1 through E7 are:

$$E1 = \overline{XL1} \cdot \overline{XL2} \qquad \text{(not XL1 and not XL2)}$$
$$E2 = XL1 + \overline{XL2} \qquad \text{(XL1 or not XL2)}$$
$$E3 = 1.0 \qquad \text{(always on)}$$
$$E4 = \overline{XL1} + XL2 \qquad \text{(not XL1 or XL2)}$$
$$E5 = E2 \qquad \text{(same as the state of E2)}$$
$$E6 = \overline{XL2} \qquad \text{(not XL2)}$$
$$E7 = XL1 \qquad \text{(the state of XL1)}$$

Number	XL1	XL2
0	0	0
1	0	1
2	1	0
3	1	1

(a) Binary values of XL1 and XL2 for given decimal numbers.

E1	E2	E3	E4	E5	E6	E7	Number
1	1	1	1	1	1	0	0
0	0	1	1	0	0	0	1
0	1	1	0	1	1	1	2
0	1	1	1	1	0	1	3

(b) Truth table for light edges.

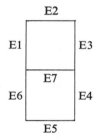

(c) Light edge definition.

Fig. 4.29 Diagram defining logic properties for Example 4.7

A logic circuit for performing the above functions is given in Fig. 4.30 where we assume the random number is available from XL1 and XL2.

The program for making this simulation will first use RNDGEN to generate a random number. The number so generated will lie between 0 and 1. The output of the random number generator will be partioned as follows:

For $0 \leq$ output $< 0.25 \longrightarrow$ Assign 0

For $0.25 \leq$ output $< 0.50 \longrightarrow$ Assign 1

For $0.50 \leq$ output $< 0.75 \longrightarrow$ Assign 2

For $0.75 \leq$ output $\leq 1.00 \longrightarrow$ Assign 3

The numbers 0, 1, 2, and 3 will be represented in binary form to give the correct values for XL1 and XL2. Next the logic functions for E1 through E7 will be expressed in terms of CSMP functional logic blocks. Finally, the output will give the random binary number and indicate the state for each edge of the light.

The program given in Fig. 4.31 will accomplish the above tasks. More than 50% of the program statements are used to create a special form of output. Note that on the TIMER card, FINTIM is given as 15.0 and DELT specified as 0.5. This means that 30 random numbers will be generated. Of course, the user has complete freedom to specify

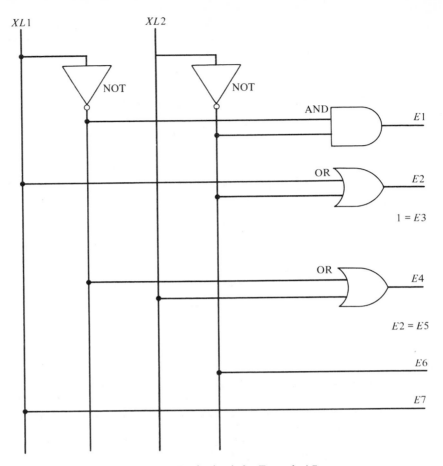

Fig. 4.30 Logic circuit for Example 4.7.

FINTIM and DELT. In this example neither the **PRINT** nor **PRTPLT** features of **CSMP** were used since all desired output is obtained from conventional **FORTRAN** statements. An example of typical output is given in Fig. 4.32 where the random number, the binary representation, and the state of each light edge are given at the top of the page. This is followed by a bold print output of the decimal equivalent.

Example 4.8

The reset-set-toggle (RST) flip-flop in the CSMP functional blocks, as given in Table 4.2, can be used for simulating many standard digital functions. This example illustrates the use of the flip-flop in constructing and testing a two-bit up-down counter. (Modulo two.)

Several comments should first be made on logic design using CSMP. First the logic statements should be placed after a NOSORT card. Logic designs using flip-flops with feedback from the output states will form algebraic loops if run as SORT statements.

```
*   PROGRAM TO TURN ON LIGHT WITH NUMBER FROM ZERO TO THREE

NOSORT
****************************************************************************
*             PART ONE - GENERATE RANDOM NUMBER                           *
****************************************************************************
      L = RNDGEN(9)
      XL2 = 0.0
      IF(L.GE.0.5) GO TO 13
      XL1 = 0.0
      IF(L.GE.0.25) XL2 = 1.0
      GO TO 14
   13 XL1 = 1.0
      IF(L.GE.0.75) XL2 = 1.0
   14 CONTINUE
****************************************************************************
*                      PART TWO - LOGIC STRUCTURE                         *
****************************************************************************
      XL1BAR = NOT(XL1)
      XL2BAR = NOT(XL2)
      E1 = AND(XL1BAR,XL2BAR)
      E2 = IOR(XL1,XL2BAR)
      E3 = 1.0
      E4 = IOR(XL1BAR,XL2)
      E5 = E2
      E6 = XL2BAR
      E7 = XL1
****************************************************************************
* PART THREE - WRITE OUT THE VALUES FOR L,XL1,XL2,E1,E2,E3,E4,E5,E6,E7*
****************************************************************************
      WRITE(6,3)
    3 FORMAT(1H1,1X,'RANDOM NUMBER    XL1   XL2    EDGE1   EDGE2   EDGE3
     $EDGE4   EDGE5   EDGE6   EDGE7')
      WRITE(6,10) L,XL1,XL2,E1,E2,E3,E4,E5,E6,E7
   10 FORMAT(3X,F10.8,7X,F3.1,2X,F3.1,4X,F3.1,4X,F3.1,4X,F3.1,
     $4X,F3.1,4X,F3.1,4X,F3.1,/////////)
****************************************************************************
*     PART FOUR - SPECIAL WRITE STATEMENTS TO FORM DECIMAL NUMBER         *
****************************************************************************
      IF(XL1.EQ.0.0.AND.XL2.EQ.0.0) GO TO 9
      IF(XL1.EQ.1.0.AND.XL2.EQ.1.0) GO TO 8
      IF(XL1.EQ.0.0.AND.XL2BAR.EQ.0.0) GO TO 11
      IF(XL1.EQ.1.0.AND.XL2BAR.EQ.1.0) GO TO 12
    9 WRITE(6,15)
*   THIS FORMAT PRINTS A ZERO
   15 FORMAT(T34,12('X')/T34,12('X')/11(T34,'XX          XX'/),
     $2(T34,'XXXXXXXXXXXX'/))
      GO TO 25
   11 WRITE(6,19)
*   THIS FORMAT PRINTS A ONE
   19 FORMAT(15(/T40,'XX'))
      GO TO 25
   12 WRITE(6,20)
*   THIS FORMAT PRINTS A TWO
   20 FORMAT(T34,12('X')/T34,12('X')/5(T44,'XX'/),T34,12('X')/
     $T34,12('X')/5(T34,'XX'/),T34,12('X')/T34,12('X')/)
      GO TO 25
    8 WRITE(6,17)
*   THIS FORMAT PRINTS A THREE
   17 FORMAT(T34,12('X')/T34,12('X')/5(T44,'XX'/),T34,12('X')/
     $T34,12('X')/5(T44,'XX'/),T34,12('X')/T34,12('X')/)
   25 CONTINUE
TIMER FINTIM =15.0, DELT = 0.5
END
STOP
```

Fig. 4.31 Program listing for NIXIE light indicator. (Example 4.7.)

RANDOM NUMBER	XL1	XL2	EDGE1	EDGE2	EDGE3	EDGE4	EDGE5	EDGE6	EDGE7
0.54687333	1.0	0.0	0.0	1.0	1.0	0.0	1.0	1.0	1.0

```
           XXXXXXXXXXXX
           XXXXXXXXXXXX
                     XX
                     XX
                     XX
                     XX
                     XX
           XXXXXXXXXXXX
           XXXXXXXXXXXX
           XX
           XX
           XX
           XX
           XX
           XXXXXXXXXXXX
           XXXXXXXXXXXX
```

Fig. 4.32   Typical program output for NIXIE light. (Example 4.7.)

Furthermore, the SORT algorithm will change the logic statements around so that the resulting simulation becomes nonsense. Second CSMP will go through the program statements twice at TIME $= 0$ even though integral functions are not used. This means that flip-flops will react to the two passes and will often give erroneous results. This problem can be avoided by starting pulses into the logic only after TIME $= 0$. Third, when integration is not used in CSMP, the program statements are executed every DELT units. Thus, users should specify a DELT on the TIMER card commensurate with the frequency desired for executing the program statements. In summary, the following guidelines are recommended when using CSMP to simulate logic containing flip-flops.

1 Always place logic statements after a NOSORT card.

2 Apply inputs to the logic *after* time $= 0$.

3 Use a DELT corresponding to the frequency desired for executing the program statements.

The logic diagram of an up-down counter using RST flip-flops is shown in Fig. 4.33. The counter counts up one unit for every X_1 pulse applied and counts down one unit for every X_2 pulse applied. Simultaneous pulses for X_1 and X_2 gives an undefined condition. A CSMP program for simulating this counter is given in Fig. 4.34. The train of pulses applied to the system start at TIME $= 0.1$. Four pulses are first applied from X_1 and followed by a continuous train of pulses from X_2. Note that variables Q1I and Q2I are introduced in the INITIAL section. This allows the states of the flip-flops to be defined so that statements for R1, S1, T1 (input to flip-flop 1) and R2, S2, T2 (input to flip-flop 2) can be written. Note also that all logic statements follow a NOSORT card and that

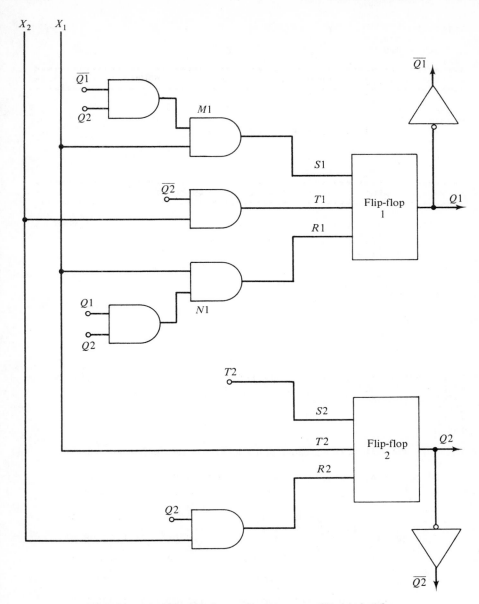

Fig. 4.33 Logic diagram for up-down counter of Example 4.8.

DELT = 0.1 is used on the TIMER card. If a DELT is not specified, the program uses
FINTIM/100 and thereby executes the program statements at this value.

The output of the program is given in Fig. 4.35. Note that Q2 is the low-order bit and
Q1 is the high-order bit. The four input pulses for X1 gives the binary count 1, 2, 3, 0
(which is correct). One will also observe that X2 causes the proper down-count.

```
INITIAL
    Q1I = 0.0
    Q2I = 0.0

*************************************************************************
*   THE FOLLOWING STATEMENTS GENERATE A TRAIN OF PULSES AS INPUT      *
*************************************************************************

DYNAMIC
NOSORT
    XX1 = IMPULS(0.1,0.1)
    YY1 = IMPULS(0.5,0.1)
    X1  = XX1 - YY1
    X2  = YY1

*************************************************************************
*   THE FOLLOWING STATEMENTS FORM AN UP-DOWN COUNTER                  *
*************************************************************************

    Q1IBAR = NOT(Q1I)
    Q2IBAR = NOT(Q2I)
    M1 = AND(Q1IBAR,Q2I)
    S1 = AND(X1,M1)
    T1 = AND(X2,Q2IBAR)
    N1 = AND(Q1I,Q2I)
    R1 = AND(X1,N1)
    R2 = AND(X2,Q2I)
    S2 = T1
    T2 = X1
    Q1 = RST(R1,S1,T1)
    Q2 = RST(R2,S2,T2)
    Q1I = Q1
    Q2I = Q2

TIMER  FINTIM = 1.5, DELT = 0.1, PRDEL = 0.1
PRINT X1, X2, Q1, Q2
TITLE OUTPUT OF A MODULO TWO COUNTER EXAMPLE 4-8
END
STOP
ENDJOB
```

Fig. 4.34 CSMP program of the up-down counter of Example 4.8.

```
OUTPUT OF A MODULO TWO COUNTER EXAMPLE 4-8
```

TIME	X1	X2	Q1	Q2
0.0	0.0	0.0	0.0	0.0
1.0000E-01	1.0000E 00	0.0	0.0	1.0000E 00
2.0000E-01	1.0000E 00	0.0	1.0000E 00	0.0
3.0000E-01	1.0000E 00	0.0	1.0000E 00	1.0000E 00
4.0000E-01	1.0000E 00	0.0	0.0	0.0
5.0000E-01	0.0	1.0000E 00	1.0000E 00	1.0000E 00
6.0000E-01	0.0	1.0000E 00	1.0000E 00	0.0
7.0000E-01	0.0	1.0000E 00	0.0	1.0000E 00
8.0000E-01	0.0	1.0000E 00	0.0	0.0
9.0000E-01	0.0	1.0000E 00	1.0000E 00	1.0000E 00
1.0000E 00	0.0	1.0000E 00	1.0000E 00	0.0
1.1000E 00	0.0	1.0000E 00	0.0	1.0000E 00
1.2000E 00	0.0	1.0000E 00	0.0	0.0
1.3000E 00	0.0	1.0000E 00	1.0000E 00	1.0000E 00
1.4000E 00	0.0	1.0000E 00	1.0000E 00	0.0
1.5000E 00	0.0	1.0000E 00	0.0	1.0000E 00

Fig. 4.35 Output results of up-down counter simulation. (Example 4.8.)

REFERENCES

1. KUO, B.C., *Automatic Control Systems*. Englewood Cliffs, N. J.: Prentice-Hall, Inc. 1967.

2. THALER, G. J., and BROWN, R. G., *Analysis and Design of Feedback Systems*. New York: McGraw-Hill Book Company, 1960.

3. D'AZZO, J. J., and HOUPIS, C. H., *Feedback Control System Analysis and Synthesis*. New York: McGraw-Hill Book Company, 1966.

4. DORF, R. C., *Modern Control Systems*. Reading, Mass.: Addison-Wesley Publishing Co., Inc., 1967.

5. CHIRLIAN, P. M., *Signals, Systems, and the Computer*. New York and London: Intext Educational Publishers, 1973.

6. DORF, R. C., *Time Domain Analysis and Design of Control Systems*. Reading, Mass.: Addison-Wesley Publishing Co., Inc., 1965.

7. SCHULTZ, D. G., and MELSA, J. L., *State Functions and Linear Control Systems*. New York: McGraw-Hill Book Company, 1967.

8. DRANSFIELD, PETER, *Engineering Systems and Automatic Control*. Englewood Cliffs, N. J.: Prentice-Hall, Inc., 1968.

9. KUO, B. C., *Analysis and Synthesis of Sampled Data Control Systems*. Englewood Cliffs, N. J.: Prentice-Hall, Inc., 1963.

10. CADZOW, J. A., and MARTENS, H. R., *Discrete-Time and Computer Control Systems*. Englewood Cliffs, N. J.: Prentice-Hall, Inc., 1970.

11. SAUCEDO, R., and SCHIRING, E. E., *Introduction to Continuous and Digital Control Systems*. New York: The MacMillan Company, 1968.

12. HYLAND, W. W., and GREEN, W. L., "Application of Real-Time Digital Automatic Gauge Control," *ISA Transactions*, Vol. 13, No. 4, 326–34, 1974.

13. STEIGLITZ, KENNETH, *An Introduction to Discrete Systems*. New York: John Wiley and Sons, 1974.

14. SMITH, C. L., *Digital Computer Process Control*. New York and London: Intext Educational Publishers, 1969.

15. WOLF, J. and GREEN, W. L., "Application of a Minicomputer for Practical Realization of Predictive Compensation." *ISA Transactions*, Vol. 12, No. 3, 1973.

16. DEBOLT, R. R., and POWELL, B. E., "A Natural 3-Mode Controller Algorithm for D.D.C." *ISA Journal* Sept., 1966, pp. 43–7.

PROBLEMS

1 The open-loop transfer function of a system is given by

$$G(s) = \frac{150}{s(s+2)(s+10)}$$

(a) Find the frequency response of the open-loop system over the range $\omega_{min} = 0.1$ to $\omega_{max} = 100.0$. Use the basic program given in Fig. 4.1 with NUM(1-6) and DEN(1-6) changed to appropriate values for this problem. Obtain 20 output points per decade. Add a statement to the program in Fig. 4.1 so that the output magnitude is expressed in decibels (dB). If VALUE is the output, this can be expressed in dB by

$$\text{VALUEX} = 20.0*\text{ALOG10(VALUE)}$$

(b) The $G(s)$ given above is used in a unity-negative feedback system as shown below. Again, use the basic program of Fig. 4.1 to obtain the closed-loop frequency response. Express the output in dB and request **PRTPLT** of output and phase for $.1 \leq \omega \leq 100.0$,

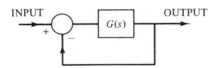

Fig. P-4.1b

(c) From the computer output of (a) and (b) give
   (i) the phase margin and gain margin of the system;
   (ii) the peak overshoot (M_p) of the closed-loop frequency response;
   (iii) the frequency at which the peak response occurs (ω_p);
   (iv) the approximate bandwidth of the closed-loop system; and, finally,
   (v) is the closed-loop system stable?

Answers:
      Part (a). At $\omega = 1.2589$, VALUEX $= 13.983$, PHASE $= -129.36$
      Part (b). At $\omega = 2.5119$, VALUEX $= 5.2604$, PHASE $= -24.841$
      Part (c). (i) Phase margin $= 10$ deg, gain margin $= 4.06$ dB
             (ii) $M_p = 15.117$ dB, (iii) $\omega_p = 3.548$, (iv) 5.3, (v) yes

2 A stereo amplifier has the following transfer function

$$G(s) = \frac{4.05 \times 10^8 s}{(s + 30)(s + 135{,}000)}$$

Use the frequency response program given in Fig. 4.1 and determine the response of the amplifier for $6.28 \leq \omega \leq 628{,}000$. Use log stepping with 10 increments per decade. Have the program plot dB magnitude and phase versus ω. What values are given by the **PRTPLT** for the low-frequency cutoff, high-frequency cutoff, bandwidth, and mid-frequency gain?

Answers:
      Low frequency cutoff $\doteq 31.5$ radians/sec
      High frequency cutoff $\doteq 1.27 \times 10^5$ radians/sec
      Bandwidth $\doteq 1.27 \times 10^5$ radians/sec
      Mid-frequency gain $\doteq 69.5$ dB

3 Make the necessary modifications to the program in Fig. 4.1 in order to find the frequency response of $V_o(s)/V_I(s)$ for the network given below. Use the **PRTPLT** statement to list the dB magnitude and phase versus ω (ω in log steps, 10 steps/per decade) over the range $0.1 \leq \omega \leq 30{,}000$. *Hint:* the transfer function is given by

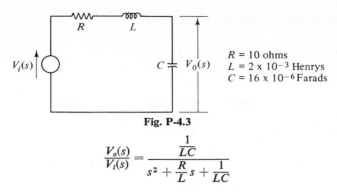

$$R = 10 \text{ ohms}$$
$$L = 2 \times 10^{-3} \text{ Henrys}$$
$$C = 16 \times 10^{-6} \text{ Farads}$$

Fig. P-4.3

$$\frac{V_o(s)}{V_i(s)} = \frac{\dfrac{1}{LC}}{s^2 + \dfrac{R}{L}s + \dfrac{1}{LC}}$$

Answer:

At $\omega = 1995.2$, $V_o(\omega) = 0.638$ dB, Phase $= -20.09°$

4 Consider the multiple-element mechanical system shown in Fig. P-4.4. The differential equations describing this system are

$$-B_3 \frac{dx_1}{dt} + \left(M_2 \frac{d^2 x_2}{dt^2} + (B_2 + B_3)\frac{dx_2}{dt} + K_2 x_2\right) = f(t)$$

$$\left(M_1 \frac{d^2 x_1}{dt^2} + (B_1 + B_3)\frac{dx_1}{dt} + K_1 x_1\right) - B_3 \frac{dx_2}{dt} = 0$$

Fig. P-4.4

By assuming zero initial conditions, taking the Laplace transform, and solving the simultaneous equations one finds

$$\frac{X_1(s)}{F(s)} = \frac{sB_3}{C_4 s^4 + C_3 s^3 + C_2 s^2 + C_1 s + C_0}$$

where: $C_4 = M_1 M_2$

$C_3 = M_1(B_2 + B_3) + M_2(B_1 + B_3)$

$C_2 = M_2 K_1 + M_1 K_2 + B_1(B_2 + B_3) + B_2 B_3$

$C_1 = K_1(B_2 + B_3) + K_2(B_1 + B_3)$

$C_0 = K_1 K_2$

Let $M_1 = M_2 = K_1 = K_2 = B_3 = 1.0$, and $B_1 = B_2 = 0.5$. Find the frequency response of the transfer function, $X_1(s)/F(s)$, over the relevant range of ω. Solve this problem by making appropriate changes of the program in Fig. 4.1.

Answers:

At $\omega = 1.0$, magnitude $= -1.938$ dB, phase $= -90°$

235

5   A particular process-transfer function is given by[16]

$$T(s) = \frac{1.25(s^2 + 0.345s + 0.32)e^{-0.5s}}{(s + 1)(s + 0.2)^2}$$

Find the frequency response over the range $.001 \le \omega \le 10$ by making appropriate changes in the program of Fig. 4.1. Use log stepping for ω with 16 steps per decade. What is the phase margin and gain margin?

Answer:

Phase margin $= 62°$, gain margin $= 0.677$ absolute

6   The transfer function of a particular band-reject filter is given by

$$G(s) = \frac{s^4 + 1.6 \times 10^3 s^3 + 8.8 \times 10^5 s^2 + 1.92 \times 10^8 s + 1.44 \times 10^{10}}{s^4 + 1.204 \times 10^4 s^3 + 3.648 \times 10^7 s^2 + 1.4448 \times 10^9 s + 1.44 \times 10^{10}}$$

Find the frequency response of this filter over the range $1 \le \omega \le 100{,}000$. Use log stepping for ω with 10 steps per decade. PRTPLT the dB magnitude of the filter response. What band of frequencies are rejected by this filter?

Answer:

Approximately 14 to 8500 radians/sec

7   A signal is given by

$$y(t) = 10.0*SIN(10.0*TIME) + 1.1*SIN(350.0*TIME)$$

Find the response of the filter in Prob. 6 to this signal over the time interval $0 \le TIME \le 1.4$ with OUTDEL $= 0.014$. Use the transfer function macro given in Example 3.12 in developing your program. Comment on the filter response for this $y(t)$.

Answer:

For $\omega = 350$, signal is attenuated by 40 dB

8   A pulse $r(t)$ defined by

$$r(t) = 1 \qquad 0.0 \le t \le 0.6$$
$$r(t) = 0 \qquad \text{all other } t$$

is applied to a linear, time-invariant, minimum-phase system. Data for the pulse response are given below.

Time	Response	Time	Response
0.000	0.000	0.645	0.5092
0.045	0.2899	0.675	0.3773
0.075	0.4221	0.720	0.2405
0.120	0.5590	0.765	0.1534
0.165	0.6464	0.855	0.0624
0.255	0.7380	0.915	0.0342
0.330	0.7710	0.975	0.0188
0.390	0.7838	1.060	0.0076
0.480	0.7934	1.200	0.0002
0.600	0.7980		

Find the frequency response of this system. *Hint:* Use the array-integrator frequency response program as shown in Fig. 4.8. Use the data above to define the AFGEN function. An approximate frequency range should start at $\omega_{min} = 0.1$ with NUM-DEC (number of decades) = 4 and XDECAD (number of points per decade) = 10. Dimensioned variables in Fig. 4.8 should be changed from 21 to 41. Allow approximately 3 min of CPU time on a 360/65 to run this problem.

Answer:

At $\omega = 19.95$ radians/sec, magnitude = -8.85 dB, phase = $-62.9°$

9 Data given in Prob. 8 were measured from the system configuration of Fig. P-4.9.

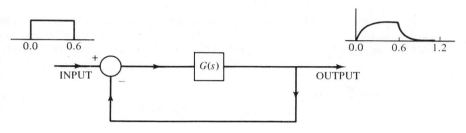

Fig. P-4.9

(a) From the frequency response of Prob. 8, determine a reasonable model for $G(s)$.

(b) Use the $G(s)$ determined from part (a) above in the configuration of Fig. P-4.9. Apply a pulse input as shown in this diagram and use CSMP to find the output response. Compare this response to the data given in Prob. 8 to determine the validity of your model. An introductory background in feedback systems is necessary for working this problem.

Answers:

$$(a) \ \ G(s) = \frac{8}{(s+2)}$$

10 The transfer function of a system is given by

$$\frac{C(s)}{R(s)} = \frac{300s^2 + 720s + 132}{s^5 + 22s^4 + 137s^3 + 496s^2 + 725s + 132}$$

(a) Express this transfer function in integration form as shown in Fig. 4.11.

(b) Write a set of state variable equations from your diagram by using the output of each integration as a state.

(c) Use the equations in part (b) and write a CSMP program for finding $c(t)$ (the output) when $r(t)$ (the input) is a unit step. *Hint:* Use variable-step integration with OUTDEL = 0.12 and FINITIM = 6.0.

Answer:

$$(c) \ \ \text{At TIME} = 0.84, \ c(t) = 1.1923$$

11 The transfer function of a system is given by

$$\frac{C(s)}{R(s)} = \frac{150}{s^3 + 12s^2 + 20s + 150}$$

(a) Express this transfer function in integration form as shown in Fig. 4.10.
(b) Write a set of state variable equations from your diagram by using the output of each integrator as a state.
(c) Use the equations in part (b) and write a CSMP program for finding $c(t)$ (the output) when $r(t)$ (the input) is a unit step. *Hint:* Use variable-step integration with OUTDEL = 0.08, FINTIM = 4.0.

Answer:

$$\text{At TIME} = 1.12, \; c(t) = 1.3443$$

12   A signal given by

$$f(t) = (4*e^{-0.1t}) + \sin(1.5t) \text{ sec}$$

is applied to the input of a sample and hold system as illustrated in Fig. P-4.12. The ideal sampler closes every $T = 0.4$ sec. Write a CSMP program which will PRTPLT $f(t)$, $f^*(t)$, and $x(t)$. Use an OUTDEL of 0.1 sec and a FINTIM of 5.0 sec.

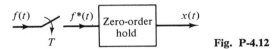

Fig. P-4.12

Answer:

$$\text{At TIME} = 2.1, \; x(t) = 3.416$$

13   The model of a continuous process is given in Fig. P-4.13.

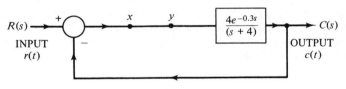

Fig. P-4.13

(a) If $r(t)$, the input, is a unit step, write a CSMP program which gives the output response, $c(t)$. Assume all initial conditions are zero.
(b) Remove the segment of the system between x and y in Fig. P-4.13 and replace it by a sample and hold as given in Prob. 12. This replacement is representative of what happens when a digital computer is used to close the process loop. Develop a CSMP simulation for this case and determine the unit-step response for $T = 0.08, 0.32$. A fixed-step integration should be used for this case with DELT being a sub-multiple of T.

Answers:

(a) At TIME = 0.96, $c(t) = 0.50984$
(b) At TIME = 1.04, T = 0.08, $c(t) = 0.65624$
   At TIME = 1.04, T = 0.32, $c(t) = 0.47167$
   Both for METHOD RKSFX, DELT = 0.008

14 The system diagram of a proposed direct digital control system using a PID control algorithm is shown in Fig. P-4.14. The digital control portion can be expressed in a

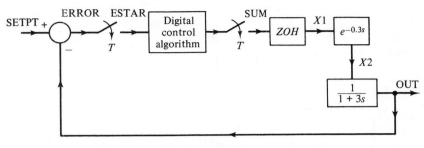

Fig. P-4.14

PROCEDURE algorithm by

```
    PROCEDURE SUM = PID(A1,SETPT,T,K,TI,TD,ESTAR,ESTAR1,INTG1)
        IF(A1.NE.1.0) GO TO 10
        IF(KEEP.NE.1.0) GO TO 10
        EPROP = ESTAR
        EDERIV = TD*(1.0/T)*(ESTAR - ESTAR1)
        INTG = INTG1 + T*ESTAR
        EINTG = (1.0/TI)*INTG
        SUM = K*(EPROP + EDERIV + EINTG)
        ESTAR1 = ESTAR
        INTG1 = INTG
10   CONTINUE
ENDPRO
```

Find the unit-step response of this system using the above control algorithm with T = 0.02, K = 9.6, TI = 1.67, TD = 0.15, INTG1 = 0.0, and ESTAR1 = 0.0. Use RKSFX with DELT = 0.02, FINTIM = 10.0, OUTDEL = 0.2. Give the PRTPLT of OUT. A review of Example 4.6 may be helpful in working this problem.

Answer:
$$\text{At TIME} = 0.8, \text{ out}(t) = 1.1431$$

15 A cascade form of a PID controller for the system in Prob. 14 is given in Fig. P-4.15.[16] Implement this controller in discrete form using a PROCEDURE function. Parameters of the controller are K = −9.6, Sampling Interval = 0.02, TI = 1.67, TD = 0.15, all initial conditions are zero. Use RKSFX fixed-step integration, FINTIM = 10.0, DELT = 0.02, OUTDEL = 0.2. Give the PRTPLT of OUT.

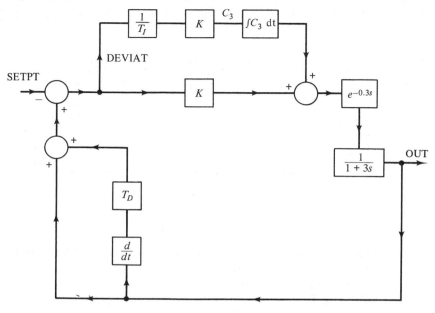

Fig. P-4.15

Answer:

$$\text{At TIME} = 1.2, \text{ out}(t) = 0.94849$$

16 Consider the system shown in Fig. P-4.16. It has been determined that a suitable controller, $D(z)$, for this system with a sampling time of $T = 0.5$ is

$$D(z) = \frac{A0 + A1*z^{-1}}{1 + B1*z^{-1}} = \frac{(0.25 - 0.24*z^{-1})}{(1.0 - 0.99*z^{-1})}$$

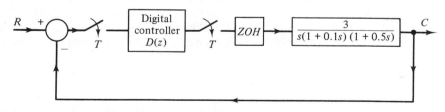

Fig. P-4.16

Use the method given in Example 4.4 for simulating the above system. Let the input, R, be a unit step. Give the PRTPLT of C and the output of the zero-order hold. Let FINTIM = 10.0, OUTDEL = 0.15, DELT = 0.025, METHOD RKSFX.

Answers:
$$\text{At TIME} = 4.2, \text{ZOH} = -0.026969, \; c(t) = 1.1697$$

17 The diagram for a digital control system is given in Fig. 4.21. Find the unit-step response of the system when the controller is expressed as a parallel realization. Use the PROCEDURE approach. See Example 4.5 for further details.

Answers:
(For $T = 0.1$, DELT $= 0.05$, OUTDEL $= 0.1$, METHOD RKSFX)
At TIME $= 1.4$, output $= 1.1532$

18 Although the subject was not discussed in the text, CSMP can be used for simulating and finding the response of digital filters. For example, the exponential filter can be expressed as

$$\frac{Y(z)}{X(z)} = \frac{(1 - \alpha)z}{z - \alpha}$$

where: $z =$ the z-transform variable

        $\alpha =$ a filter parameter

   $Y(z) = z$-transform of the filter output

   $X(z) = z$-transform of the filter input

A digital algorithm for this filter is

$$Y(nT) = (1 - \alpha)X(nT) + \alpha Y[(n - 1)T]$$

where: $Y(nT) = n$th value of the filter output

   $Y((n - 1)T) = (n - 1)$st value of the filter output

      $X(nT) = n$th value of the filter input

          $T =$ filter sampling time

The frequency response of the discrete filter for several values of α is shown in Fig. P-4.18(a).

Let an input signal to the filter be given by

$$y(t) = \underset{\text{term A}}{\sin (0.05t)} + \underset{\text{term B}}{0.5 \sin (0.44t)}$$

We wish to design a filter which will attenuate term B by 15 dB but pass term A. To achieve this we select $\alpha = 0.9$ and $T = 1.43$ sec. The program in Fig. P-4.18(b) can be used to simulate this filter. Run this program and observe the output. Comment on your results.

Answer:
$$\text{At TIME} = 80.08, \; x(t) = -1.0729, \; y(t) = -0.21010$$

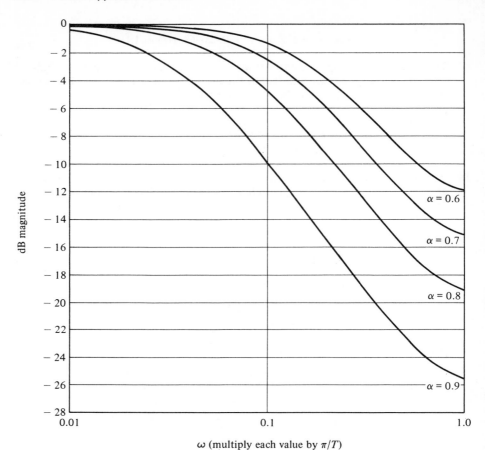

Fig. P-4.18(a) Magnitude response of the digital exponential filter.

```
            ***PROBLEM INPUT STATEMENTS***

PARAM  ALPHA = 0.9, Y1 = 0.0
*   SAMPLING TIME = 1.43
DYNAMIC
      NOSORT
      X = SIN(0.05*TIME) + 0.5*SIN(0.44*TIME)
      Y = (1.0 - ALPHA)*X + ALPHA*Y1
      Y1 = Y
*  WHEN USING THIS PROGRAM BE SURE THAT DELT ON THE
*  TIMER CARD IS SET EQUAL TO THE SAMPLING TIME
*  MAKE SURE Y1 IS UPDATED AS SHOWN
TIMER    FINTIM = 250.0,  OUTDEL = 2.86, DELT = 1.43
PRTPLT  X, Y
LABEL ILLUSTRATICN OF EXPONENTIAL FILTER,  PROB 4-18
END
STOP
ENDJCB
```

Fig. P-4.18(b) Program listing for Problem 4.18.

19 The algorithm of a third-order discrete Butterworth filter is given by

$$Y(n) = 0.0039*X(n) + 0.0116*X(n-1) + 0.0116*X(n-2) + 0.0039*X(n-3)$$
$$+ 2.3051*Y(n-1) - 1.8317*Y(n-2) + 0.4957*Y(n-3)$$

where the coefficients have been selected to give a bandwidth of $\pi/9T$. Let the input to the filter be identical to the input signal given in Prob. 18. Use a sampling time of $T = 2$ sec. This should attenuate term B by 30 dB. Write a CSMP program to simulate this filter. Use a FINTIM $= 125$, DELT $= 2.0$, and OUTDEL $= 2.0$. Compare the results to Prob. 18.

Answer: At TIME $= 68$, $x(t) = 0.29501$, $y(t) = -0.14972$

5
CSMP III

The preceding chapters specifically dealt with the form of CSMP called System/360 Continuous System Modeling Program (S/360 CSMP). An extension of S/360 CSMP called CSMP III is available. It incorporates the following major changes and improvements.

1 Additional output versatility including extensive x–y plotting capability is available.

2 Several new functional blocks have been added.

3 Double precision calculations can be used to improve accuracy.

4 The model size has been increased.

There is very little difference in using the two forms of CSMP. Most programs written for S/360 CSMP are completely compatible with CSMP III and will run without modifications. The following is a brief check list of items that should be considered when running a program on a different form of CSMP.

1 The S/360 CSMP PRTPLT statement is treated as an OUTPUT instruction in CSMP III. The OUTPUT statement cannot be used in S/360 CSMP.

2 Subscripted variables are allowed in CSMP III PRINT and OUTPUT statements, but are not permitted in any S/360 CSMP output instructions.

3 The S/360 CSMP CONTINUE statement is equivalent to the END CONTINUE instruction used in CSMP III.

4 In S/360 CSMP data cards are placed between the labels DATA and ENDDATA. The equivalent labels in CSMP III are INPUT and ENDINPUT.

5 The use of the array form of the INTGRL function is considerably different in the two forms of CSMP.

6 There are several function blocks that can only be used in CSMP III. They are listed in this chapter and in Appendix I.

7 The JCL (job control language) cards are different.

It is the intent of this chapter to point out the major differences in using the two forms of the CSMP program and to describe the additional features of CSMP III. This material does not contain a detailed description of all the features and capabilities of CSMP III. The reader should refer to the IBM manual[1] for additional information.

This chapter is divided into three sections: output, functions, and control statements. The greatest difference between the two forms of the CSMP program exists in output capability. Output capabilities are covered in the first section. The second section contains descriptions of the additional CSMP III functions. A few control statements are slightly different in CSMP III. This is covered in the last section.

Output Statements

Additional output statements are provided in CSMP III as well as methods for controlling the format of the output. The beginning user should start with the standard output statements and gradually work up to using the various available options.

The four major output statements are PRINT, OUTPUT, PREPARE, and RANGE: Each of these statements can be used alone or with auxiliary instructions to alter the standard output format. FORTRAN output capability is available in both forms of CSMP.

PRINT

The PRINT statement is used exactly as in S/360 CSMP. The changes listed below provide additional flexibility.

1 Up to 55 variables can be listed on a PRINT card.

2 The PRINT statement can be used for subscripted variables. For example, the following card will print the values of A, B, X(1), Y(6), Y(7), Y(8), and Y(9).

PRINT A, B, X(1), Y(6–9)

TITLE

As in S/360 CSMP, the TITLE statement is used for specifying the heading of PRINT output. Each TITLE statement, which can be continued to one additional card, provides up to 120 characters in one line of output. A maximum of six cards can be included in a program. This means that three TITLE statements, each containing one continuation card; six TITLE cards with no continuations; or an equivalent number of cards can be used.

For example, the following two TITLE statements provide a total of two lines of heading for PRINT output.

TITLE SIMULATION OF THE CONTROL SYSTEM FOR NATURAL GAS $\cdots$
TRANSMISSION
TITLE ALL UNITS ARE GIVEN IN THE METRIC SYSTEM

RERUN

The RERUN card is used to print in the heading the values of listed parameters. For example, when several runs are made using the CALL RERUN statement, the current values of X, Y, and Z are printed as part of the heading by using the following statement.

RERUN X, Y, Z

PRINTPAGE

This statement allows the user to specify the size of pages used for PRINT output. If the size is not specified by the HEIGHT and WIDTH variables, the height will be the standard sixty lines and the width will be 132 characters. An example of a PRINTPAGE statement is shown below.

PRINTPAGE HEIGHT = 50, WIDTH = 104

This instruction will provide a page 50 lines high and 104 characters wide. Page width determines the maximum number of variables that can be listed in column form as summarized below.

Page Width	Maximum Number of Output Variables in Column Form
132	9
120	8
104	7
91	6
78	5

OUTPUT

The OUTPUT statement provides an additional means for both printing and plotting output. It is used in place of the S/360 CSMP PRTPLT statement. If PRTPLT is used in CSMP III, it is automatically treated as an OUTPUT statement. Up to fifty-five variables can be included in one statement. An example of a valid OUTPUT card is shown below.

OUTPUT X, W, T(3), Q(3–12), P

The output format is automatically determined by the number of variables that are listed on the OUTPUT statement. For five or fewer variables, the output is both plotted and printed on the same plot. When there are more than five variables the output is only printed. The following example illustrates the use of the OUTPUT statement with the three different formats.

Example 5.1

Consider the problem of calculating the steady-state error of an accelerometer for a sinusoidal input of frequency ω. Fig. 5.1 shows the basic elements of an accelerometer.

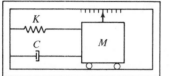

$$\zeta = \frac{C}{2\sqrt{KM}}$$

$$\omega_N = \sqrt{\frac{K}{M}}$$

Fig. 5.1   Elements of an accelerometer.

The steady-state percent error is given by the following expression

$$\text{ERROR} = \frac{100}{\left[\left(1 - \left(\frac{\omega}{\omega_n}\right)^2\right)^2 + \left(2\zeta\frac{\omega}{\omega_n}\right)^2\right]^{1/2}} - 100 \qquad (5.1)$$

Symbols used in program

where ζ = damping factor   ZETA

$\frac{\omega}{\omega_n}$ = ratio of exciting   FREQ
       frequency to natural
       frequency

The program shown in Fig. 5.2 calculates the percent error as a function of frequency ratio and damping factor. The independent variable is the frequency ratio (FREQ) and each subscripted variable (ERROR(I)) represents the error for a particular value of ζ. The ERROR(I) variable corresponds to a damping factor of I/20. The printing format depends on the number of variables listed in the OUTPUT statement. For 1–5 variables, both a printer-plot and tabular listing are given on the same page. The plot for each variable is independently scaled and identified by a unique symbol. The output for the following instruction containing four variables is given in Fig. 5.3.

OUTPUT ERROR(1), ERROR(8–9), ERROR(14)

As with the S/360 CSMP PRTPLT instruction, multiple OUTPUT statements can be used to provide separate plots. For example, the following three cards will provide three separate plots of X, Y, and Z.

OUTPUT X
OUTPUT Y
OUTPUT Z

```
LABEL   STEADY-STATE ERROR OF AN ACCELEROMETER AS A FUNCTION OF FREQUE...
NCY AND DAMPING FACTOR
  RENAME   TIME = FREQ
  FIXED  I
  STORAGE   ERROR(20)
NOSORT
  DO 1  I = 1,20
  ZETA = FLOAT(I)/20.0
1 ERROR(I) = 100.0/SQRT((1.0 - FREQ*FREQ)**2 + (2.0*ZETA*FREQ)**2)    ...
  - 1CC.0
  OUTPUT   ERROR(1), ERROR(8-9), ERROR(14)
LABEL   TYPICAL OUTPUT FOR 1-5 VARIABLES
  OUTPUT   ERROR(4), ERROR(6-10)
LABEL   TYPICAL OUTPUT FOR 6-9 VARIABLES
  OUTPUT   ERROR(1-20)
LABEL   TYPICAL OUTPUT FOR 10-55 VARIABLES
  TIMER   FINTIM = 1.0, OUTDEL = 0.02
END
STOP
ENDJOB
```

Fig. 5.2 Program to calculate the steady-state error of an accelerometer.

In S/360 CSMP, three separate PRTPLT cards or the following single statement will also give three separate plots.

PRTPLT X, Y, Z

For 6–9 variables, a column listing is used. It is similar to the format used by the PRINT statement. The output from the following instruction containing six variables is given in Fig. 5.4.

OUTPUT ERROR(4), ERROR(6–10)

For 10–55 variables a row format is used. The names of the variables are listed in the left column and each successive column contains the magnitude of the variables for increasing values of the independent variable (TIME). Figure 5.5 shows an example of this type of output for the following statement.

OUTPUT ERROR(1–20)

Similar to the PRTPLT instruction, the print interval for the OUTPUT statement is specified by OUTDEL.

LABEL

This statement specifies the heading that appears at the top of each page of OUTPUT printout. Only one continuation card is allowed for each LABEL instruction. LABEL statements provide headings for all OUTPUT instructions that follow the LABEL cards. LABEL instructions that follow an OUTPUT statement refer only to the statement they follow. Examples are shown by the program of Fig. 5.2 and outputs in Figs. 5.3, 5.4, and 5.5. A typical LABEL statement is shown below.

LABEL SIMULATION OF THE INDUSTRIAL DYNAMICS OF A ...
MANUFACTURING PLANT

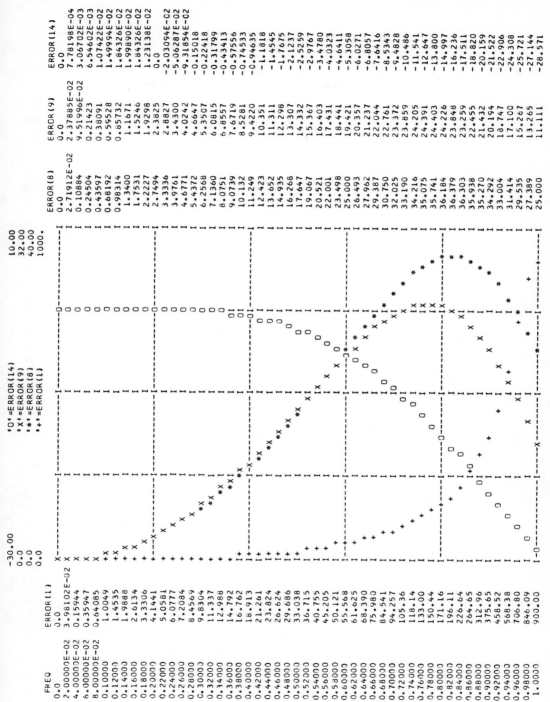

Fig. 5.3 Typical printer-plot from OUTPUT statement for 1–5 variables.

249

STEADY-STATE ERROR OF AN ACCELEROMETER AS A FUNCTION OF FREQUENCY AND DAMPING FACTOR
TYPICAL OUTPUT FOR 6-9 VARIABLES

FREQ	ERROR(4)	ERROR(6)	ERROR(7)	ERROR(8)	ERROR(9)	ERROR(10)
0.0	0.0	0.0	0.0	0.0	0.0	0.0
2.00000E-02	3.68042E-02	3.27911E-02	3.01971E-02	2.71912E-02	2.37885E-02	1.99890E-0
4.00000E-02	0.14738	0.13132	0.12088	0.10884	9.51996E-02	7.99561E-0
6.00000E-02	0.33218	0.29585	0.27225	0.24504	0.21423	0.17982
8.00000E-02	0.59196	0.52689	0.48465	0.43597	0.38091	0.31946
0.10000	0.92775	0.82510	0.75854	0.68192	0.59528	0.49870
0.12000	1.3409	1.1914	1.0945	0.98314	0.85732	0.71727
0.14000	1.8332	1.6268	1.4934	1.3400	1.1671	0.97485
0.16000	2.4066	2.1328	1.9559	1.7531	1.5246	1.2711
0.18000	3.0636	2.7107	2.4832	2.2227	1.9298	1.6053
0.20000	3.8068	3.3623	3.0763	2.7494	2.3825	1.9771
0.22000	4.6397	4.0895	3.7365	3.3336	2.8827	2.3856
0.24000	5.5657	4.8946	4.4651	3.9761	3.4300	2.8298
0.26000	6.5891	5.7798	5.2635	4.6771	4.0242	3.3088
0.28000	7.7144	6.7478	6.1333	5.4372	4.6647	3.8210
0.30000	8.9469	7.8014	7.0759	6.2568	5.3507	4.3650
0.32000	10.292	8.9437	8.0932	7.1360	6.0815	4.9389
0.34000	11.757	10.178	9.1865	8.0751	6.8557	5.5405
0.36000	13.349	11.507	10.358	9.0739	7.6719	6.1671
0.38000	15.075	12.936	11.608	10.132	8.5281	6.8159
0.40000	16.945	14.467	12.938	11.249	9.4220	7.4834
0.42000	18.968	16.105	14.351	12.423	10.351	8.1655
0.44000	21.156	17.853	15.846	13.652	11.311	8.8578
0.46000	23.520	19.715	17.423	14.935	12.298	9.5549
0.48000	26.073	21.696	19.083	16.268	13.307	10.251
0.50000	28.831	23.797	20.824	17.647	14.332	10.940
0.52000	31.810	26.022	22.645	19.067	15.367	11.614
0.54000	35.026	28.373	24.542	20.521	16.403	12.265
0.56000	38.499	30.851	26.510	22.001	17.431	12.885
0.58000	42.250	33.456	28.544	23.498	18.441	13.463
0.60000	46.301	36.184	30.632	25.000	19.421	13.990
0.62000	50.676	39.030	32.766	26.493	20.357	14.455
0.64000	55.397	41.986	34.929	27.962	21.237	14.846
0.66000	60.490	45.040	37.103	29.387	22.044	15.152
0.68000	65.977	48.172	39.267	30.750	22.761	15.361
0.70000	71.878	51.359	41.393	32.025	23.372	15.462
0.72000	78.207	54.568	43.451	33.190	23.859	15.444
0.74000	84.969	57.759	45.403	34.216	24.205	15.296
0.76000	92.152	60.881	47.211	35.075	24.391	15.009
0.78000	99.723	63.874	48.829	35.741	24.403	14.576
0.80000	107.61	66.667	50.210	36.184	24.226	13.990
0.82000	115.71	69.179	51.307	36.379	23.848	13.248
0.84000	123.85	71.326	52.072	36.303	23.259	12.347
0.86000	131.78	73.016	52.461	35.938	22.455	11.289
0.88000	139.18	74.162	52.436	35.270	21.432	10.077
0.90000	145.66	74.688	51.969	34.292	20.194	8.7149
0.92000	150.77	74.529	51.043	33.004	18.747	7.2117
0.94000	154.06	73.645	49.652	31.414	17.100	5.5766
0.96000	155.15	72.025	47.807	29.535	15.267	3.8210
0.98000	153.81	69.684	45.530	27.389	13.265	1.9576
1.0000	150.00	66.667	42.857	25.000	11.111	0.0

Fig. 5.4 Typical print-out from OUTPUT statement for 6–9 variables.

STEADY-STATE ERROR OF AN ACCELEROMETER AS A FUNCTION OF FREQUENCY AND DAMPING FACTOR
TYPICAL OUTPUT FOR 10-55 VARIABLES

FREQ	0.0	2.00000E-02	4.00000E-02	6.00000E-02	8.00000E-02	0.10000	0.12000	0.14000	0.16000
ERROR(1)	0.0	3.98102E-02	0.15944	0.35947	0.64085	1.0049	1.4535	1.9888	2.6134
ERROR(2)	0.0	3.91998E-02	0.15703	0.35400	0.63107	0.98949	1.4310	1.9576	2.5719
ERROR(3)	0.0	3.82080E-02	0.15302	0.34492	0.61476	0.96375	1.3934	1.9057	2.5029
ERROR(4)	0.0	3.68042E-02	0.14738	0.33218	0.59196	0.92775	1.3409	1.8332	2.4066
ERROR(5)	0.0	3.50037E-02	0.14015	0.31583	0.56265	0.88152	1.2735	1.7402	2.2831
ERROR(6)	0.0	3.27911E-02	0.13132	0.29585	0.52689	0.82510	1.1914	1.6268	2.1328
ERROR(7)	0.0	3.01971E-02	0.12088	0.27225	0.48465	0.75854	1.0945	1.4934	1.9559
ERROR(8)	0.0	2.71912E-02	0.10884	0.24504	0.43597	0.68192	0.98314	1.3400	1.7531
ERROR(9)	0.0	2.37885E-02	9.51996E-02	0.21423	0.38091	0.59528	0.85732	1.1671	1.5246
ERROR(10)	0.0	1.99890E-02	7.99561E-02	0.17982	0.31946	0.49870	0.71727	0.97485	1.2711
ERROR(11)	0.0	1.57776E-02	6.31256E-02	0.14185	0.25169	0.39229	0.56314	0.76364	0.99303
ERROR(12)	0.0	1.11847E-02	4.46930E-02	0.10028	0.17761	0.27614	0.39516	0.53383	0.69113
ERROR(13)	0.0	6.17981E-03	2.46735E-02	5.51910E-02	9.72900E-02	0.15033	0.21350	0.28581	0.36603
ERROR(14)	0.0	7.78198E-04	3.06702E-03	6.54602E-03	1.07422E-02	1.49999E-02	1.84326E-02	1.99850E-02	1.84326E-02
ERROR(15)	0.0	-5.06592E-03	-2.01263E-02	-4.55780E-02	-8.19550E-02	-0.12973	-0.18980	-0.26311	-0.35097
ERROR(16)	0.0	-1.12610E-02	-4.49066E-02	-0.10129	-0.18079	-0.28378	-0.41106	-0.56320	-0.74120
ERROR(17)	0.0	-1.77460E-02	-7.11975E-02	-0.16045	-0.28558	-0.44699	-0.64481	-0.87968	-1.1517
ERROR(18)	0.0	-2.47955E-02	-9.18216E-02	-0.22305	-0.39650	-0.61919	-0.89111	-1.2120	-1.5816
ERROR(19)	0.0	-3.21350E-02	-0.12868	-0.28918	-0.51329	-0.80026	-1.1495	-1.5597	-2.0300
ERROR(20)	0.0	-3.99475E-02	-0.15977	-0.35873	-0.63596	-0.99013	-1.4195	-1.9223	-2.4961

FREQ	0.18000	0.20000	0.22000	0.24000	0.26000	0.28000	0.30000	0.32000	0.34000
ERROR(1)	3.3306	4.1441	5.0581	6.0777	7.2084	8.4569	9.8304	11.337	12.988
ERROR(2)	3.2770	4.0764	4.9740	5.9747	7.0837	8.3072	9.6520	11.126	12.738
ERROR(3)	3.1879	3.9638	4.8343	5.8037	6.8768	8.0590	9.3566	10.776	12.326
ERROR(4)	3.0636	3.8068	4.6397	5.5657	6.5891	7.7144	8.9469	10.292	11.757
ERROR(5)	2.9043	3.6061	4.3910	5.2621	6.2226	7.2762	8.4270	9.6793	11.038
ERROR(6)	2.7107	3.3623	4.0895	4.8946	5.7798	6.7478	7.8014	8.9437	10.178
ERROR(7)	2.4832	3.0763	3.7365	4.4651	5.2635	6.1333	7.0759	8.0932	9.1865
ERROR(8)	2.2227	2.7494	3.3336	3.9761	4.6771	5.4372	6.2568	7.1360	8.0751
ERROR(9)	1.9298	2.3825	2.8827	3.4300	4.0242	4.6647	5.3507	6.0815	6.8557
ERROR(10)	1.6053	1.9771	2.3856	2.8298	3.3088	3.8210	4.3650	4.9389	5.5405
ERROR(11)	1.2504	1.5346	1.8445	2.1785	2.5349	2.9119	3.3071	3.7180	4.1418
ERROR(12)	0.86581	1.0565	1.2616	1.4791	1.7070	1.9431	2.1845	2.4286	2.6720
ERROR(13)	0.45276	0.54440	0.63914	0.73494	0.82953	0.92043	1.0049	1.0802	1.1429
ERROR(14)	1.23138E-02	0.0	-2.03094E-02	-5.06287E-02	-9.31854E-02	-0.15018	-0.22418	-0.31799	-0.43413
ERROR(15)	-0.45436	-0.57500	-0.71442	-0.87425	-1.0565	-1.2630	-1.4957	-1.7568	-2.0485
ERROR(16)	-0.94608	-1.1788	-1.4406	-1.7326	-2.0561	-2.4123	-2.8026	-3.2282	-3.6901
ERROR(17)	-1.4616	-1.8096	-2.1964	-2.6222	-3.0876	-3.5928	-4.1382	-4.7239	-5.3502
ERROR(18)	-1.9997	-2.4657	-2.9793	-3.5398	-4.1467	-4.7991	-5.4962	-6.2369	-7.0203
ERROR(19)	-2.5589	-3.1451	-3.7868	-4.4822	-5.2295	-6.0264	-6.8709	-7.7606	-8.6931
ERROR(20)	-3.1383	-3.8462	-4.6166	-5.4463	-6.3320	-7.2700	-8.2569	-9.2888	-10.362

Fig. 5.5 Typical print-out from OUTPUT statement for 10-55 variables.

PAGE

The PAGE card provides a number of options for selecting the format and page size for OUTPUT printout. The following parameters which appear on a PAGE card can be used to provide logarithmic scaling, contoured and shaded printer-plots, merged printer-plots, reduced size pages, and a method for interfacing with x–y plotting devices.

HEIGHT and WIDTH
XYPLOT
MERGE
CONTOR
SHADE
NTAB and NPLOT
GROUP
LOG
SYMBOL

As in the use of the LABEL statement, a PAGE card which is placed before a group of OUTPUT statements applies to them all. PAGE statements which are placed after an OUTPUT statement refer only to the statement they follow. If instructions are not specified on a PAGE statement, the standard format is always automatically used.

HEIGHT and WIDTH.   These variables are used to control page height and width for both printer and x–y plotter output. If the HEIGHT variable is not specified, the standard 60 lines per page is used for printer-plot output. For x–y plotter output, the default option is either 8 in. or 8 cm depending on the local plotter system. The WIDTH variable can be used to reduce the size from the standard width of 132 characters to the minimum width of 52 characters. For x–y plotter output, the WIDTH variable refers to the width of the plot. The standard width is 10 in. or 10 cm. An example of specifying a printer-plot 120 characters wide and 50 lines high is shown below.

PAGE HEIGHT = 50, WIDTH = 120

XYPLOT.   For computer installations where a plotting subroutine has been incorporated into the CSMP III System, the XYPLOT symbol can be used to initiate off-line x–y plotting. Because of the various makes and models of digital plotters, each computer facility needs to develop its own plotting subroutine.[1,2] When the XYPLOT symbol is listed on the PAGE card, the variables contained in the OUTPUT statement are printed by an x–y plotter rather than a line printer. The first variable listed in the OUTPUT statement is plotted on the abscissa, and all the remaining variables are plotted on the ordinate. LABEL statements are automatically transferred to the plotting routines. If the scale is not specified, automatic scaling is used. An example of the method used to specify the scale is shown by the following statements.

OUTPUT X(4.0, 12.0), Y(0.0, 25.0)
PAGE XYPLOT

For this example, the variable X is plotted on the abscissa with a scale ranging from 4.0 to 12.0, and the variable Y is plotted on the ordinate with the scale of 0.0 to 25.0.

If automatic scaling is desired, the following statements should be used.

OUTPUT X, Y, Z
PAGE XYPLOT

The above statements will plot X on the abscissa versus the variables Y and Z on the ordinate. Note that the OUTPUT statement used with XYPLOT allows complete flexibility in the choice of the independent variable.

The PREPARE instruction described in Chap. 3 can also be used to produce a data set for off-line plotting.

MERGE.   This symbol is used to merge printer-plots from several runs to one plot for OUTPUT statements containing only one variable. As an example, consider the problem of plotting on one graph the dynamic response of a spring-mass-damper system for three values of damping.

Example 5.2

Fig. 5.6 shows a linear spring-mass-damper system with the given parameters and initial conditions. The equation of motion is given by Eq. (5.2).

$$m\ddot{x} + c\dot{x} + kx = f \qquad\qquad (5.2)$$

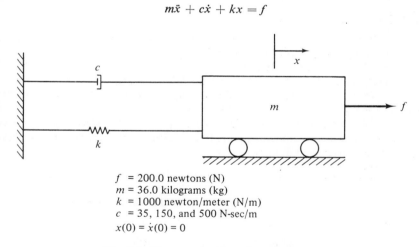

f = 200.0 newtons (N)
m = 36.0 kilograms (kg)
k = 1000 newton/meter (N/m)
c = 35, 150, and 500 N-sec/m
$x(0) = \dot{x}(0) = 0$

Fig. 5.6   Linear spring-mass-damper system.

The program of Fig. 5.7 uses a standard PARAMETER statement to make three runs for specific values of c. The MERGE parameter is used to combine the time-history of displacement for all three runs in one printer-plot.

Fig. 5.8 shows the resulting OUTPUT.

An alternate form of print-plot is available by specifying the CONTOR and SHADE parameters. They can effectively illustrate the transient response of distributed systems. For example, the transient and spatial temperature distribution in one-dimensional heat flow can be illustrated by using the CONTOR or SHADE parameter.

```
*    CSMP III PROGRAM TO SIMULATE SPINT—MASS—DAMPER SYSTEM AND TO
*       ILLLSTRATE THE USE OF THE MERGE STATEMENT
    CONSTANT  F = 200.0,  M = 36.0,  K = 1000.0
    PARAMETER  C = (35.0,  150.0,  500.0)

*    XDC = ACCELERATION            XD = VELOCITY

    XDD = (F - C*XD - K*X)/M
    XD = INTGRL(0.0,XDD)
    X = INTGRL(0.0,XD)
      OUTPUT  X
      PAGE  MERGE
      TIMER  FINTIM = 1.96,  CUTDEL = 0.04
   END
   STOP
   ENDJOB
```

Fig. 5.7 Program to simulate spring-mass-damper system and to illustrate MERGE output.

CONTOR. This parameter provides a contoured form of a printer-plot. Scaling for the plot is automatic when the following statement is used.

PAGE CONTOR

A second form of the CONTOR parameter can be used to specify the scale. In the following statement, the scaled range of the plot is between 3.0 and 45.0.

PAGE CONTOR = (3.0, 45.0)

Example 5.3

The problem of simulating the transient-temperature distribution in a rod illustrates printer-plots using both CONTOR and SHADE parameters. The copper rod shown in Fig. 5.9 is insulated on the outside surface to prevent heat flow in the radial direction. The initial temperature of the rod is 300°K and the left-hand end is connected to an infinite heat source which maintains this end at a temperature of 300°K. At the start of the simulation, the right-hand end is brought in contact with a surface having a constant temperature of 500°K. As in Example 3.3, the copper rod is divided into 20 equal-size elements. Assuming a constant temperature in each element, the transient temperature of the ith element is given by Eq. (5.3).

$$\frac{dT(i)}{dt} = Q[T(i-1) - 2.0*T(i) + T(i+1)] \qquad i = 1, 2, \ldots, 20 \qquad (5.3)$$

Symbols used in computer program

where $Q = k/(\rho c L^2)$ Q

        ρ = density of copper,   RO
            8890 kg/m³

        c = specific heat,   C
            398 joule/kg-°K

        k = thermal conductivity,   K
            386 watt/m-°K

        L = length of element,   L
            0.02 m

Figure 5.10 shows the program for solving the set of 20 equations.

MERGED OUTPUT PRESENTATION FOR X

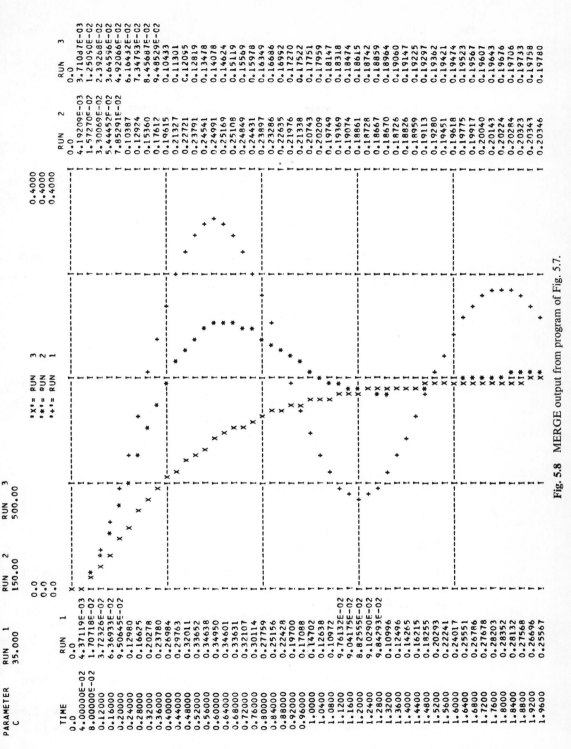

Fig. 5.8 MERGE output from program of Fig. 5.7.

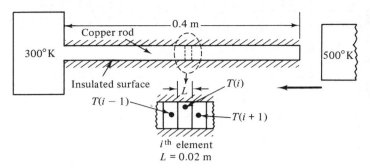

Fig. 5.9 Insulated copper rod.

Note in the CONTOR printer-plot of Fig. 5.11 there are 11 variable groups each covering a specific range of temperatures.

The following listing shows the symbols for each range:

Symbols	Percent range	Temperature values
L	less than 2.5%	300 to 305 °K
1	7.5% to 12.5%	315 to 325 °K
2	17.5% to 22.5%	335 to 345 °K
3	27.5% to 32.5%	355 to 365 °K
4	37.5% to 42.5%	375 to 385 °K
5	47.5% to 52.5%	395 to 405 °K
6	57.5% to 62.5%	415 to 425 °K
7	67.5% to 72.5%	435 to 445 °K
8	77.5% to 82.5%	455 to 465 °K
9	87.5% to 92.5%	475 to 485 °K
H	more than 97.5%	495 to 500 °K

The percent numbers in the above listing are valid for all CONTOR plots while the temperature values apply only to Fig. 5.11.

SHADE.   This parameter provides a shaded print-plot and is used exactly the same as the CONTOR parameter. Scaling is automatic when the following statement is used.

PAGE SHADE

The scale range of −5.0 to 4.5 is specified by the following form.

PAGE SHADE = (−5.0, 4.5)

The output from the SHADE parameter of the program of Fig. 5.10 is shown in Fig. 5.12.

Note that there are ten levels of intensity. Variables having values near the high side of the scale are represented by dark plotting. Variables in the lowest 10%

```
LABEL   CSMP III PROGRAM TO SIMULATE THE TRANSIENT TEMPERATURE
LABEL   DISTRIBUTION OF A COPPER ROD
INITIAL
  CONSTANT   RO = 8890.0, C = 398.0, K = 386.0, L = 0.02,              ...
  TLEFT = 300.0, TRIGHT = 500.C
  FIXED   I
  Q = K/(RO*C*L*L)
  TABLE  TI(1-20) = 20*300.0
DYNAMIC
NOSORT
  TD(1) = Q*(TLEFT - 2.0*T(1) + T(2))
  DO 1  I = 2,19
1   TD(I) = Q*(T(I-1) - 2.0*T(I) + T(I+1))
  TD(20) = Q*(T(19) - 2.0*T(20) + TRIGHT)
SORT
  T = INTGRL(TI, TD, 20)
  OUTPUT  T(1-20)
LABEL   EXAMPLE OF A CONTCR PRINTER-PLOT
  PAGE  CONTOR
  OUTPUT  T(1-20)
LABEL   EXAMPLE OF A SHADE PRINTER-PLOT
  PAGE  SHADE
  TIMER  FINTIM = 450.0,   OUTDEL = 10.0
END
STOP
ENDJOB
```

Fig. 5.10 Program to simulate the temperature of copper rod.

of the scale range are represented by blanks. At each 10% increase in magnitude of the output variables, the printing becomes one shade darker.

The remaining five PAGE parameters (NTAB, NPLOT, GROUP, LOG, and SYMBOL) are only applicable to printer-plot output. This occurs when OUTPUT statements contain less than six variables.

NTAB. This parameter specifies the number of variables that will be both print-plotted and tabulated. When NTAB is used on the PAGE card as shown below,

PAGE NTAB = 2

only the first two variables in the OUTPUT statement will be both print-plotted and tabulated. The remaining variables on the OUTPUT statement will only be print-plotted.

NPLOT. This parameter specifies the number of variables that will be both print-plotted and tabulated. All remaining variables listed on the OUTPUT statement will only be tabulated.

GROUP. This parameter specifies the number of variables that are to be print-plotted with the same scale. The common scale is determined by the entire group of variables. The following statement specifies that the first three variables listed on the OUTPUT card will be print-plotted with the same scale.

PAGE GROUP = 3

LOG. This parameter specifies that number of variables on the OUTPUT card that are to be print-plotted with a logarithmic ordinate scale.

SYMBOL. This parameter allows the user to specify the identification symbols used in printer-plots. To illustrate the use of several PAGE parameters, consider the output from the following statements used in the program of Fig. 5.2.

CSMP III PROGRAM TO SIMULATE THE TRANSIENT TEMPERATURE
DISTRIBUTION OF A COPPER ROD
EXAMPLE OF A CONTOR PRINTER-PLOT

CONTOUR PRESENTATION FOR T(I)

'1'= 320.	'2'= 340.	'3'= 360.	SCALE VALUES '4'= 380.	'5'= 400.	'L'= 300.
'6'= 420.	'7'= 440.	'8'= 460.	'9'= 480.	'H'= 500.	

TIME
0.0
10.000
20.000
30.000
40.000
50.000
60.000
70.000
80.000
90.000
100.00
110.00
120.00
130.00
140.00
150.00
160.00
170.00
180.00
190.00
200.00
210.00
220.00
230.00
240.00
250.00
260.00
270.00
280.00
290.00
300.00
310.00
320.00
330.00
340.00
350.00
360.00
370.00
380.00
390.00
400.00
410.00
420.00
430.00
440.00
450.00

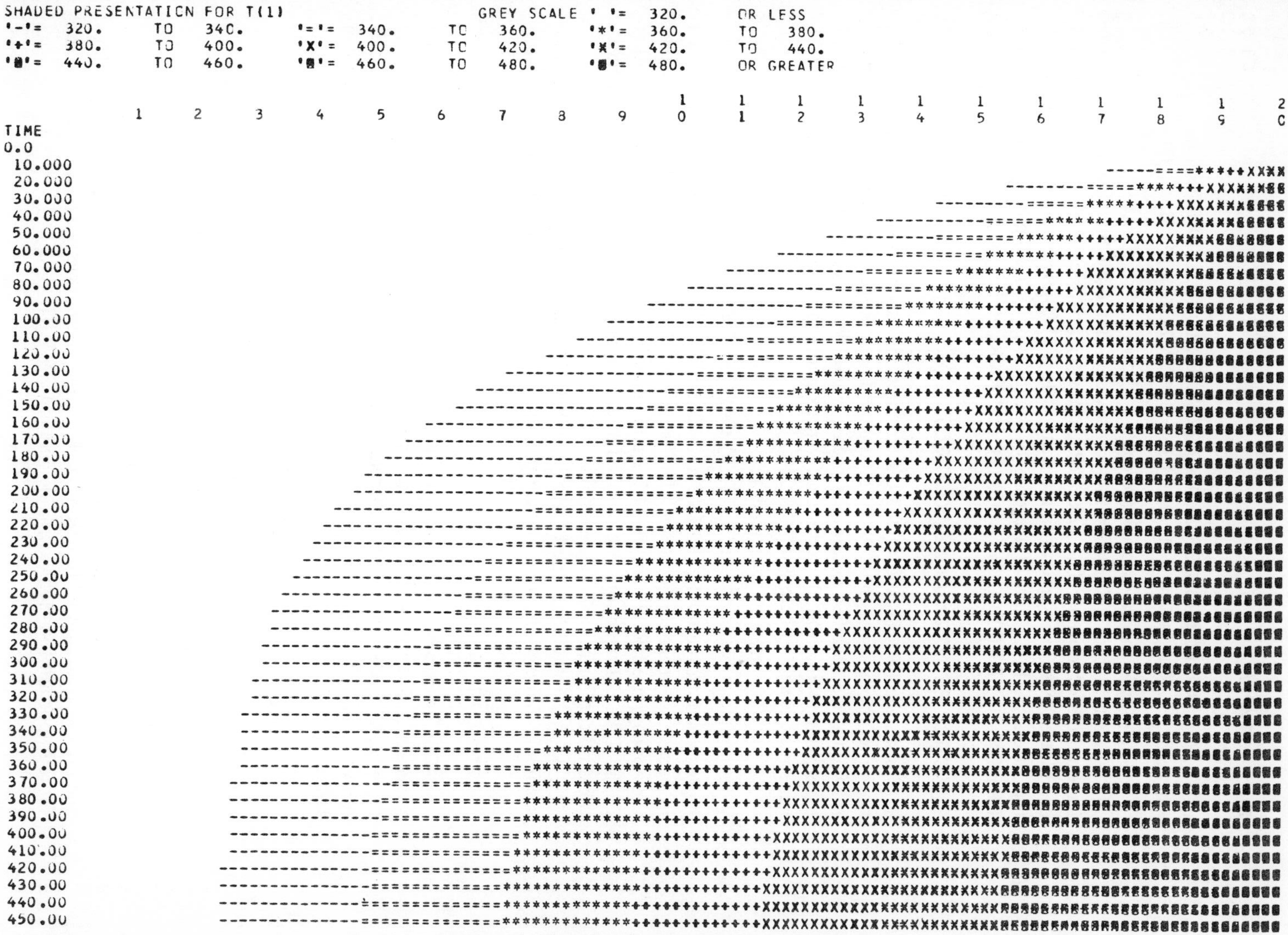

SHADED PRESENTATION FOR T(1)

			GREY SCALE		
'-'= 320. TO 340.	'='= 340. TO 360.		'-'= 320. TO 360. OR LESS	
'*'= 360. TO 380.	'+'= 380. TO 400.		'*'= 360. TO 380.	
'X'= 400. TO 420.	'■'= 420. TO 440.		'X'= 420. TO 440.	
'■'= 440. TO 460.	'■'= 460. TO 480.		'■'= 480. OR GREATER	

Column scale: 1 2 3 4 5 6 7 8 9 10 11 12 13 14 15 16 17 18 19 20

TIME
0.0
10.000
20.000
30.000
40.000
50.000
60.000
70.000
80.000
90.000
100.00
110.00
120.00
130.00
140.00
150.00
160.00
170.00
180.00
190.00
200.00
210.00
220.00
230.00
240.00
250.00
260.00
270.00
280.00
290.00
300.00
310.00
320.00
330.00
340.00
350.00
360.00
370.00
380.00
390.00
400.00
410.00
420.00
430.00
440.00
450.00

Fig. 5.12 Example of SHADE printer-plot output.

OUTPUT ERROR(8), ERROR(9), ERROR(12), ERROR(15)
PAGE NTAB = 2, GROUP = 4, SYMBOL = (A, B, C, D)

The output for the above statements is shown in Fig. 5.13.

Note that the NTAB parameter provides that only the first two variables have tabulated output. The GROUP parameter provides that all four variables are plotted with the same ordinate scale. The SYMBOL parameter specifies that the letters A, B, C, and D are used to identify the printer plots of the variables ERROR(8), ERROR(9), ERROR(12), and ERROR(15), respectively.

CALL PRINT and CALL OUTPUT

Both the PRINT and OUTPUT instructions can be invoked at any point in the program by using the CALL statement. These output statements are normally invoked when specified conditions occur, where the IF instruction is usually used to branch to the CALL statement. An example of using the CALL OUTPUT statement is shown below.

NOSORT
IF(X.GT.6.0) CALL OUTPUT

Note that both the CALL PRINT and CALL OUTPUT statements must be used in nosort or procedure sections. The use of the CALL instruction assumes that the appropriate PRINT or OUTPUT statement is defined in the program. If either of the CALL instructions are invoked at a normal PRDEL or OUTDEL interval, the duplicate printing is suppressed.

As in S/360 CSMP, the FORTRAN WRITE statement in nosort or procedure sections is allowed. In some circumstances, this type output may be desirable.

RANGE

This instruction is used to list the maximum and minimum values of selected variables and the time of occurrence. Its function in CSMP III is exactly the same as in S/360 CSMP. The only difference is that the maximum number of variables allowed in a RANGE statement is 110. An example of a valid instruction is shown below.

RANGE G, H, X1, Y1, T, COST, Q, . . .
L, J, R

Subscripted variables should not be used in a RANGE statement.

Functions

In addition to the functions available in S/360 CSMP, CSMP III provides the user with additional function blocks. This section briefly describes some of the more important functions available *only* in CSMP III.

'C'=ERROR(12)
'B'=ERROR(9)
'A'=ERROR(8)

-40.00 40.00
-40.00 40.00
-40.00 40.00
-40.00 40.00

FREQ	ERROR(8)	ERROR(9)
0.0	0.0	0.0
2.00000E-02	2.71912E-02	2.37885E-02
4.00000E-02	0.10884	9.51996E-02
6.00000E-02	0.24504	0.21423
8.00000E-02	0.43597	0.38091
0.10000	0.68192	0.59528
0.12000	0.98314	0.85732
0.14000	1.3400	1.1671
0.16000	1.7531	1.5246
0.18000	2.2227	1.9298
0.20000	2.7494	2.3825
0.22000	3.3336	2.8827
0.24000	3.9761	3.4300
0.26000	4.6771	4.0242
0.28000	5.4372	4.6647
0.30000	6.2568	5.3507
0.32000	7.1360	6.0815
0.34000	8.0751	6.8557
0.36000	9.0739	7.6719
0.38000	10.132	8.5281
0.40000	11.249	9.4220
0.42000	12.423	10.351
0.44000	13.652	11.311
0.46000	14.935	12.298
0.48000	16.268	13.307
0.50000	17.647	14.332
0.52000	19.067	15.367
0.54000	20.521	16.403
0.56000	22.001	17.431
0.58000	23.498	18.441
0.60000	25.000	19.421
0.62000	26.493	20.357
0.64000	27.962	21.237
0.66000	29.387	22.044
0.68000	30.750	22.761
0.70000	32.025	23.372
0.72000	33.190	23.859
0.74000	34.216	24.205
0.76000	35.075	24.391
0.78000	35.741	24.403
0.80000	36.184	24.226
0.82000	36.379	23.848
0.84000	36.303	23.259
0.86000	35.938	22.455
0.88000	35.270	21.432
0.90000	34.292	20.194
0.92000	33.004	18.747
0.94000	31.414	17.100
0.96000	29.535	15.267
0.98000	27.389	13.265
1.0000	25.000	11.111

Fig. 5.13 Printer-plot to illustrate the use of NTAB, GROUP, and SYMBOL.

261

Arbitrary Function Generators

The AFGEN and NLFGEN function generators used in S/360 CSMP are also available in CSMP III. There are two additional arbitrary function generators available for CSMP III users, FUNGEN and TWOVAR.

FUNGEN

This arbitrary-function generator allows the user to choose the degree of interpolation. A typical statement is shown below.

$$Y = FUNGEN(ABC, N, X)$$

where   Y = dependent variable (output)

   ABC = function name, it is defined on a FUNCTION card

   N = degree of interpolation, may be 1, 2, 3, 4, or 5. N must be an integer constant or integer variable.

   X = independent variable (input)

When N = 1, FUNGEN is identical to AFGEN; and when N = 2, FUNGEN uses second-degree interpolation which is the same as NLFGEN.

CALL FGLOAD

In addition to specifying *x–y* data in a FUNCTION statement as illustrated in Example 2.4, a CALL FGLOAD instruction using subscripted variables can be used to load the data. A typical set of statements follows.

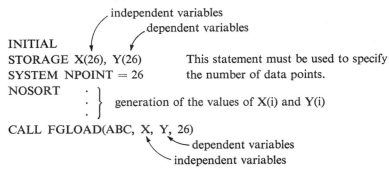

In the above example, 26 pairs of subscripted variables are read in or generated in the Initial segment. This data is then entered into the function ABC by the CALL FGLOAD instruction. The values of the independent variables X must be monitonically increasing and single-valued. A typical set of instruction for using the data entered by the CALL FGLOAD statement is shown below.

COST = AFGEN (ABC, UNITS)

FUNCTION ABC
                        dummy statement to instruct the CSMP translator
                        that the name ABC is an arbitrary function

This function loading capability is useful when the *x–y* relationship results from complex calculations. Computer time can be saved by performing the calculations only once in the Initial segment and then using an arbitrary function generator to return the values in the Dynamic segment.

TWOVAR

This arbitrary function generator can be used for functions of the form $Z = F(X, Y)$. An example of this statement is shown below.

$$Z = \text{TWOVAR(ABC, Y, X)}$$

where Z = dependent variable

       ABC = function name

    X & Y = independent variables

A graphical representation of Z as a function of X and Y is given by Fig. 5.14.

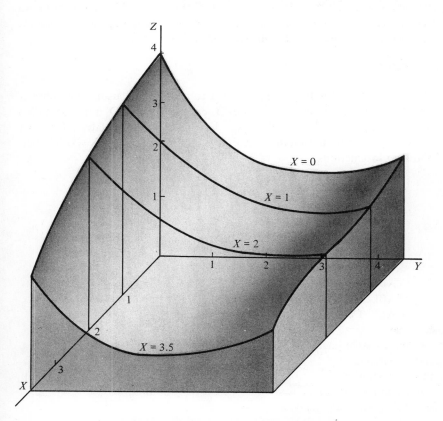

Fig. 5.14 Functional relationship between *X, Y,* and *Z*.

Note in Fig. 5.14 that the three-dimensional surface is represented by four lines each having a constant value of X. The functional relationship of Y and Z for constant values of X is the format used to specify the function Z = F(X, Y). Using the function name ABC, the following statements can be used to define the functional relationship shown in Fig. 5.14.

Value of X   (Y, Z) Pairs of data

FUNCTION ABC, 0.0 = (0.0, 4.0), (0.5, 3.0), (1.5, 2.0), (2.5, 1.7), ...
                     (3.5, 1.7), (4.5, 2.0)

FUNCTION ABC, 1.0 = (0.0, 3.5), (1.0, 2.8), (2.0, 1.9), (3.0, 1.5), ...
                     (4.5, 1.8)

FUNCTION ABC, 2.0 = (0.0, 3.0), (1.0, 2.3), (2.0, 1.8), (3.2, 1.6), ...
                     (3.8, 1.6), (4.5, 1.6)

FUNCTION ABC, 3.5 = (0.0, 2.2), (1.0, 1.2), (2.0, 0.7), (2.8, 0.75), ...
                     (3.5, 0.9), (4.0, 1.05), (4.5, 1.2)

Each FUNCTION statement contains the pairs of (Y, Z) values for a constant value of X. The cards must be arranged such that the values of X be monotonically increasing in each FUNCTION statement. If the TWOVAR function should have to extrapolate for values of either X or Y outside the specified range, a diagnostic message will appear and computation will proceed.

CALL TVLOAD

Data for the TWOVAR function generator can be loaded using the subroutine TVLOAD. Its use is similar to FGLOAD, except TVLOAD must be called to load the data for each family of curves having a constant value of X. An example of loading the data contained in Fig. 5.14 for the curve X = 0 is given by the following instructions.

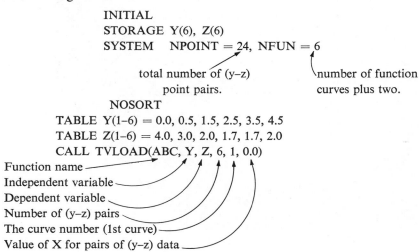

```
                INITIAL
                STORAGE  Y(6), Z(6)
                SYSTEM    NPOINT = 24, NFUN = 6
```

           total number of (y–z)   number of function
                point pairs.   curves plus two.

```
                NOSORT
              TABLE Y(1–6) = 0.0, 0.5, 1.5, 2.5, 3.5, 4.5
              TABLE Z(1–6) = 4.0, 3.0, 2.0, 1.7, 1.7, 2.0
              CALL  TVLOAD(ABC, Y, Z, 6, 1, 0.0)
```

Function name

Independent variable

Dependent variable

Number of (y–z) pairs

The curve number (1st curve)

Value of X for pairs of (y–z) data

The CALL TVLOAD statement is used three more times to load the y–z data for the curves $X = 1.0$, $X = 2.0$, and $X = 3.5$.

An example of the second CALL TVLOAD statement is shown below.

$$\text{CALL TVLOAD(ABC, Y, Z, 5, 2, 1.0)}$$

The value of X must be monotonically increasing for the successive CALL TVLOAD instructions. The range of Y should be the same for each call instruction and must be monotonically increasing.

A typical set of instructions for using the TWOVAR arbitrary function which is loaded by the subroutine TVLOAD is shown below.

$$Q = \text{TWOVAR(ABC, R, S)}$$

FUNCTION ABC ⎱
FUNCTION ABC ⎰ ◄————— Two dummy FUNCTION statements are required.

SLOPE

This function is used to calculate the slope of a function specified by data points in a FUNCTION statement. It is not the time derivative as given by the DERIV function. The standard form is given below.

$$Y = \text{SLOPE(ABC, N, X)}$$

where   Y = the slope of the function ABC at X.

$$Y = \frac{df(X)}{dX}$$

ABC = function name containing the x–y data. The data may also be loaded by the CALL FGLOAD statement.

N = an integer constant or variable giving the degree of the polynomial that is used to fit the data. The slope is computed from a polynomial curve fit. N may be 1, 2, or 3.

Example 3.12 in Chap. 3 illustrates the use of a macro to handle a general transfer function. CSMP III has a function block for this purpose.

TRANSF

Consider the following general block diagram written in Laplace transform notation.

$$r(s) \longrightarrow \boxed{\dfrac{a_m s^m + a_{m-1} s^{m-1} + \cdots + a_1 s + a_{m+1}}{b_n s^n + b_{n-1} s^{n-1} + \cdots + b_1 s + b_{n+1}}} \longrightarrow c(s)$$

The output for the above block can be obtained by using the following statement.

$$C = \text{TRANSF}(n, \text{B}, m, \text{A}, \text{R})$$

where n & m = integers giving the order of the denominator and numerator of the transfer function, respectively

  B & A = subscripted variables defining the transfer function of the denominator and numerator, respectively

  R = input

The following statements should be used to declare A and B as subscripted variables and to assign values to $A(i)$ and $B(i)$.

$$\text{STORAGE } A(m+1), \ B(n+1)$$
$$\text{TABLE } A(1-(m+1)) = a_1, a_2, a_3, \ldots, a_{m+1}$$
$$\text{TABLE } B(1-(n+1)) = b_1, b_2, b_3, \ldots, b_{n+1}$$

Consider the example of simulating the following block diagram by the use of the TRANSF function.

$$R(s) \longrightarrow \boxed{\dfrac{s^2 + 5s^2 + 11.2}{1.1s^4 + 4s^3 + 7s^2 + 2s + 0.5}} \longrightarrow C(s)$$
$$r(t) = 12.0$$

The program for determining the response is shown below.

```
STORAGE A(3), B(5)
TABLE A(1 − 3) = 5.0, 1.0, 11.2
TABLE B(1 − 5) = 2.0, 7.0, 4.0, 1.1, 0.5
C = TRANSF (4, B, 2, A, 12.0)
PRINT C
TIMER FINTIM = 4.0, PRDEL = 0.08
END
STOP
ENDJOB
```

Control Statements

Additional control capabilities are available in CSMP III. They include branching instructions, integration techniques, and double-precision calculations. This section describes the additional control capabilities as well as the changes in control statements.

END CONTINUE

This statement is used in CSMP III as the CONTINUE instruction is used in S/360 CSMP. It allows a run to be interrupted so that data or control statements can be changed during the simulation. A detailed explanation of the CSMP CONTINUE statement is contained in Chap. 3 in the section on translation control statements.

INPUT and ENDINPUT

These labels are used to identify the set of cards containing data to be entered into the program by the FORTRAN READ(5, XYZ) instruction. The INPUT card must immediately follow the END statement. All data cards must be contained between INPUT and ENDINPUT. The ENDINPUT label must be punched in card columns 1–8.

```
                    END
                    INPUT
                      .
                      .    } data cards
                      .
                    ENDINPUT
                    STOP
                    ENDJOB
```

In S/360 CSMP, the DATA and ENDDATA labels are used in place of INPUT and ENDINPUT. This is described in Chap. 3 in the section on data statements.

In CSMP III it is possible by the use of the CALL FINISH statement to terminate a run by branching to the Terminal segment. Also new parameters and control variables can be introduced in the simulation by employing the CALL CONTIN card.

CALL FINISH

This statement can be used in nosort or procedure sections to branch to the Terminal segment. If there is no Terminal segment, the program will terminate when the CALL FINISH instruction is encountered. The CALL FINISH instruction is normally used in conjunction with an IF statement as shown below.

IF(FORCE .GT. FMAX) CALL FINISH

CALL CONTIN

This statement is similar to the CALL RERUN instruction described in Chap. 2. When the program encounters a CALL CONTIN statement, a new run is initiated starting with the value of TIME from the previous run. The CALL CONTIN card does not reset initial conditions. As with the CALL RERUN instruction, the CALL CONTIN statement should only be used in the Terminal segment and is normally used with an IF statement.

Integration Techniques

There are two additional integration methods available in CSMP III; RKSDP and STIFF. As in S/360 CSMP, if the integration technique is other than the variable-step Runge-Kutta, it must be specified on a METHOD card as shown below.

METHOD STIFF

RKSDP. This is the double-precision version of the variable-step Runge-Kutta method. To take advantage of the increased accuracy, the ABSERR and RELERR parameters should be used to decrease the error-bound on the integrator output. Additional information for using double-precision is contained in a following section.

STIFF. This variable-step integration method should be used for the class of problems represented by so-called "stiff" equations. These equations have solutions which have exponents which are greatly different as typified by Eq. (5.4).

$$x = Ae^{-t} + Be^{-100t} \tag{5.4}$$

STIFF integration is used because numerical solutions using a method such as Runge-Kutta sometimes exhibit what is called partial-induced-instability when solving a stiff equation.[3]

Array Integration

The array or specification form of the INTGRL function is also available in CSMP III as illustrated by the following statement.

$$Y = INTGRL(XO, X, 40)$$

The above instruction specifies an array of 40 integrators: Y = output, XO = initial condition, and X = integrand.

Since the above INTGRL statement in CSMP III automatically specifies that Y, XO, and X are subscripted variables, they must not be included in a STORAGE or DIMENSION statement. Unlike S/360 CSMP, the initial conditions can be loaded using a TABLE statement. The EQUIVALENCE card that is required in S/360 CSMP is not used in CSMP III. Examples of programs using a CSMP III array integrator are shown in Figs. 5.10 and 5.15. The program of Fig. 5.15 shows the double-precision calculations used in the program of Fig. 5.10.

There are two additional variables that can be included on the TIMER card, DELMAX and TIME.

DELMAX

This symbol specifies the maximum allowable integration step size for all variable-step integration methods. If not specified, the smaller of PRDEL or OUTDEL is used.

TIME

This variable specifies the value of TIME at the beginning of the run. If not included on the TIMER card, it is set equal to zero.

Double-Precision Operations

In some simulation problems the round-off errors with single-precision arithmetic are too large for satisfactory solutions. Double-precision calculation can improve accuracy and is available in CSMP III. The following describes the various aspects of using double-precision.

Integration. The following card provides double precision variable-step Runge-Kutta integration.

METHOD RKSDP

The numerical integration calculations are performed in double-precision but the output is returned in single-precision which is rounded from the double-precision results.

If double-precision output is required, the specification form of the INTGRL statement must be used for all integration. The following two control cards should be included.

SYSTEM DPINTG
METHOD RKSDP

Double-precision initial conditions may be set using the TABLE instruction.

Calculations. There are several rules that must be followed when using double-precision numbers.

1 All double-precision variables must be subscripted and all of these variables not appearing in the specification form of the INTGRL function must be declared on a REAL*8 card.

/   REAL*8   Q(15), P(30), S(60)
              └─Column 7
            └─ Virgule in column 1

The above instruction must appear before the first structure statement.

2 Double-precision numbers can be rounded and set equal to single-precision numbers using the following statement.

Y = ZZRND(X(4))

In the above instruction, the double-precision variable X(4) is rounded and set equal to the single-precision variable Y.

3 Double-precision symbols starting with the letters I, J, K, L, M, and N must be included on a FIXED card.
4 The double-precision symbol for TIME is ZZTIME.
5 The exponential form using D instead of E can be used for assigning values to double-precision constants.

Y(3) = 7.987654D5 (7.987654×10^6)
PI(1) = 3.1415926535898D0 (3.1415926535898)

Output. Double-precision variables cannot be used in standard CSMP III output statements. They must be equated, or rounded using the ZZRND instruction before they can appear in CSMP output statements.

$$X = XDP(1) \qquad \text{(equated)}$$
$$Y = ZZRND(YDP(1)) \quad \text{(rounded)}$$
$$\text{OUTPUT X, Y}$$

If double-precision output is desired, the FORTRAN WRITE statement should be used as covered in Chap. 3 in the section on data output.

The program of Fig. 5.15 illustrates double-precision integration for calculating the transient temperature distribution of the copper rod of Example 5.3. A FORTRAN WRITE instruction is used to print the double-precision result which is shown in Fig. 5.16.

Note the differences between the double-precision program of Fig. 5.15 and the single-precision program of Fig. 5.10.

```
*   CSMP III PROGRAM TO SIMULATE THE TRANSIENT TEMPERATURE DISTRIBUTION
*   OF A COPPER ROD USING DOUBLE PRECISION CALCULATIONS
INITIAL
/       REAL*8 DP(7)
   FIXED   I
   TABLE   TI(1-20) = 20*3.0D2
   SYSTEM  DPINTG
   METHOD  RKSDP
   RELERR  T(1-20) = 1.0E-8
   ABSERR  T(1-20) = 1.0E-8
NOSORT

*       THE DOUBLE PRECISION SUBSCRIPTED VARIABLE DP(I) IS USED
*    TO INPUT THE VALUES OF THE PARAMETERS OF THE COPPER ROD OF EXAMPLE 5.3

   DP(1) = 8.89D3
   DP(2) = 3.98D2
   DP(3) = 3.86D2
   DP(4) = 2.0D-2
   DP(5) = 3.0D2
   DP(6) = 5.0D2
   DP(7) = DP(3)/(DP(1)*DP(2)*DP(4)*DP(4))
100   FORMAT(5G22.12)
101     FORMAT(/,6X,'TIME',18X,'T(4)',18X,'T(8)',17X,'T(10)',17X,'T(16)')
   WRITE(6,101)
DYNAMIC
NOSORT
   TD(1) = DP(7)*(DP(5) - 2.0*T(1) + T(2))
   DO 1 I = 2,19
1   TD(I) = DP(7)*(T(I-1) - 2.0*T(I) + T(I+1))
   TD(20) = DP(7)*(T(19) - 2.0*T(20) + DP(6))
SORT
   T = INTGRL(TI, TD, 20)
NOSORT

*    THE FOLLOWING FORTRAN WRITE STATEMENT PROVIDES OUTPUT AT THE END OF
*    EACH TENTH VALID INTEGRATION STEP

   IF(KEEP.NE.1)   GO TO 2
   CT = CT + 1.0
   IF(CT.LT.9.5)   GO TO 2
   WRITE(6,100)(TIME, T(4), T(8), T(10), T(16))
   CT = 0.0
2 CONTINUE
   TIMER   FINTIM = 450.0
END
STOP
ENDJOB
```

Fig. 5.15 Simulation of the temperature of copper rod using double precision.

TIME	T(4)	T(8)	T(10)	T(16)
1.26562500000	300.000000000	300.000000000	300.000000000	300.004681049
3.09375000000	300.000000000	300.000000001	300.000000174	300.194794311
5.90625000000	300.000000000	300.000000950	300.000061033	301.805008651
10.4062500000	300.000000205	300.000221716	300.004986118	308.069261558
17.7187500000	300.000089036	300.014688666	300.135408143	322.048061889
30.0937500000	300.010141710	300.336381644	301.497811928	343.563464148
51.4687500000	300.306926535	302.971772836	307.765962064	369.055400686
89.7187500000	303.023022391	312.727054459	323.260691628	394.918449206
148.218750000	310.683180099	329.427550268	344.141355339	415.596172422
199.968750000	317.385209377	341.543210126	357.834603171	426.321822957
253.968750000	322.984081578	351.104582398	368.286939071	433.850591096
303.468750000	326.863540646	357.595660186	375.294859587	438.733635794
355.218750000	329.884090965	362.611840784	380.685218191	442.442048677
406.968750000	332.095773735	366.280281305	384.619834664	445.134832194
450.000000000	333.481153192	368.564808895	387.068383299	446.807280527

$$$ SIMULATION HALTED FOR FINISH CONDITION TIME   450.00

Fig. 5.16 Double precision output from program of Fig. 5.15.

REFERENCES

1. *Continuous System Modeling Program III* (*CSMP III*) Program Reference Manual, SH19-7001-2, Program Number 5734-XS9, IBM Corporation, Data Processing Division, White Plains, N. Y.

2. *Continuous System Modeling Program,* Program Logic Manual, LY19-7000, IBM Corporation, Data Processing Division, White Plains, N. Y.

3. Fox, Leslie and D. F. Mayers, *Computing Methods for Scientists and Engineers,* Clarendon Press at Oxford, 1968, p. 215.

PROBLEMS

1 The equations which defined the concentrations of xenon 135 (X) and iodine 135 (I) in a nuclear reactor which has been suddenly shut down are given below.

$$\dot{X} = -2.09 \times 10^{-5}X + 2.88 \times 10^{-5}I$$
$$\dot{I} = -2.88 \times 10^{-5}I$$

The initial concentrations are

$$X(0) = 3 \times 10^{15} \text{ atoms per unit volume}$$
$$I(0) = 7 \times 10^{16} \text{ atoms per unit volume}$$

Find the concentrations at $t = 100,000$ sec. Plot the results using a logarithmic ordinate.

Answer:

At TIME $= 100,000$ $X = 1.761 \times 10^{16}$ $I = 3.929 \times 10^{15}$

2 In an epidemic of a contagious disease, the following Kermack-McKendrick set of equations model the number of the population who are susceptible to the disease S, the number of infectious carriers C, and the number of individuals who have recovered and are immune R.

$$\dot{S} = -A \cdot S \cdot C$$
$$\dot{C} = A \cdot S \cdot C - B \cdot C$$
$$\dot{R} = B \cdot C$$

A and B are constants which characterize the epidemic. Time is measured in days.

Using the following initial conditions for a population of 1000,

$$S(0) = 900$$
$$C(0) = 10$$
$$R(0) = 90$$

solve the Kermack-McKendrick equation using $A = 0.001$ and $B = 0.072$.

Include S, C, and R in a common printer-plot using the same scale for all variables.

Answer:

At TIME = 10 days, $S = 27.4$, $C = 631.2$, $R = 341.4$

3 Solve the following equation using both the RKSFX and STIFF methods.

$$\ddot{y} + 201\dot{y} + 200y = 0$$
$$y(0) = 1.0$$
$$y(0) = 0$$

Compare the numerical solution with the exact solution.

$$y = \frac{200}{199}e^{-t} - \frac{1}{199}e^{-200t}$$

Answer:

The difference between the exact solution and the numerical solution using RKSFX and STIFF is less than 10^{-4}. However, several program interrupts due to underflow occur when using RKSFX.

4 Use the TRANSF function to solve for the dynamic response of the system represented by the following block diagram. A unit-step input occurs at $t = 0$.

$$r(s) \longrightarrow \boxed{\frac{s^2 + 5s + 7}{s^6 + 8s^5 + 23s^4 + 44s^3 + 41s^2 + 14s + 4}} \longrightarrow c(s)$$

Answer:

At TIME = 10.0 $c = 2.1671$

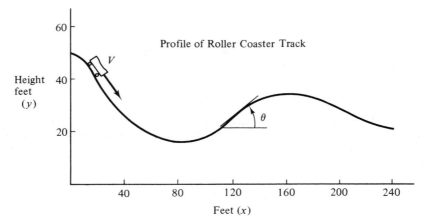

Fig. P-5.5

5 A 640 lb gravity-powered roller coaster travels on a track shown in the above profile. Aerodynamic drag equal to $0.005v^2$ and a rolling resistance of 10 lb oppose motion. The velocity at the start of the run is 15 ft/sec. Use the SLOPE function to evaluate dy/dx in simulating the motion of the vehicle.

The equation of motion is

$$\frac{w}{g}\dot{v} = -w \sin \theta - 0.005v^2 - 10.0$$

where $\quad \sin \theta = \dfrac{\dfrac{dy}{dx}}{\sqrt{1 + \left(\dfrac{dy}{dx}\right)^2}}$

$$x = \int_0^t \frac{v\, dt}{\sqrt{1 + \left(\dfrac{dy}{dx}\right)^2}}$$

w = weight

g = acceleration of gravity, 32.17 ft/sec^2

Answer:

At TIME = 6.0 sec: v = 35.1 ft/sec and x = 197.0 ft.

6 A 27 lb weight moving downward at a velocity of 200 in/sec strikes and sticks to a lead cylinder.

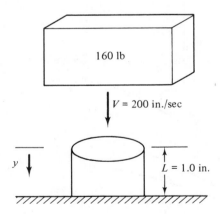

160 lb

V = 200 in./sec

y

L = 1.0 in.

Fig. P-5.6

The dynamic resisting-force of the lead cylinder for large strains is approximated by the following formula.

$$F = 620\left[1 + 0.004\left(\frac{\dot{y}}{L-y}\right)^{2.02}\right]\left(\frac{y}{L-y}\right)^{0.44}$$

The equation that describes the motion of the mass is given below.

$$\frac{27}{386}\ddot{y} = 27 - F$$

Generate data for the resisting-force F in the Initial segment using the following range for y and $\dot{y}$.

$$0 \lesssim y \lesssim 0.25 \text{ in.}$$

$$0 \lesssim \dot{y} \lesssim 200 \text{ in./sec}$$

Load the generated data by using the TVLOAD subroutine and then use the TWOVAR function to calculate F to determine the maximum deflection of the load cylinder.

Answer:

$$y_{\max} = 0.150 \text{ in. at TIME} = 0.00343 \text{ sec}$$

7   The device below is used to control the water level in a tank. The flow characteristics of the valve are given in the figure below. The outflow from the tank is given by the following expression.

$$q_o = 5.05\sqrt{h}$$

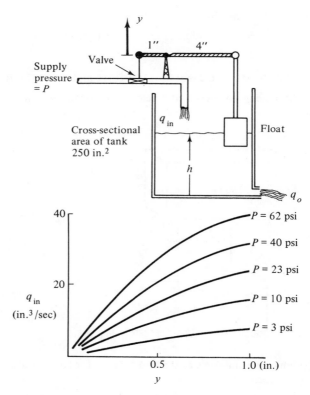

Fig. P-5.7

At the start of the simulation, $h = 10$, $P = 23$ psi, and $y = 0.5$. Use the TWOVAR function to represent the valve characteristics and solve for the dynamics of the system when P is suddenly increased to 60 psi.

The equation describing the water level is

$$250 \frac{dh}{dt} = q_{in} - q_o.$$

Answer:

At TIME $= 40$ sec, $h = 10.69$ in.

8 The curve below represents an odd periodic function. Using the FUNGEN function with 5th-order interpolation, find the Fourier coefficients for the first five terms of the periodic function. The odd Fourier coefficients are given by the following expression.

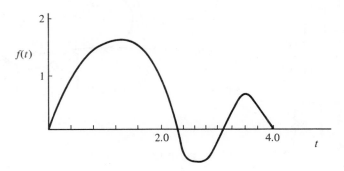

Fig. P-5.8

$$b_n = \frac{2}{p} \int_0^p f(t) \sin\left(\frac{n\pi t}{p}\right) dt$$

$$2p = \text{period} = 8.0$$

Answer:

$$b_1 = 0.84, \; b_2 = 0.91, \; b_3 = 0.32, \; b_4 = -0.40, \; b_5 = 0.27$$

9 The road-holding ability of an automobile can generally be improved by decreasing the unsprung mass. Use the below model of one corner of an automobile to find the time average ratio of tire force to vehicle weight as the car passes over a depression in the road. Do the simulation for unsprung masses of 80 and 150 lb.
Equations of motion are

$$m_1 \ddot{y}_1 = k_t(y - y_1) + k(y_2 - y_1) + c(\dot{y}_2 - \dot{y}_1) + m_1 g$$
$$m_2 \ddot{y}_2 = k(y_1 - y_2) + c(\dot{y}_1 - \dot{y}_2) + m_2 g$$

initial conditions:

$$y_1(0) = \frac{(m_1 + m_2)g}{k_t} \qquad y_2(0) = y_1(0) + \frac{m_2 g}{k}$$

$$\dot{y}_1(0) = \dot{y}_2(0) = 0$$

$$\frac{\text{Tire force}}{\text{weight}} = \frac{k_t(y - y_1)}{(m_1 + m_2)g}$$

Answer:

Unsprung mass 80 lb: average force/weight $= 0.6937$
Unsprung mass 150 lb: average force/weight $= 0.6733$

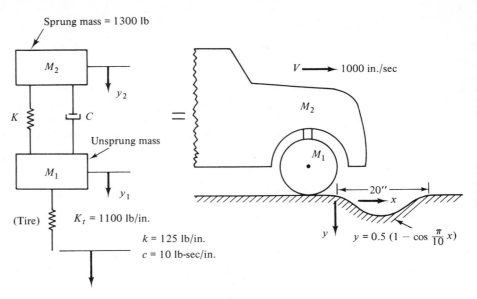

Fig. P-5.9

10   A control system is represented by the following block diagram. For the following input, use the TRANSF and REALPL functions to find the time response of the output.

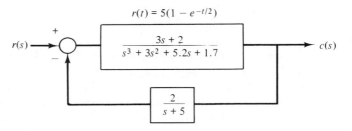

Fig. P-5.10

Answer:

$$\text{At TIME} = 6.0, \ c = 3.5697$$

11   Motorcycle helmets are tested by dropping a helmet containing a headform on a rigid surface. The following data for two different helmets was taken at the Southwest Research Institute and is the acceleration measured in the headform during a Z-90 test. The helmet has a velocity at impact of 19.6 ft/sec. Using the OVERLAY statement, find the rebound velocities of both helmets using one program.

   The following expression can be used to estimate the injury to a human head from an impact. A value greater than 1000 is usually fatal.

$$\text{HIC} = \int_0^t a^{2.5} \, dt$$

a is the acceleration in g's.

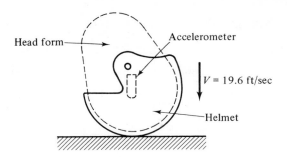

Fig. P-5.11

Calculate the value of HIC for both helmets.

Test Results of a Good Helmet		*Test Results of a Poor Helmet*	
Time (sec)	*Acceleration (g's)*	*Time (sec)*	*Acceleration (g's)*
0	0	0	0
0.001	24	0.001	100
0.002	45	0.0015	180
0.003	84	0.002	235
0.004	125	0.0025	220
0.005	144	0.003	258
0.006	135	0.0035	246
0.007	120	0.004	205
0.008	71	0.0045	175
0.009	52	0.005	110
0.01	20	0.0055	75
0.011	12	0.006	48
0.012	0	0.0065	14
		0.007	0

Answer:

Good helmet: rebound velocity = 7.34 ft/sec
HIC = 935.7

Poor helmet: rebound velocity = 10.85 ft/sec
HIC = 2653

12 Each of the three tanks initially contain 500 gal of brine with 50 lb, 100 lb, and 150 lb of salt dissolved in tanks 1, 2, and 3, respectively. The flow between tanks is shown in Fig. P-5.12. By assuming the brine is kept well stirred, the following equations describe the amount of salt in each tank.

$$\frac{dw_1}{dt} = \frac{5w_2}{500 - t} - \frac{3w_1}{500 + 2t}$$

$$\frac{dw_2}{dt} = \frac{3w_1}{500 + 2t} + \frac{4w_3}{500 - t} - \frac{8w_2}{500 - t}$$

$$\frac{dw_3}{dt} = \frac{3w_2}{500 - t} - \frac{4w_3}{500 - t}$$

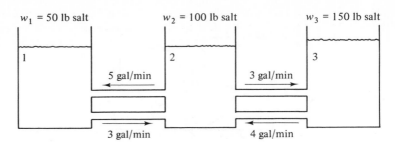

Fig. P-5.12

Find the amount of salt in each tank after 4 hours. Use the GROUP parameter to plot the time-history of w_1, w_2, and w_3 with the same scale.

Answer:

$$w_1 = 186.77 \text{ lb of salt}$$
$$w_2 = 54.927 \text{ lb of salt}$$
$$w_3 = 58.308 \text{ lb of salt}$$

13 When a perfectly flexible and homogeneous cable is hung between two support points, the equation of its shape is called the *catenary*. The exact shape is given by the solution to the following differential equation.

$$\frac{d^2y}{dx^2} = \frac{w}{h}\sqrt{1 + \left(\frac{dy}{dx}\right)^2}$$

where: $w = 0.17$ lb/ft, weight per unit length of cable

$h =$ tension in cable at the lowest point

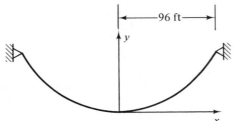

x   **Fig. P-5.13**

Find the shape of the free hanging cable for the following values of h.

$$h = 10, 40, \text{ and } 300 \text{ lb}$$

Use the MERGE parameter to plot all three curves on one page.

Answer:

$$\text{At: } x = 96 \text{ ft}$$
$$h = 10 \text{ lb, } y = 97.341 \text{ ft}$$
$$h = 40 \text{ lb, } \quad y = 19.857 \text{ ft}$$
$$h = 300 \text{ lb, } y = 2.6118 \text{ ft}$$

14   An aluminum circular fin on a 0.5 ft. diameter pipe is used to transfer energy from the surface of the pipe to the ambient air. One fin can transfer 3200 Btu/hour. For a thin circular fin having only a radial temperature distribution, the following equation gives the steady-state temperature for natural convection.

$$\frac{d^2T}{dr^2} + \frac{1}{r}\frac{dT}{dr} - \frac{2h}{kL}(T - T_a) = 0$$

where: T = temperature of fin

T_a = ambient temperature = 80°F

$h = 0.19(T - T_a)^{0.33}$ Btu/hr-ft²-°F

$k = 118$ Btu/hr-ft-°F

L = thickness of fin = 0.0085 ft

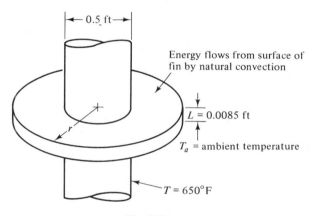

Fig. P-5.14

Find the temperature at a radius of 0.75 ft. Note that the above equation requires the integration to start at $r = 0.25$ ft. An initial condition for the temperature gradient on the inside surface of the fin can be calculated from the following relationship.

$$q = 3200 = -kA\frac{dT}{dr}\bigg|_{r=0.25}$$

$$A = \text{area} = 0.5\pi L$$

Answer:

$$T(r = 0.75) = 171.55°F$$

15   Solve problem 14 using the following boundary conditions

$$T(r = 0.25) = 650°F$$

$$\frac{dT}{dr}(r = 0.75) = -\frac{h}{k}[T(r = 0.75) - T_a]$$

In the second boundary condition, the heat flow from the outer edge of the fin is set equal to the energy removed from the same surface by convection. Since the temperature gradient at the outer edge is specified, the CALL RERUN statement must be used in an iterative procedure to calculate the temperature gradient at the inside surface: $dT(r = 0.25)/dr$.

Answer:

$$\frac{dT(r = 0.25)}{dr} = -1146°\text{F/IN}$$
$$T(r = 0.75) = 451.7°\text{F}$$

I

LISTING AND

DEFINITION OF ALL

CSMP FUNCTIONS

The information in this section gives a summary of the functional blocks in both CSMP/360 and CSMP III. The first portion contains those functions which are identical for both versions of CSMP. The second portion gives those functions which have been added to the newer version, CSMP III.

Material in this Appendix can be a valuable asset to the user in that a brief description of the function is given along with the example numbers in the text which use the function.

Table A-1
Functions Involving Integration And Differentiation

Program Statement	*Mathematical/Transformation Equivalent*
INTEGRATION	
$Y = INTGRL(IC, X)$ $IC = y(t)\|_{t=t_0}$ $X =$ the input variable or function (a) For CSMP/360, $t_0 = 0$. (b) For CSMP III, t_0 is specified on the TIMER card by TIME $= t_0$. If not specified, $t_0 = 0$. These comments apply for all blocks that contain a t_0.	$$y = \int_{t_0}^{t} x \, dt + y(t_0)$$ $t_0 =$ starting time $t =$ independent variable Equivalent Laplace transform $$\frac{Y(s)}{X(s)} = \frac{1}{s}$$

Comments

This is probably one of the most frequently used functional blocks of CSMP. If the user does not specify a method of integration, then RKS variable step will automatically be used in the program. However, six other methods of integration are available and any one of these can be selected by listing the particular integration form on a METHOD card (see pages 81 to 93). Users may define their own integration through the use of a FORTRAN subroutine named CENTRL (pages 93 to 95).

A vector or array integration form is also available by using the specification form

$$Y = INTGRL(IC, X, N)$$

where Y = an output array

IC = array of initial conditions

X = array of inputs or integrands

N = the number of elements in the input array

Before using the specification form of integration, the reader should review the material starting on page 103. Application of this form of integration for S/360 CSMP is given in Example 3.3, page 104 and Example 4.2, page 200. Examples of the use of array integration in CSMP III are shown in Chapter 5.

Application of the ordinary integration form, that is Y = INTGRL(IC,X), can be found in the following examples: Ex. 2.1, p. 14; Ex. 2.2, p. 18; Ex. 2.3, p. 25; Ex. 2.4, p. 30; Ex. 2.5, p. 35; Ex. 2.6, p. 39; Ex. 2.7, p. 42; Ex. 2.8, p. 52; Ex. 2.9, p. 58; Ex. 2.10, p. 60; Ex. 3.1, p. 90; Ex. 3.6, p. 114; Ex. 3.8, p. 126; Ex. 3.9, p. 134; Ex. 3.12, p. 151; Ex. 3.13, p. 158; Ex. 3.14, p. 167; Ex. 3.16, p. 177; Ex. 4.3, p. 210; Ex. 4.4, p. 211; Ex. 4.5, p. 215; Ex. 4.6, p. 221; Ex. 5.2, p. 253; Ex. 5.3, p. 254.

Program Statement	*Mathematical/Transformation Equivalent*
DIFFERENTIATION $Y = DERIV(IC, X)$ $IC = \dfrac{dx(t)}{dt}\bigg|_{t=t_0}$ $X = $ input variable	$y(t) = \dfrac{dx(t)}{dt}$ Equivalent Laplace transform $\dfrac{Y(s)}{X(s)} = s$

Comments

Generally, this function is not used to any considerable extent in simulation. DERIV is not a MEMORY or HISTORY function and therefore cannot be used alone in a closed loop except when the loop is broken by an implicit function. An example using the DERIV function is given on p. 48.

Program Statement	*Mathematical/Transformation Equivalent*
REAL POLE or FIRST-ORDER LAG $\text{Y} = \text{REALPL(IC, P, X)}$ $\text{IC} = y(t)\|_{t=t_0}$ $\quad TC = 1/p$ $\text{X} = $ the forcing function $\quad$ or input	$p\dfrac{dy}{dt} + y = x$ Equivalent Laplace transform $\dfrac{Y(s)}{X(s)} = \dfrac{1}{ps+1}$

Comments

This function is often used in control system simulation. The terms real pole and first-order lag are frequently used in control systems and engineering; hence these terms are used as descriptive titles. However, the equation

$$p\frac{dy}{dt} + y = x$$

is simply a first-order differential equation. When the transfer function is expressed as

$$\frac{Y(s)}{X(s)} = \frac{A}{s+B}$$

the form must be changed to

$$\frac{Y(s)}{X(s)} = \frac{\dfrac{A}{B}}{1 + \dfrac{s}{B}}$$

See pp. 21 to 22 for further comments.

Applications using the REALPL function are given in Ex. 2.2, p. 18; Ex. 2.6, p. 39; Ex. 2.8, p. 52; Ex. 4.2, p. 200; Ex. 4.3, p. 210; Ex. 4.4, p. 211; Ex. 4.6, p. 221.

Program Statement	*Mathematical/Transformation Equivalent*
LEAD-LAG $\text{Y} = \text{LEDLAG(P1, P2, X)}$ *Note:* P1 and P2 are $\quad$ parameters; initial $\quad$ conditions are not used $\quad$ with this function.	$p_2\dfrac{dy}{dt} + y = p_1\dfrac{dx}{dt} + x$ Equivalent Laplace transform $\dfrac{Y(s)}{X(s)} = \dfrac{p_1 s + 1}{p_2 s + 1}$

Comments

This is another function often used in control system simulation. The term lead-lag is descriptive of compensation in control systems. A lead compensator is of the form given above when $p_1 > p_2$. A lag compensator is of the same form

Comments (*continued*).

but with $p_2 > p_1$. Further comments and application of this function are given in Example 2.6, p. 39.

Program Statement	*Mathematical/Transformation Equivalent*
COMPLEX POLES Y = CMPXPL(IC1, IC2, P1, P2, X) $IC1 = y(t)\|_{t=t_0}$ $IC2 = \dot{y}(t)\|_{t=t_0}$ P1 = ζ P2 = ω_n	$$\frac{d^2y}{dt^2} + 2P_1P_2\frac{dy}{dt} + P_2^2 y = x$$ Equivalent Laplace transform $$\frac{Y(s)}{X(s)} = \frac{1}{s^2 + 2P_1P_2 s + P_2^2}$$

Comments

This function is particularly designed for second-order polynomials with complex roots. If the roots are real, one may employ REALPL twice. Nevertheless, CMPXPL will also handle the case when roots of $s^2 + 2P_1P_2s + P_2^2$ are real. Ordinarily, one encounters this polynomial in the form $s^2 + 2\zeta\omega_n s + \omega_n^2$. Obviously, P_2 is equivalent to ω_n and P_1 equivalent to ζ. If one desires to simulate the transfer function

$$\frac{Y(s)}{X(s)} = \frac{1}{s^2 + 1.6s + 4}$$

then $P_2 = 2$ and $P_1 = 0.4$.

Application of this function is given in Example 2.7, p. <del>42</del> 45.

Program Statement	*Mathematical/Transformation Equivalent*
MODE-CONTROLLED INTEGRATOR Y = MODINT(IC, X1, X2, X3) $IC = y(t)\|_{t=t_0}$	$y(t) = \int_{t_0}^{t} x_3\, dt + IC;$ for $x_1 > 0$, any x_2 $y(t) = IC$; for $x_1 \leq 0$, $x_2 > 0$ $y(t) = $ last output ; for $x_1 \leq 0$, $x_2 \leq 0$ from integrator

Comments

This function allows the user to start an integration, integrate for prescribed conditions, then hold the value of y, restart the integration for other prescribed conditions. For some simulations, the value of y may have large step-changes. If this occurs while using a variable-integration step-size the condition on DELMIN may not be satisfied and the simulation will halt. The user can correct the situation by changing to a fixed-step integration with appropriate step-size.

See Example 3.13, p. 158, for application of the mode-controlled integrator.

Table A-2

Special CSMP Functions

Program Statement	*Mathematical/Transformation Equivalent*
ZERO-ORDER HOLD $Y = ZHOLD(X1, X2)$ X1 = a trigger signal X2 = input to the hold	$y(t) = x_2$; for $x_1 > 0$ $y(t) = $ last value of x_2; for $x_1 \leq 0$ $y(t)\vert_{t=t_0} = 0$ Equivalent Laplace transform $$\frac{Y(s)}{X(s)} = \frac{1 - e^{-sT}}{s}$$ $T = $ the sampling interval

Comments

This function has important application in the design and simulation of discrete data systems. The physical property of the zero-order hold is illustrated in the following sketches.

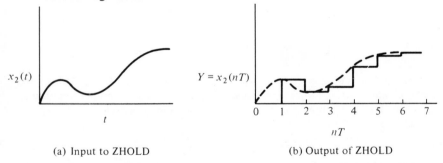

(a) Input to ZHOLD	(b) Output of ZHOLD

A convenient signal for X1 is the IMPULS function. Thus, to sample and hold a sine function of a frequency of 30 radians/sec, every 1 sec, one may write

$$X2 = SIN(30.0*TIME)$$
$$X1 = IMPULS(0.0, 1.0)$$
$$Y = ZHOLD(X1, X2)$$

It is important to recognize that if a variable-step integration method is used in a problem, integration will generally not occur when IMPULS is applied. This means that sampling will not occur at nT but rather at those values of TIME for which an integration is performed. This problem can be overcome by using a fixed-step integration in which T is a multiple of DELT.

Application of the zero-order hold can be found in Ex. 4.3, p. 210; Ex. 4.4, p. 211; Ex. 4.5, p. 215; Ex. 4.6, p. 221.

Program Statement	*Mathematical/Transformation Equivalent*

DEAD TIME (DELAY)
 	
Y = DELAY(N, P, X)
 	
P = the ideal delay time
 	
N = number of points of X
    sampled during P.
    N must be an integer
    constant ≥ 3 but $\leq 16{,}378$
	$y = x(t - p)$; for $t \geq p$
	$y = 0 \qquad$; for $t < p$
	
	Equivalent Laplace transform
	
	$$\frac{Y(s)}{X(s)} = e^{-ps}$$

Comments

The role of this function is shown in the following sketches. As a rule of thumb one can select N as a number in the range of expected integrations during P. The program uses P/N as the sample interval of X when $P/N \geq$ DELT, otherwise DELT is used.

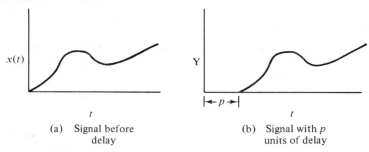

    (a) Signal before   (b) Signal with p
          delay   units of delay

Dead time is often present in process control systems and hence **DELAY** is a convenient function to use when simulating these systems. Another important use of the **DELAY** function stems from the simulation of discrete systems represented by z-transforms. Thus, if

$$D(z) = \frac{M(z)}{N(z)} = \frac{A + Bz^{-1}}{1 + Cz^{-1}}$$

the Delay function can be used for z^{-1} to simulate $D(z)$ much in the same way that INTGRL is used to simulate s^{-1}.

Application of the Delay function is given in Ex. 2.8, p. 52; and Ex. 4.4, p. 211.

Problem Statement	*Mathematical/Transformation Equivalent*
IMPLICIT FUNCTION Y = IMPL(IG, ERROR, FOFY) IG = an initial guess for Y ERROR = an acceptable error in finding a solution for Y FOFY = the function of y, $f(y)$	$y = f(y)$ $\lvert f(y) - y \rvert \leq$ error

. *Comments*

The implicit function is used to solve a problem of the form

$$F(y) = 0$$

This function can be expressed (usually in several different forms) as

$$y = f(y)$$

$$\text{where } F(y) = f(y) - y = 0$$

The expression for $f(y)$ must be the last statement used in the implicit loop.

Further comments on the implicit function and how it can be used to break algebraic loops are given on pp. 176 to 178. For application see Ex. 3.15, p. 176; and Ex. 3.16, p. 177.

Table A-3

Signal Sources

Program Statement	Function Definition

RAMP FUNCTION
 	
$Y = RAMP(T)$
 	
T = starting time of ramp
	$y = t - T$; for $t \geq T$
	$y = 0$; for $t < T$
	
	Y 45°
	           T
	          t

Comments

This function is often used for special test-signal purposes. Note that the slope of the ramp is 1. This can be changed to any value by

$$Y = Q*RAMP(T)$$

Also, combinations of the ramp can be used to generate other special signals. For example,

$$X1 = RAMP(0.0)$$
$$X2 = RAMP(2.0)$$
$$X3 = X1 - X2$$

produces the signal shown below.

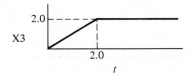

Program Statement	*Function Definition*
STEP FUNCTION Y = STEP(T) T = starting time of the step	$y = 0$; for $t < T$ $y = 1$; for $t \geq T$

Comments

The step input is one of the most frequently used signals for investigating the transient response of a system. The amplitude of the step can be changed to values other than one by using the statement,

$$Y = A*STEP(T)$$

where A is the desired amplitude. Single-shot pulses can easily be generated by multiple use of the step function. For example, the statements

$$X1 = STEP(0.5)$$
$$X2 = STEP(1.5)$$
$$X3 = X1 - X2$$

produces the pulse sketched below. Application of the step function can be found in the following examples. Ex. 2.2, p. 18; Ex. 2.6, p. 39; Ex. 2.7, p. 42; Ex. 2.8, p. 52; Ex. 3.12, p. 151; Ex. 3.13, p. 158; Ex. 3.15, p. 176; Ex. 3.16, p. 177; Ex. 4.2, p. 200; Ex. 4.3, p. 210; Ex. 4.4, p. 211; Ex. 4.5, p. 215; Ex. 4.6, p. 221.

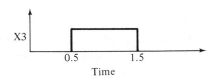

Program Statement	*Function Definition*
PULSE FUNCTION Y = PULSE(P, TRIG) P = pulse width TRIG = trigger starting each pulse	$y = 1$; for time duration P following the application of TRIG. $y = 0$; for all other time

Comments

This function offers a convenient way for generating a train of pulses. Periodic pulses can be generated by using a periodic trigger. The spacing between pulses will be random provided the trigger is random (with P less than the duration between trigger applications). For an illustration of generating a pulse train see the comment notes of the Impulse function.

The pulse function is used in the following examples. Ex. 2.7, p. 42; and Ex. 3.13, p. 158.

Program Statement	*Function Definition*
IMPULSE FUNCTION $\quad$ Y = IMPULS(T, P) T = time when first impulse is applied P = spacing between impulses	$y = 0$; for $t < T$ $y = 1$; for $t = T + NP$ $y = 0$; for $t \neq T + NP$ $\quad N = 0, 1, 2, \ldots$![diagram of impulse train with Y axis, pulses spaced P apart, starting at T]

Comments

This function has an amplitude of 1 at discrete points of time. In most cases a unit impulse is considered to have infinite amplitude and an area of 1 (Dirac function). The user should note this difference.

The IMPULS is particularly useful as a trigger function. As an example, a train of unit pulses can be generated as follows.

$$X1 = IMPULS(0.0, 1.0)$$
$$X2 = PULSE(0.2, X1)$$

The signal X2 generated by these two statements is shown below. The impulse function is also useful in simulating the ideal sampler in discrete data systems (see Example 4.3, p. 210). The use of the impulse function is also given in Ex. 2.7, p. 42; Ex. 3.13, p. 158; Ex. 4.4, p. 211; Ex. 4.5, p. 215; Ex. 4.6, p. 221.

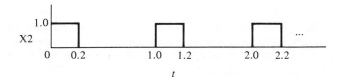

Program Statement	*Function Definition*
VARIABLE SINE FUNCTION Y = SINE(DELAY,OMEGA,PHASE) DELAY = delay time for starting the function OMEGA = frequency of the function in radians/unit time PHASE = phase shift in radians	$y = \sin(\text{omega}(t\text{-delay}) + \text{phase})$; for $t \geq$ delay $y = 0$; for $t <$ delay

Comments

    This function offers the user a convenient way of displacing a sine function. If the user is not concerned about either phase shift or delay, the ordinary FORTRAN SIN can be used as in the following statement

$$Y = \text{SIN(OMEGA} \ast \text{TIME)}$$

where OMEGA is a specified value of radian frequency and TIME is the CSMP variable for time. See Example 3.10, p. 145 for an illustration of the CSMP SINE function.

Problem Statement	*Function Definition*
UNIFORM DISTRIBUTION RANDOM NUMBER GENERATOR Y = RNDGEN(N) N = any odd integer which is used as a "seed" value	Uniform distribution of Y $P(y)$ is the probability density function.

Comments

    This function generates numbers between 0 and 1 with a uniform distribution. For a given seed value, the random sequence will be unchanged from one computer run to the next. Details for using RNDGEN are given on p. 119.

    Application of this function is given in Ex. 3.13, p. 158; Ex. 4.7, p. 224.

Program Statement	Function Definition
NORMAL DISTRIBUTION RANDOM NUMBER GENERATOR $\quad$ Y = GAUSS(N, P1, P2) $\quad$ N = an odd integer seed value $\quad$ P1 = mean $\quad$ P2 = standard deviation	Normal distribution of Y $P(y)$ is the probability density function

Comments

$\quad$ This function produces random values of y which statistically have a Gaussian distribution with a specified mean and standard deviation. Both random number generators return numbers only at valid integration steps (KEEP = 1).

$\quad$ Additional comments concerning the use of GAUSS are given on p. 118. See Example 3.7, p. 118 for application.

Table A-4

Function Generation

Problem Statement	Function Definition
ARBITRARY FUNCTION GENERATOR $\quad$ (linear interpolation) $\quad\quad$ Y = AFGEN(FUNCT, X) $\quad\quad$ X = an independent variable FUNCT = name of a function defined by a FUNCTION card. Any valid FORTRAN symbol can be used.	 Linear interpolation between data points (A,B,C,D,E,F)

Comments

$\quad$ AFGEN is used to generate a function specified by the user. Linear or straight line interpolation is used between given values of x. Data points for x and y are listed on a FUNCTION card (placed at the beginning of the program) as below

$$x_1 \quad y_1 \quad x_2 \quad y_2 \quad \ldots \text{etc.}$$
$$\downarrow \quad \downarrow \quad \downarrow \quad \downarrow$$

FUNCTION LUCK = (0.0, 1.2),(0.3, 1.8),(0.7, 2.3), (1.1, 0.9),(1.7, $-$0.3)

Comments (continued).

The function is called by a statement in the program such as

$$Y = AFGEN(LUCK, X)$$

Further explanation of AFGEN is given on p. 31. Application of this function is given in Ex. 2.4, p. 30; Ex. 3.13, p. 158; Ex. 4.2, p. 200.

Problem Statement	*Function Definition*
ARBITRARY FUNCTION GENERATOR (quadratic interpolation) $Y = NLFGEN(FUNCT, X)$ X = an independent variable FUNCT = name of a function defined by a FUNCTION card.	 Quadratic interpolation between points (A,B,C,D)

Comments

The use of this function is identical to AFGEN except that quadratic interpolation is used between data points. See the comments given for AFGEN and also refer to p. 31.

Application of NLFGEN can be found in Ex. 2.10, p. 60; Ex. 2.11, p. 65; Ex. 3.5, p. 110.

Table A-5

Nonlinear System Characteristics

Program Statement	*Definition of Characteristic*
DEAD SPACE $Y = DEADSP(P1, P2, X)$ X = input function P1, P2 = parameters defining the range of the dead space	 $y = 0 \quad$; for $P_1 \leq x \leq P_2$ $y = x - P_2$; for $x > P_2$ $y = x - P_1$; for $x < P_1$

Comments

This function can be used for simulating dead space in a system nonlinearity such as a valve. The slope on the active portion is not limited to 45° and can be changed to a specified value by

$$Y = Q*DEADSP(P1, P2, X)$$
$$(Q = \text{specified slope})$$

Application of this function is given in Example 2.9, p. 58.

Program Statement	*Definition of Characteristic*
LIMITER $Y = LIMIT(P1,\ P2,\ X)$ $X = \text{an input}$ $P1, P2 = \text{parameters defining}$ $\text{the range of limiting}$	 $y = P_1 ; \text{ for } x < P_1$ $y = P_2 ; \text{ for } x > P_2$ $y = x \ ; \text{ for } P_1 \leq x \leq P_2$

Comments

Used for defining limiting or saturation in a system. The slope of the function can be changed from 45° by the statement

$$Y = Q*LIMIT(P1, P2, X)$$
$$(Q = \text{user defined slope})$$

A typical application of this function is given in Example 2.10, p. 60.

Program Statement	*Definition of Characteristic*
QUANTIZER $Y = QNTZR(P,\ X)$ $P = \text{width of quantized signal}$ $X = \text{independent variable}$	 $y = KP; \text{ for } (K - \tfrac{1}{2})P < x \leq (K + \tfrac{1}{2})P$ $K = 0, \pm 1, \pm 2, \pm 3, \ldots$

Comments

Useful for quantizing a function or signal. For example, a sin function can be quantized as follows.

$$X1 = SIN(10.0*TIME)$$
$$X2 = X1*QNTZR(0.05, TIME)$$

This function can be used to simulate the output of a digital to analog converter.

Program Statement	*Definition of Characteristic*
HYSTERESIS $$Y = HSTRSS(IC, P1, P2, X)$$ P1, P2 = parameters defining the range of the hysteresis IC = value of Y at t_o. X = independent variable	$$y = x - P_2$$ for $\quad [x(t) - x(t - \Delta t)] > 0$ and $\quad y(t - \Delta t) \le (x - P_2)$ $$y = x - P_1$$ for $\quad [x(t) - x(t - \Delta t)] < 0$ and $\quad y(t - \Delta t) \ge (x - P_1)$ otherwise $\quad y = y(t - \Delta t)$

Table A-6

Switching Functions

Program Statement	*Description of Switch*
OUTPUT SWITCH $$Y1, Y2 = OUTSW(X1, X2)$$	$$y_1 = x_2, y_2 = 0 \; ; \text{for } x_1 < 0$$ $$y_1 = 0 \;, y_2 = x_2; \text{for } x_1 \ge 0$$

Program Statement	Description of Switch
INPUT SWITCH $Y = INSW(X1, X2, X3)$	 $y = x_2$; for $x_1 < 0$ $y = x_3$; for $x_1 \geq 0$

Program Statement	Description of Switch
COMPARATOR $Y = COMPAR(X1, X2)$	 $y = 0$; for $x_1 < x_2$ $y = 1$; for $x_1 \geq x_2$

Program Statement	Description of Switch
FUNCTION SWITCH $Y = FCNSW(X1, X2, X3, X4)$	 $y = x_2$; for $x_1 < 0$ $y = x_3$; for $x_1 = 0$ $y = x_4$; for $x_1 > 0$

Program Statement	Description of Switch
RESETTABLE FLIP-FLOP $Q = RST(R, S, T)$	

$$Q = 0; \text{ for } R > 0 \text{ (regardless of } S \text{ and } T)$$
$$Q = 1; \text{ for } R \leq 0, \text{ and } S > 0 \text{ (regardless of } T)$$

$$\left.\begin{array}{l} Q = 0; \\ Q = 1; \\ Q = 0; \\ Q = 1; \end{array}\right\} \text{for } \begin{array}{l} R \leq 0 \\ S \leq 0 \end{array} \text{ and } \left\{\begin{array}{l} T > 0 \\ T > 0 \\ T \leq 0 \\ T \leq 0 \end{array}\right. \text{ when } \left\{\begin{array}{l} Q(t - \Delta t) = 1 \\ Q(t - \Delta t) = 0 \\ Q(t - \Delta t) = 0 \\ Q(t - \Delta t) = 1 \end{array}\right.$$

Comments

This switch has the operating characteristics of a reset-set-toggle flip-flop. $Q(t - \Delta t)$ is the previous state of the flip-flop. At the start of a simulation, $Q(t - \Delta t) = 0$ for R and $S \leq 0$ with either state of T.

Application of this flip-flop is given in Example 4.8, p. 228.

Table A-7
Logic Functions

Program Statement	Logic Function Description
AND $Y = AND(X1, X2)$ Applications given in Ex. 3.13, p. 158 Ex. 4.7, p. 224 Ex. 4.8, p. 228	 $y = 1; \text{ for } x_1 > 0, x_2 > 0$ $y = 0; \text{ otherwise}$
NAND (not and) $Y = NAND(X1, X2)$	 $y = 0; \text{ for } x_1 > 0, x_2 > 0$ $y = 1; \text{ otherwise}$

Program Statement	*Logic Function Description*
NOT $Y = NOT(X)$ Application given in Ex. 4.7, p. 224 Ex. 4.8, p. 228	 $y = 1$; for $x \leq 0$ $y = 0$; for $x > 0$
INCLUSIVE OR $Y = IOR(X1, X2)$ Application given in Ex. 4.7, p. 224	 $y = 0$; for $x_1 \leq 0$, $x_2 \leq 0$ $y = 1$; otherwise
NOR (not or) $Y = NOR(X1, X2)$	 $y = 1$; for $x_1 \leq 0$, $x_2 \leq 0$ $y = 0$; otherwise
EXCLUSIVE OR $Y = EOR(X1, X2)$	 $y = 1$; for $x_1 \leq 0$, $x_2 > 0$ $y = 1$; for $x_1 > 0$, $x_2 \leq 0$ $y = 0$; otherwise
EQUIVALENT $Y = EQUIV(X1, X2)$	 $y = 1$; for $x_1 \leq 0$, $x_2 \leq 0$ $y = 1$; for $x_1 > 0$, $x_2 > 0$ $y = 0$; otherwise

FORTRAN Functions. All FORTRAN functions are available for use in CSMP simulation. Also, the majority of FORTRAN techniques can be used with

CSMP including WRITE statements, DO loops, GO TO, and IF. In some cases the FORTRAN statements require special handling procedures within the body of a CSMP program. Several examples are given in the text in which FORTRAN is included with the CSMP program.

In the following table some of the more common FORTRAN functions are listed for convenience. Users should refer to any standard FORTRAN text for a more complete listing.

Table A-8

FORTRAN Functions

Program Statement	*Mathematical Description*
EXPONENTIAL Y = EXP(X)	$y = e^x$
NATURAL LOGARITHM Y = ALOG(X)	$y = \ln(x)$
COMMON LOGARITHM Y = ALOG10(X)	$y = \log_{10}(x)$
ARCTANGENT Y = ATAN(X)	$y = \tan^{-1}(x)$
TRIGONOMETRIC SINE Y = SIN(X)	$y = \sin(x)$
TRIGONOMETRIC COSINE Y = COS(X)	$y = \cos(x)$
SQUARE ROOT Y = SQRT(X)	$y = \sqrt{x}$
HYPERBOLIC TANGENT Y = TANH(X)	$y = \tanh(x)$
ABSOLUTE VALUE (Real argument and output) Y = ABS(X)	$y = |x|$

Program Statement	*Mathematical Description*
ABSOLUTE VALUE (Integer argument and output) Y = IABS(X)	$y = \lvert x \rvert$
TRANSFER OF SIGN X = SIGN(A, B)	$X = \dfrac{A \cdot B}{\lvert B \rvert}$
LARGEST VALUE (Integer arguments and real output) Y = AMAX0(X1, X2, ... , XN)	$y = \max{(x_1, x_2, \ldots, x_n)}$
LARGEST VALUE (Real arguments and output) Y = AMAX1(X1, X2, ... , XN)	$y = \max{(x_1, x_2, \ldots, x_n)}$
LARGEST VALUE (Integer arguments and output) Y = MAX0(X1, X2, ... , XN)	$y = \max{(x_1, x_2, \ldots, x_n)}$
LARGEST VALUE (Real arguments and integer output) Y = MAX1(X1, X2, ... , XN)	$y = \max{(x_1, x_2, \ldots, x_n)}$
SMALLEST VALUE (Integer arguments and real output) Y = AMIN0(X1, X2, ... , XN)	$y = \min{(x_1, x_2, \ldots, x_n)}$
SMALLEST VALUE (Real arguments and output) Y = AMIN1(X1, X2, ... , XN)	$y = \min{(x_1, x_2, \ldots, x_n)}$
SMALLEST VALUE (Integer arguments and output) Y = MIN0(X1, X2, ... , XN)	$y = \min{(x_1, x_2, \ldots, x_n)}$

Program Statement	*Mathematical Description*
SMALLEST VALUE (Real arguments and integer output) Y = MIN1(X1, X2, . . . , XN)	$y = \min(x_1, x_2, \ldots, x_n)$

Additional CSMP Functions. The following functions have been added to CSMP III. Refer to Chap. 5 for addition comments concerning CSMP III.

Table A-9

Functions Added to CSMP III.

Program Statement	*Transformation Equivalent*

GENERAL LAPLACE TRANSFORM
(Available only in CSMP III)
  Y = TRANSF(N, B, M, A, X)
Format For Entering Data:
  STORAGE
    B(N + 1), A(M + 1)
TABLE
    B(1 − (N + 1)) = B(1),
    B(2), . . . , B(N + 1),
    A(1 − (M + 1)) = A(1),
    A(2), . . . , A(M + 1)
	$X(s) \quad \boxed{\dfrac{a_m s^m + a_{m-1} s^{m-1} + \cdots + a_1 s + a_{m+1}}{b_n s^n + b_{n-1} s^{n-1} + \cdots + b_1 s + b_{n+1}}} \quad Y(s)$
	
	where $m \le n$

Comments

The TRANSF function can be used to find the response of a transfer function with m zeros and n poles ($m \le n$) due to an arbitrary input, x. The user should note with caution that a_{m+1} and b_{n+1} are the last coefficients of the numerator and denominator respectively while a_m and b_n are the first coefficients.

Suppose one desires to find the step response of the following transfer function using TRANSF.

$$G(s) = \frac{Y(s)}{X(s)} = \frac{3s^2 + 4s + 2}{s^3 + 6s^2 + 9s + 4}$$

The statements given below can be used in the program for this purpose.

STORAGE DENCOF(4), NUMCOF(3)
TABLE   DENCOF(1-4) = 9.0, 6.0, 1.0, 4.0, . . .
    NUMCOF(1-3) = 4.0, 3.0, 2.0

to set to unity   DENCOF( ) = 0., 1., 0.
            NUMCOF( ) = 0., 1., 0.

Comments (*continued*).

$$IN = STEP(0.0)$$
$$OUT = TRANSF(3, DENCOF, 2, NUMCOF, IN)$$

For further comments see Example 3.12, p. 151; and p. 265.

Program Statement	Equivalent Mathematical Expression
VARIABLE FLOW TRANSPORT DELAY (Available only in **CSMP III**) Y = PIPE(N, IC, P, X1, X2, ND) N = the number of intervals required to define the delay of P. IC = the initial condition of entire pipeline P = the holdup quantity X1 = flow rate X2 = delayed characteristic ND = degree of interpolation for retrieving delayed characteristic. This may be 1 or 2. This functional block is used to simulate the transient flow of an incompressible fluid in a system having a time delay.	$$f_v = \int_{t_0}^{t} x_1 \, dt$$ $$q = \int_{t_0}^{t} x_1 x_2 \, dt$$ $$y = IC; \text{ for } f_v < P$$ $$y = \frac{q(f_v - P) - q(f_v(t - \Delta t) - P)}{f_v - f_v(t - \Delta t)};$$ $$\text{for } f_v \geq P$$ where f_v = flow volume q = weighted volume

Program Statement	Equivalent Mathematical Expression
FUNCTION GENERATOR WITH DEGREE OF INTERPOLATION CHOSEN BY USER (Available only in **CSMP III**) Y = FUNGEN(FUNCT, N, X) FUNCT = function name N = degree of interpolation to be used. User may select from 1, 2, 3, 4, or 5. X = value of abscissa	$$y = f(x)$$ Degree of interpolation between A, B, C, D, E, and F depends upon selection of N.

Comments

Use of this function is very similar to AFGEN and NLFGEN. See p. 262.

Program Statement	*Function Definition*
SLOPE OF A CURVE (Available only in CSMP III) $\quad$ Y = SLOPE(FUNCT, N, X) FUNCT = user's name of the curve $\quad$ N = the degree of interpolation $\quad\quad$ to be used, N = 1, 2, or 3. $\quad$ X = value of abscissa	$y = \dfrac{df}{dx}\Big|_x$ See p. 265.
ARBITRARY FUNCTION OF 2 VARIABLES (Available only in CSMP III) $\quad$ Y = TWOVAR(FUNCT, Z, X) See p. 263.	 $y = f(x, z)$
SAMPLING INTERVAL SWITCH (Available only in CSMP III) $\quad$ Y = SAMPLE$\left[\text{P1, P2,} \begin{pmatrix} P3 \\ N \end{pmatrix} \right]$ P1 = the starting time for sampling $\quad\quad$ to occur P2 = the last time for sampling to $\quad\quad$ occur P3 = the time interval between $\quad\quad$ samples if entered as a $\quad\quad$ floating-point number $\quad$ N = the number of sampling $\quad\quad$ intervals if entered as a $\quad\quad$ fixed-point number	$y = 1$ for TIME $= p_1 + k p_3 \le p_2$ or $y = 1$ for TIME $= p_1 + \dfrac{k(p_2 - p_1)}{n} \le p_2$ $k = 0, 1, 2, 3, \ldots$ for both the above $\quad\quad$ conditions $y = 0$ otherwise

Program Statement	*Equivalent Mathematical Expression*
SCALAR-TO-ARRAY CONVERTOR (Available only in CSMP III) CALL ARRAY (V1, V2, . . . , VN, X) This statement should only be used in nosort or procedural sections. Storage locations for the subscripted variable must be allocated by eithe a STORAGE of DIMENSION statement.	$x(1) = v_1$ $x(2) = v_2$ $\quad\quad \cdot$ $\quad\quad \cdot$ $\quad\quad \cdot$ $x(n) = v_n$

Program Statement	*Equivalent Mathematical Expression*
ARRAY-TO-SCALAR CONVERTOR (Available only in CSMP III) Y1, Y2, ..., YM = SCALAR(X(2))	$y_1 = x(2)$ $y_2 = x(3)$ . . . $y_m = x(m + 1)$
DOUBLE PRECISION FLOATING-POINT TO SINGLE PRECISION (Available only in CSMP III) Y = ZZRND(X(2))	The ZZRND function transforms by rounding the double precision variable X(2) to the single precision variable Y.

```
MACRO FTOT=FRICT(SR,FSLIDE,FEXT,V)
          FSTAT=FSLIDE*SR
          FZ=DEADSP(-FSTAT,FSTAT,FEXT)
          FN=FEXT+FSLIDE
          FP=FEXT-FSLIDE
          FTOT=FCNSW(V,FN,FZ,FP)
ENDMAC
*THIS MACRO DESCRIBES A LIMITED INTEGRATOR BY D.A.SMITH
MACRO X=LIMINT(XIC,LOWLIM,UPLIM,XP)
PROCED XPC=LIN(XP,X,LOWLIM,UPLIM)
        XPC=XP
        IF((X.GE.UPLIM).AND.(XP.GT.0.)) XPC=0.
        IF((X.LE.LOWLIM).AND.(XP.LT.0.)) XPC=0.
ENDPRO
        XC=INTGRL(XIC,XPC)
        X=LIMIT(LOWLIM,UPLIM,XC)
ENDMAC
```

II

DIAGNOSTIC MESSAGES, PROGRAM RESTRICTIONS, AND RESERVED WORDS

This material involving diagnostic messages, restrictions, and reserved words was primarily taken from IBM Manuals.[1,2]

Diagnostic Messages

Diagnostic messages may occur during both the translation and execution phases of the program and are designed to be self-explanatory. Some of the diagnostic checks detect illegal characters or incorrect syntax; the symbol "$" is printed below the detected error prior to the associated diagnostic message. A "warning only" message is printed when an error is not wholly discernible in translation or does not destroy the "validity" of simulation. Some examples of these errors are:

Control variable name not a systems variable

Parameter value not specified

Variable used as input to a section not available from any prior section

Some examples of errors causing a run halt at the end of translation are:

Incorrect structure or data statement format

Invalid data card type

Unspecified implicit loop

RELERR specification on other than an integrator output name

Examples of errors causing a run halt during execution are:

Failure of an integration or implicit function to meet the error criterion

A misspelled subroutine name

The following is an alphabetical list of diagnostic messages with their explanations and suggested corrections. The messages with the label "S/360" apply only to S/360 CSMP, the messages with the label "III" apply only to CSMP III, and messages designated with "Both" apply to both forms of CSMP.

III   CALL CONTIN CAN ONLY BE USED IN A TERMINAL SEGMENT
        "CALL CONTIN" can be used only in a Terminal segment. If it is used elsewhere, the run terminates.

Both   CALL RERUN CAN ONLY BE USED IN A TERMINAL SEGMENT
        "CALL RERUN" can be used only in a Terminal segment. If it is used elsewhere, the run terminates.

III   CENTRAL INTEGRATION ROUTINE NOT SUPPLIED
        On the METHOD execution control card, the user has used the word, CENTRL, to specify his integration method. However, he has not supplied the integration deck to the program. The run will be terminated.

Both   CSMP STATEMENT INCORRECTLY WRITTEN
        The translation phase has detected an error in the statement printed before this message. The statement should be checked carefully, including parentheses and commas. Although translation of the source statements will continue, the run will be terminated before the execution phase.

Both   CSMP STATEMENT OUT OF SEQUENCE
        The sequence of input statements cannot be processed and the run will be terminated before the execution phase. The statement should be checked for sequence in the input deck to see if it has been misplaced. MACRO definitions must precede all structure statements. An INITIAL segment, when used, must precede the DYNAMIC segment. If used, the TERMINAL segment must follow the DYNAMIC segment.

S/360   DATA HAS NOT BEEN SPECIFIED FOR AN AFGEN FUNCTION
S/360   DATA HAS NOT BEEN SPECIFIED FOR AN NLFGEN FUNCTION
        An AFGEN (or NLFGEN) function generator has been used in a structure statement but the corresponding data has not been specified using the FUNCTION statement. The run will be terminated.

III   DATA HAS NOT BEEN SUPPLIED FOR A TWOVAR FUNCTION
        A TWOVAR function generator has been used in a structure statement, but the data has not been specified on FUNCTION statements. The run will be terminated.

III   DATA NOT SUPPLIED FOR FUNCTION
        An AFGEN, NLFGEN, FUNGEN, or SLOPE function generator has been used in a structure statement, but a FUNCTION statement has not specified the corresponding data. The run will be terminated.

S/360   DYNAMIC STORAGE EXCEEDED. THIS CASE CANNOT BE RUN
The 8000-word limitation on simulator data storage has been exceeded. The storage in this array includes the current values of model variables, function and error tables, central integration history, and subscripted variable values. The problem should be analyzed to determine where equations can be combined to reduce the number of required entries in the array.

S/360   ERROR—CENTRAL INTEGRATION ROUTINE NOT SUPPLIED
The user has used the word CENTRL for his integration method on the METHOD execution control card; however, he has not supplied the integration deck to the program. The run will be terminated.

Both   ERROR IN COORDINATE ENTRIES
An error has been detected in the previously printed FUNCTION data statement. There is either an odd number of entries in the data table or an improper sequence of X-coordinate values. The run will be terminated.

S/360   ERROR IN PRINT-PLOT STATEMENT
An error has been detected in the PRTPLT output control statement. The statement should be checked for a correct number of parentheses and commas for specifying lower and upper limits, particularly if one or the other is missing, and commas are used to indicate this. Although the run continues, everything on the card after the error is disregarded.

Both   ERROR IN TABLE ENTRY
In the previously printed TABLE data statement, an error has been detected. Although reading of the data statements will continue, the run will be terminated before execution.

Both   EXCEEDED MAXIMUM ITERATIONS ON IMPLICIT LOOP
One hundred iterations of the implicit loop have been run and convergence has not yet occurred. The run has been terminated. One possibility is to change the error condition, so that the convergence criteria can be met.

III   FAMILY OF PARAMETER VALUES AFTER END CONTINUE DELETED
Following an END CONTINUE statement, a multiple-value parameter has been defined. These values will be deleted and the program will use the previous value of the parameter.

III   FINTIM HAS NOT BEEN SET. RUN DELETED
FINTIM either has not been specified or was made equal to zero.

S/360   FINTIM IS ZERO. THIS CASE CANNOT BE EXECUTED
FINTIM either has not been specified or has been specified as being equal to zero.

III   FUNCTION DATA EXCEEDS STORAGE
Function data supplied in the execution input exceeds the core storage allocated. The run is terminated.

III   FUNCTION DATA FOR LOAD EXCEEDS STORAGE
The FGLOAD or TVLOAD subprogram has been called and the available storage for the function as specified by the SYSTEM NPOINT=n-statement is not sufficient.

III   FUNCTION DATA FOR LOAD INCORRECT
The FGLOAD or TVLOAD subprogram has been called and the input data (X values) are not monotonically increasing.

III   FUNCTION NAME xxxxxx HAS BEEN SPECIFIED PREVIOUSLY
The name assigned to function data on a FUNCTION statement has been already assigned to another function, parameter, or variable.

Both   GENERATED STATEMENT NO. xx
The Translator has detected an error during generation of statement xx of an invoked macro in the structure of the model. Carefully check the corresponding statement of the macro definition for proper spelling and punctuation.

Both   ILLEGAL CHARACTER OR DOUBLE OPERATOR
In the previously printed statement, an illegal character or double operator has been detected. Although translation of the source statements will continue, the run will be terminated before the Execution phase.

III   ILLEGAL SPECIFICATION ON RERUN
Following an END RERUN statement or END CONTINUE, there is an OUTPUT or PREPARE statement or a TIMER statement with an OUTDEL parameter. This is an illegal specification.

Both   INCORRECT IMPLICIT STATEMENT
The Translation phase has detected an error in the IMPL structure statement printed before this message. The statement should be checked to see that the third argument is the output name of the last statement in the definition and that the block output appears at least once to the right of an equal sign. Although translation of the source statements will continue, the run will be terminated before the Execution phase.

Both   INCORRECT MACRO STATEMENT
The Translation phase has detected an error in the macro use statement printed before this message. The statement should be checked to ensure that the number of arguments and outputs is correct and that the argument list ends with a parenthesis. Although translation of the source statements will continue, the run will be terminated before the Execution phase.

S/360   INCORRECT TIMER VAR NAME**WARNING ONLY
One of the system variable names (FINTIM, DELT, PRDEL, OUTDEL, or DELMIN) has been misspelled on the TIMER execution control card. The user should also check the possibility that the system variable has been renamed. Although the run will continue, the system variable misspelled will be unchanged.

Both   INPUT NAME SAME AS OUTPUT NAME
The output variable name to the left of the equal sign has also been used as an input name on the right side of the equal sign. Except as output of a memory type functional element, such usage is not permissible in a parallel, sorted section. The run will be terminated.

III   INPUT P IS LESS THAN ZERO FOR DELAY
Delay time P of the DELAY function is found to be less than zero, which is an invalid condition. The execution of this run will be terminated.

III   INPUT P2 IS LESS THAN P1 FOR DEADSP
Parameter P2 is found to be less than parameter P1 for the DEADSP function and this is an invalid condition.

III   INPUT P2 IS LESS THAN ZERO FOR IMPULS
The time between impulses P2 is found to be less than zero for the IMPULS function, which is an invalid condition.

S/360   INPUT TO FUNCTION GENERATOR nnnnnn ABOVE SPECIFIED RANGE
INPUT = xxxx.xxxx
The input (xxxx.xxxx) to the function generator named nnnnnn is above the maximum specified range. The program will take the value for the maximum specified and continue. This message will be printed only once, even though the condition is reached several times.

S/360   INPUT TO FUNCTION GENERATOR nnnnnn BELOW SPECIFIED
RANGE INPUT = xxxx.xxxx
The input (xxxx.xxxx) to the function generator named nnnnnn is below the minimum specified range. The program will take the value for the minimum specified and continue. This message will be printed only once, even though the condition is reached several times.

III   INPUT TO FUNCTION name $\begin{bmatrix} \text{ABOVE} \\ \text{BELOW} \end{bmatrix}$ INPUT DATA

[CURVE NO. nnnn] CALL mmmm INPUT = w at v

The input to the function generator (interpolating for the function named in the message) is above the maximum or below the minimum input specified. The output of the function generator is set equal to the output value of the function corresponding to the input extreme which was violated. This message is printed only once even though the condition may be reached several times. The curve number nnnn indicates (for functions of two variables) for which curve the first input was in error. For example, if the model contains the following function of two variables:

$$\text{FUNCTION F, } -5. = (0., 3.5), (1., 6.1), (3., 8.6)$$
$$\text{FUNCTION F, } 3.2 \;\; = (0., 3.8), (1., 7.5), (3.2, 9.6)$$
$$\text{FUNCTION F, } 10.2 = (0., 4.2), (1.2, 8.3), (2.8, 9.3)$$

and the diagnostic message specifies CURVE NO. 2, then the error is in the first input to the function curve

$$\text{FUNCTION F, } 3.2 \;\; = (0., 3.8), (1., 7.5), (3.2, 9.6)$$

The call number mmmm is the first argument of the interpolation function in the UPDATE FORTRAN subroutine generated by the Translator. For example, if the model contains the statement

$$\text{OP} = \text{AFGEN (FUNCT, XIN)}$$

the UPDATE subroutine would contain the corresponding statement

$$\text{OP} = \text{AFGEN (n, FUNCT, XIN)}$$

The first argument n is an integer constant and would be the call number in the diagnostic message for this statement if the value of XIN is outside the range of the function FUNCT. The input value out of range is specified by INPUT $= w$. The value of the independent variable for this out-of-range condition is given by AT v.

SECOND INPUT TO FUNCTION name $\begin{bmatrix} \text{ABOVE} \\ \text{BELOW} \end{bmatrix}$ INPUT DATA

CALL mmmm INPUT $= w$ AT v

The second input to the function generator TWOVAR (interpolating for the function of two variables named in the message) is above the maximum or below the minimum input specified. The output of the function generator is set equal to the output value of the function corresponding to the input extreme which was violated. This message is printed only once even though this condition may be reached several times. The call number mmmm is the first argument of the interpolation function block in the UPDATE subroutine generated by the Translator. For example, if the model contains the statement

OUT $=$ TWOVAR (F, XIN, ZIN)

the UPDATE subroutine would contain the corresponding statement

OUT $=$ TWOVAR (n, F, XIN, ZIN)

The first argument n is an integer constant and would be the call number in the diagnostic message for this statement if the value of ZIN is outside the range of the function F. The second input value which is out of range is specified by INPUT $= w$. The value of the independent variable for this out-of-range condition is given by AT v.

Both LABEL INCORRECTLY WRITTEN
The label used in the preceding statement cannot be recognized by the program. Check for proper spelling. The statement will be disregarded; the run will continue.

III MACRO WITHIN MACRO USED IN A PROCEDURAL
Macros, separately defined, can be invoked within the definition of other macros if overall parallel structure is implied. Invocation of a macro within a procedure within a macro definition is therefore not permissible. Similarly, a macro containing other macros in its definition may not be invoked from a procedure or from a procedural section.

S/360 MACRO xxxxxx WITHIN MACRO yyyyyy USED IN A PROCEDURAL SECTION
Macros, separately defined, may be invoked within the definition of other macros if overall parallel structure is implied. Invocation of a macro within a procedure within a macro definition is therefore not permissible. Similarly, a macro containing other macros in its definition may not be invoked from a procedure or from a procedural section.

III METHOD SELECTED EXCEEDS STORAGE
The integration method specified in the execution input exceeds the core storage allocated. The run is terminated. The integration methods available, in order of increasing size, are: RECT, TRAPZ, SIMP, ADAMS, RKSFX, RKS, STIFF, RKSDP, and MILNE.

S/360 MORE THAN 10 PRTPLT STATEMENTS
More than ten **PRTPLT** output control statements have been specified. Only the first ten will be used.

Both NUMBER EXCEEDS 12 CHARACTERS
In the previously printed statement, a number exceeding twelve characters in a macro argument or integrator block initial condition has been detected. Although translation of the source statements will continue, the run will be terminated before the execution phase.

Both NUMBER INCORRECTLY WRITTEN
In the previously printed statement, a number written incorrectly has been detected. If detected during the translation phase, translation of the source statements will continue; however, the run will be terminated before the execution phase.

S/360 ONLY FIRST 10 CONDITIONS FOR JOB END WILL BE TESTED
More than ten specifications have been given with the FINISH execution control statement. Although the run will continue, only the first ten specifications will be used.

III ONLY FIRST 10 CONDITIONS WILL BE TESTED
More than ten specifications have been given through the FINISH execution control statement. Although the run will continue, only the first ten specifications will be used.

Both ONLY FIRST 50 VALUES WILL BE USED
The multiple-value form of the PARAMETER data statement specifies more than 50 values for the parameter. A sequence of runs will be performed using only the first 50 values.

S/360 ONLY FIRST 50 VARIABLES WILL BE PREPARED
More than 50 variables (including TIME) have been specified with PREPARE or PRTPLT output control statements. Although the run will continue, only the first 50 variables will be used.

III ONLY FIRST 220 VARIABLES WILL BE PREPARED
More than 220 variables (including TIME) have been specified through PRE-PARE or OUTPUT print control statements. Although the run will continue, only the first 220 variables will be used.

S/360 ONLY FIRST 50 VARIABLES WILL BE PRINTED
More than 50 variables (including TIME) have been requested with PRINT execution control statements. Only the first 50 will be printed; others will be ignored.

III ONLY FIRST 55 VARIABLES WILL BE PRINTED
More than 55 variables (including TIME) have been requested through PRINT execution control statements. Only the first 55 will be printed; others will be ignored.

S/360 ONLY FIRST 100 VARIABLES WILL BE RANGED
More than 100 variables (including TIME) have been specified with the RANGE output control statement. Although the run will continue, only the first 100 variables will be used.

III   ONLY FIRST 110 VARIABLES WILL BE RANGED
More than 110 variables (including TIME) have been specified through the
RANGE output control statement. Although the run will continue, only the
first 110 variables will be used.

Both   ONLY LAST VALUE OF FAMILY USED FOR CONTINUE RUN
A multiple-value parameter has been used in a run that is to be continued. The
continue control feature will be implemented only with the last value of the parameter. This is a warning message. The run will continue.

Both   OUTPUT NAME HAS ALREADY BEEN SPECIFIED
In the previously printed statement, the output variable name to the left of the
equal sign has been used before as an output variable name; that is, it has occurred to the left of the equal sign in a preceding section. The run will be continued.

Both   PARAMETERS NOT INPUT OR OUTPUTS NOT AVAILABLE TO SORT
SECTION ***SET TO ZERO***
A list of variable names will be printed following this heading. The run is con-
tinued. Variables that are not parameters specified on data cards are set to zero.
Output variable names that are not available to this sort section are initially set
to zero, but may change as the problem is run.

Both   PROBLEM CANNOT BE EXECUTED
At least one diagnostic message will have been printed among the source statements indicating the reason why the problem cannot be executed. The run will
be terminated.

S/360   PROBLEM INPUT EXCEEDS TRANSLATION TABLE nn
During translation of the problem, a table has been exceeded and the run will
terminate. The specific table is identified by nn in the following list:

nn	*Translation Table*
1	More than 500 statement output names.
2	More than 1400 statement input names (temporary count during trans-lation).
3	More than 400 parameter names
4	More than 300 INTGRL or MEMORY outputs
5	More than 1400 input names and unique block names
6	More than 20 FIXED variable names
7	More than 100 non-zero initial condition numeric values
8	More than 10 FORTRAN specification cards
9	More than 100 unique block names and symbolic names with first letter I, J, K, L, M, or N but not appearing on FIXED statements
10	More than 25 STORAGE variable names
11	More than 15 sections (SORT or NOSORT)
12	More than 100 MACRO arguments, outputs, and statement numbers for one MACRO

nn	*Translation Table*
13	More than 50 MACRO functions
14	More than 120 MACRO definition cards
15	More than 50 HISTORY or MEMORY functions
16	More than 15 MEMORY functions
17	More than 85 variables that are neither parameters specified on data cards nor outputs of a following SORT section
18	More than 150 duplicate names in COMMON (outputs or inputs to INTGRL blocks)
19	More than 100 parameters in one SORT sequence
20	More than 600 structure statements in a single SORT section
21	More than 180 characters in a single macro-generated statement. This restriction is violated by a procedural macro with more than 25 names for input and output variables or by a macro which includes a statement of excessive complexity. To circumvent the restriction, simplify the macro definition.

III   PROBLEM INPUT EXCEEDS TRANSLATION TABLE nn

A table has been exceeded during translation of the problem, and the run will terminate. The specific table is identified by nn, which refers to the following list:

nn	*Translation Table*
1	More than 600 macro and statement output names
2	More than 1900 macro and statement input names
3	More than 300 INTGRL blocks
4	More than 400 parameter and function names
5	More than 50 STORAGE or 100 integrator array specifications
6	More than 50 user-defined history and memory functions
7	More than 50 macro functions defined
8	More than 125 MACRO definition statements
9	More than 100 literal constants (does not include zero values)
10	More than 25 FIXED variable names
11	More than 20 sort sections
12	Sum of arguments, outputs and statement numbers of a macro block definition is more than 100
13	More than 300 duplicate names used with INTGRL blocks
14	More than 600 structure statements in a single sort section

nn	*Translation Table*

15   More than 100 variables in a single SORT sequence

16   More than 180 characters in a single macro-generated statement. This restriction is violated by a procedural macro with more than 25 names for input and output variables or by a macro which includes a statement of excessive complexity. To circumvent the restriction, simplify the macro definition.

S/360   PRTPLT, PREPARE, AND RANGE VARIABLES EXCEED 100. ALL RANGE VARIABLES STARTING WITH xxxxxx HAVE BEEN DELETED
More than 100 variables (including TIME) have been specified with PRTPLT, PREPARE, and RANGE output control statements. Although the run will continue, only the first 100 variables will be used for this run.

III   RANGE VARIABLES DELETED STARTING WITH NAME
Too many variables have been specified on RANGE, OUTPUT, and PREPARE statements. "Name" and all those following are deleted from RANGE area.

S/360   RERUN FROM TERMIN CANCELED FOR CONTIN RUN
The TERMINAL segment cannot be used to cause a rerun when a CONTINUE translation control statement started the run. The TERMINAL computation statements will be executed but any CALL RERUN will be ignored.

S/360   SIMULATION HALTED
The run was terminated because a FINISH condition was satisfied. The variable name and its value are printed.

III   SIMULATION HALTED FOR FINISH CONDITION
The run was terminated because a finish condition was satisfied. The variable name and its value are printed.

Both   SIMULATION INVOLVES AN ALGEBRAIC LOOP CONTAINING THE FOLLOWING ELEMENTS
A list of output variable names will be printed following this diagnostic. The sort subprogram has been unable to find an integration or memory block in the loop involving these variables. The run will be terminated before the execution phase.

S/360   SYMBOLIC NAME xxxxxx NOT DEFINED
An error has been detected on the PARAMETER, INCON, CONSTANT, or TIMER card printed before this message. Although input to the execution phase will continue, the simulation will not be run.

III   SYMBOLIC NAME NOT DEFINED IN MODEL
An error has been detected on the PARAMETER, INCON, CONSTANT, or TIMER card printed before this message. Although input to the Execution phase will continue, the simulation will not be run.

Both   SYMBOLIC NAME EXCEEDS SIX CHARACTERS
In the previously printed statement, a symbolic name exceeding six characters has been detected. Although translation of source statements will continue, the run will be terminated before Execution.

Both   SYMBOLIC NAME INCORRECTLY WRITTEN
In the previous statement, a symbolic name has been written incorrectly. The job will be terminated.

III   TOO MANY CONTINUATION CARDS
The previously printed statement has been continued on too many cards. If a MACRO label statement has over three continuation statements or if a structure statement has over eight continuation statements, the user should make multiple statements or use more columns on individual cards. Although translation of the source statements will continue, the run will be terminated before the Execution phase.

S/360   TOO MANY CONTINUATION CARDS. MAX=n
The previously printed statement has been continued on too many cards. If N = 3, a MACRO label statements has over three continuation statements. If N = 8, a structure statement has over eight continuation statements. The user should make multiple statements or use more columns on individual cards. Although translation of the source statements will continue, the run will be terminated before the execution phase.

Both   TOO MANY LEFT PARENTHESES

Both   TOO MANY RIGHT PARENTHESES
Too many left (or right) parentheses have been detected in the statement printed before this diagnostic. Although the translation of the source statements will continue, the run will be terminated before the execution phase.

Both   VARIABLE STEP DELT LESS THAN DELMIN. SIMULATION HALT
The simulation will not be continued because the specified DELT is less than the specified DELMIN.

Programs Restrictions

Size limitations of the CSMP programs may be exceeded when solving very large problems. The following table gives some of the restrictions for both S/360 CSMP and CSMP III. Additional restrictions appear in this Appendix in the section on diagnostic messages under the message, "PROBLEM INPUT EXCEEDS TRANSLATION TABLE nn."

	S/360 CSMP	CSMP III
Number of statement and MACRO output names	500	600
Numbers of statement and MACRO input names	1400	1900
Number of parameter and function names	400	400
Number of integrators plus statements with memory and history	300	300
Number of structure statements in a single sort section	600	600
Number of sort sections	15	20
Number of parameters in a single sort sequence	100	100

	S/360 CSMP	CSMP III
Number of user-supplied memory and history functions	50	50
Number of memory functions	15	
Number of MACRO functions	50	50
Sum of MACRO arguments, outputs, and statements number for one MACRO	100	100
Number of FINISH specifications	10	10
Number of RANGE and PREPARE variables	†	‡
Number of STORAGE variables	25	50
Number of statement sent directly to FORTRAN (identified by a / in column 1)	10	
Number of FIXED variables	20	25

†The total number of RANGE, PRTPLT, and PREPARE variables must be 100 or less.

‡Two times the number of RANGE variables plus the number of PREPARE and OUTPUT variables must be equal to or less than 220.

Reserved Words

The words listed below are reserved for special use in CSMP. They should not be used as variable or subprograms names.

ABS	GO
BACKSPACE	GOTO
CALL	IABS
COMMON	IDIM
CONTINUE	IF
DABS	IFIX
DBLE	INTEGER
DEFINE	ISIGN
DIM	PAUSE
DIMENSION	READ
DFLOAT	REAL
DO	RETURN
DOUBLE	REWIND
DSIGN	SIGN
END	SNGL
ENDFILE	STOP
EQUIVALENCE	SUBROUTINE
EXIT	WRITE
EXTERNAL	
FIND	
FLOAT	
FORMAT	
FUNCTION	

In addition to the above words, the following restrictions must be followed for variable names.

1 Certain variable names are reserved for use by the system, and cannot appear in a CSMP structure statement. These names are NALARM, IZxxxx, and ZZxxxx, where x is any digit.

2 KEEP is a COMMON variable and it must be used consistently with its intended purpose.

3 DELT, DELMIN, DELMAX, FINTIM, TIME, PRDEL, and OUTDEL are system reserved names and, unless renamed, must appear only in their intended context.

4 TIME is the name for the independent variable and should be used only for that purpose.

5 CSMP subroutines, unless renamed, must be used only as intended. These names are MAINEX, CENTRL, NUMER, ALPHA, DEBUG, UPDATE, CSTORE, and the standard CSMP functional block names.

6 The statement numbers 30000 to 39999, inclusive, are reserved for system use in the UPDATE subprogram.

MATHEMATICS OF INTEGRATION METHODS

There are six different integration methods available in S/360 CSMP. CSMP III provides an additional method for solving stiff equations. All methods use centralized integration. This means that integration is performed after all structure statements have been evaluated.

In the Milne and Runge-Kutta methods, the step-size is automatically adjusted to user-specified error-bounds during the problem execution. The following gives the mathematics of the integration methods as listed in the IBM programs reference manuals.[1,2]

Milne Fifth-Order Predictor-Corrector (MILNE)

Predictor: $Y_{t+\Delta t}^{p} = Y_{t-\Delta t} + \dfrac{\Delta t}{3}(8X_t - 5X_{t-\Delta t} + 4X_{t-2\Delta t} - X_{t-3\Delta t})$

Corrector: $Y_{t+\Delta t}^{c} = \dfrac{1}{8}(Y_t + 7Y_{t-\Delta t}) + \dfrac{\Delta t}{192}(65X_{t+\Delta t} + 243X_t + 51X_{t-\Delta t} + X_{t-2\Delta t})$

Estimate: $Y_{t+\Delta t} = 0.96116 Y_{t+\Delta t}^{c} + 0.03884 Y_{t+\Delta t}^{p}$

Integration interval control is based on the following criteria:

$$\frac{|Y^c - Y^p|}{A + R|Y^c|} \simeq \frac{\text{Error}}{A + R|Y^c|} \leq 1$$

Runge-Kutta Fourth-Order (RKS)

$$Y_{t+\Delta t} = Y_{\bullet} + \frac{1}{6}(K_1 + 2K_2 + 2K_3 + K_4)$$

$$K_1 = \Delta t\, f(t, Y_t)$$

$$K_2 = \Delta t\, f\left(t + \frac{\Delta t}{2}, Y_t + \frac{K_1}{2}\right)$$

$$K_3 = \Delta t\, f\left(t + \frac{\Delta t}{2}, Y_t + \frac{K_2}{2}\right)$$

$$K_4 = \Delta t\, f(t + \Delta t, Y_t + K_3)$$

The interval (Δt) for both variable step integrations will be reduced to satisfy the following criterion:

$$\frac{|Y_{t+\Delta t} - Y^s|}{A + R|Y_{t+\Delta t}|} \cong \frac{\text{Error}}{A + R|Y_{t+\Delta t}|} \leq 1$$

In both the MILNE and RKS methods, Y^s is $Y_{t+\Delta t}$ calculated by Simpson's rule. A and R are the absolute and relative errors corresponding to the values specified by ABSERR and RELERR.

A form of the fourth-order Runge-Kutta for a fixed-step-size (RKSFX) is also available. The mathematics are the same as shown above with the exception that the error criterion is not used.

In addition to RKSFX method, there are four fixed-step integration methods. Adams Second-Order (ADAMS)

$$Y_{t+\Delta t} = Y_t + \frac{\Delta t}{2}(3\dot{Y}_t - \dot{Y}_{t-\Delta t})$$

Simpson's Rule (SIMP)

Predictor: $\quad Y^p_{t+\Delta t/2} = Y_t + \frac{\Delta t}{2}X_t$

$$Y^p_{t+\Delta t} = Y^p_{t+\Delta t/2} + \frac{\Delta t}{2}X_{t+\Delta t/2}$$

Corrector: $\quad Y^c_{t+\Delta t} = Y_t + \frac{\Delta t}{6}(X_t + 4X_{t+\Delta t/2} + X_{t+\Delta t})$

Trapezoidal (TRAPZ)

Predictor: $\quad Y^p_{t+\Delta t} = Y_t + \Delta t\, X_t$

Estimate: $\quad Y_{t+\Delta t} = Y_t + \frac{\Delta t}{2}(X_t + X_{t+\Delta t})$

Rectangular (RECT)

$$Y_{t+\Delta t} = Y_t + \Delta t\, X_t$$

In all of the above equations, the common terminology is $X_t \equiv \dot{Y}_t \equiv f(t)$. The value of $X_{t+\Delta t}$ used in the estimate is based on the prediction $Y^p_{t+\Delta t}$.

A method for handling stiff equations[3,4] (STIFF) is available in CSMP III. The solution is computed by the following step by step procedure.

Given the differential equation $\dot{y} = f(t, y)$, $y(t_0) = y_0$

(1) $\dot{y}_A(t) = [y(t) - y(t - h_0)]/h_0$

(2) $d_1 = \dot{y}(t) - \dot{y}_A(t)$

(3) $y_p(t + \delta) = y(t) + \delta\dot{y}(t)$ where $\delta \leq h/4$

(4) $\dot{y}_p(t + \delta) = f[t + \delta, y_p(t + \delta)]$

(5) $d_2 = [\dot{y}_p(t + \delta) - \dot{y}(t)]/\delta$

(6) $\lambda = \begin{cases} d_2/d_1 & d_1 \neq 0, \\ 0 & d_1 = 0, \end{cases}$ $c_1 = \begin{cases} (e^{\lambda h} - 1)/\lambda h & \lambda < 0, \\ 1 + \lambda h/2 & \lambda \geq 0 \end{cases}$

   $c_0 = \begin{cases} e^{\lambda h} & \lambda < 0 \\ 1 + \lambda h & \lambda \geq 0 \end{cases}$

(7) $y_c(t + h) = y(t) + h\dot{y}_A(t) + hc_1 d_1$

(8) $\dot{y}_c(t + h) = f[t + h, y_c(t + h)]$

(9) $E = h[\dot{y}_c(t + h) - (\dot{y}_A(t) + c_0 d_1)]$

(10) $E_e = 2\delta[\dot{y}_p(t + \delta) - \dot{y}(t)]$

Note that $y, y_c, y_p, y_A, y_{pe}, d_1, d_2, c_0, c_1$, and λ are vector quantities.

Step 1	To start up
	$h_0 = 0$
	$\dot{y}_A(0) = \dot{y}(0)$
Step 2	y is assumed to be the sum of an asymptotic part and a perturbation from the asymptote.
Steps 3 and 4	These steps constitute Euler integration with step size δ.
Step 5	$d_2 \approx \ddot{y}(t)$. It is assumed that the form of $f(t, y)$ makes calculation of $\ddot{y}$ difficult.
Step 6	y_c is computed solution at end of current step; h is step-size.
Step 7	E is used to monitor step-size h.
Step 8	E_e is used to monitor Euler step-size.

The section on integration methods in Chap. 3 contains details for using the various methods. Information on accuracy and computer time is also included in this section.

In the event that none of the methods satisfies the user's requirements, the user can supply his own integration method. This method is entered in the system by the name of CENTRL. Example 3.2 illustrates the procedure for using a user-supplied integration method.

REFERENCES

1. *System/360 Continuous System Modeling Program*, User's Manual GH20-0367-4, Program Number 360A-CX-16X. IBM Corporation, Technical Publications, White Plains, N. Y.

2. *Continuous System Modeling Program III (CSMP III)*, Program Reference Manual SH19-7001-2, Program Number 5734-XS9, IBM Corporation, Data Processing Division, White Plains, N. Y.

3. Fox, LESLIE and D. F. MAYERS, *Computing Methods for Scientists and Engineers*, Clarendon Press at Oxford, 1968, p. 215.

4. FOWLER, M. E., and R. M. WARTEN, "A Numerical Integration Technique for Ordinary Differential Equations with Widely Separated Eigenvalues," *IBM Journal of Research and Development*, II, No. 5, Sept. 1967, 537-43.

INDEX

MACRO FTOT = FRICT (SR, FSLIDE, FEXT, V)

ESTAT = FSLIDE * SR

$$FEXT = TORQUE\ POWER - SPRING\ LOAD = IN \cdot LBF$$
$$\left(IN \cdot LB / RAD \right)$$

FZ =

DEADSP

- ESTAT
  FEXT
  ESTAT

FSLIDE = COULOMB FRICTION COEFFICIENT

SR = GAIN FACTOR TIMES FSLIDE

FN = IN·LB ERROR + COULOMB FRICTION

FP = " -

" "

V = INPUT LOGIC SWITCH

FTOT = FCNSW (V, FN, FZ, FP)

where

$$FTOT = \begin{cases} FN & , V < 0 \\ FZ & , V = 0 \\ FP & , V > 0 \end{cases}$$